DETERMINANTS AND CONTROLS OF SCIENTIFIC DEVELOPMENT

THEORY AND DECISION LIBRARY

AN INTERNATIONAL SERIES
IN THE PHILOSOPHY AND METHODOLOGY OF THE
SOCIAL AND BEHAVIORAL SCIENCES

VOLUME 10

DETERMINANTS AND CONTROLS OF SCIENTIFIC DEVELOPMENT

Edited by

KARIN D. KNORR, HERMANN STRASSER, AND
HANS GEORG ZILIAN
Institute for Advanced Studies, Vienna, Austria

D. REIDEL PUBLISHING COMPANY
DORDRECHT-HOLLAND / BOSTON-U.S.A.

Library of Congress Cataloging in Publication Data
Main entry under title:

Determinants and controls of scientific development.

(Theory and decision library ; v. 10)
Proceedings of a conference initiated by a project group at the Institute for Advanced Studies, Vienna, and held at Otto-Möbes-Volkswirtschaftsschule, Graz-Stifting, Austria, June 16–22, 1974.
Includes bibliographies and indexes.
1. Social sciences—Methodology—Congresses. 2. Science—History—Congresses. 3. Science—Philosophy—Congresses. I. Knorr, Karin D., 1944– II. Strasser, Hermann, 1941– III. Zilian, Hans Georg, 1945– IV. Institut für Höhere Studien und Wissenschaftliche Forschung, Vienna.
HM38.D48 300'.1'5 75-16419
ISBN 90-277-0600-X

Published by D. Reidel Publishing Company,
P.O. Box 17, Dordrecht, Holland

Sold and distributed in the U.S.A., Canada and Mexico
by D. Reidel Publishing Company, Inc.
Lincoln Building, 106 Old Derby Street,
Hingham, Mass. 02043, U.S.A.

Printed in The Netherlands by D. Reidel, Dordrecht

TABLE OF CONTENTS

SECTION C / SOCIETAL COMPONENTS IN SCIENTIFIC DEVELOPMENT

PREFACE

This book constitutes the outcome of an international conference held at the Otto-Möbes-Volkswirtschaftsschule, Graz-Stifting (Austria), from June 16 to 22, 1974. The conference was initiated by a project group working on determinants and controls of social science development at the Institute for Advanced Studies and Scientific Research in Vienna and organized by the editors of this volume. It was held under the auspices of the Austrian Ministry of Science and Research.

The main topics of the conference were those at the forefront of the 'state of argumentation' (to borrow from one of our contributors) in philosophy and sociology of science ever since the controversy between Thomas S. Kuhn and Sir Karl R. Popper has sharpened our awareness for the methodological and substantial presuppositions involved with questions of growth and development in science. Let us give two examples of those topics. The borderline between sociology of science and philosophy of science has become more and more unclear; while the work of at least some philosophers of science comes to have an empirical flavour, sociologists of science are increasingly apt to include logical and methodological components of the research process as their objects of examination. Papers included in this volume testify to both tendencies. It has been the concern of the conference to raise the problem of how historical and sociological findings become significant for philosophy of science issues, a problem we have to cope with when trying to assess the relevance of philosophical claims for the issue of growth in science.

The second example refers to the debate on 'external' (i.e. societal) as opposed to 'internal' (i.e. methodological and community-specific) determinants of scientific development, a debate which is now in the process of being replaced rather than resolved. Indeed, if social facts are conceived of as nothing but accounts generated by our communal rhetoric, there is no sense in disclosing the alleged external influence on scientific results. Equally, if science in being 'faithful to the facts' (to borrow from yet another of the participants) represents nothing but its own mode of discourse, there is no use in searching for 'external' correlates. Rather, science has to recollect the standards which ground its version of good speech and to grasp its activity as a socially and historically located mode of production of sense. To conclude, the 'external-internal' dichotomy gives way to one which contrasts the standards of good speech with their communal grounds, or, to introduce a position recently advanced, to one which considers the question of 'cognitive' versus 'social' constitution of scientific knowledge as much more central to the idea of progress than the first mentioned concepts. While 'internal' and 'external' determinants each include both social and cognitive aspects but relate it to different systems, that of science and that of society, the 'social' versus 'cognitive' dichotomy blurs the distinction in favor of one based on the conceptual type of variables used.

In face of the complexity of the issues involved and the diversity of viewpoints encountered it has been our editorial policy to offer a selective reconstruction rather than a faithful report of the actual discussions at the conference. The volume is built around the topics which have been at the focus of three different study groups set up at the conference. The topics are those of methodological presuppositions of sociology of science and metascience, of cognition and communication in science development, and of societal components as reflected in scientific thinking and in the structure of the community of scientists. The papers presented in this volume have been selected according to their primary relevance for the discussion in the anglo-saxon scene. There is a second volume of papers presented at the Graz conference which will be edited by H. Strasser, K. Knorr, and H. G. Zilian and published under the title *Beiträge zur Theorie der Wissenschaftssteuerung* in 1976. Some of those papers are included here; others are more specifically suited to the state of discussion in the German-speaking countries. The effort we have made by editing the papers is not an isolated one. There are several groups in Western Europe and the United States working on science development right now, one of them (PAREX) preparing a collection of case studies on social and cognitive factors in the emergence and growth of new disciplines (G. Lemaine, R. MacLeod, M. Mulkay, and P. Weingart, eds.).

One of the nice things about the topics we are dealing with is their reflexive character – a factor which added significantly to the spirit of the conference. The authors of the papers presented here had all the opportunity to engage in a 'two-pronged' strategy, that of doing sociology of science and that of at the same time applying it to themselves. We hope that part of the spirit of reflectedness has been preserved in this volume. Conscious role enactment on the basis of empirical results about the behavior of scientists is something intriguing and awkward at the same time. Quite naturally, there is a constant ambivalence between the (often unconscious) tendency to play the game and the conscious attempt to step out of it – if only so far as to prove one's sophistication. Participants at the conference admirably succeeded in coping with this aspect of presentation of self. We take the opportunity to thank them for their kind cooperation.

To Werner Leinfellner and Gerald L. Eberlein, who acted as our editorial advisors, we can only express our deep indebtedness. The editors also wish to thank Mrs. Hertha Firnberg, Head of the Ministry of Science and Research, for her interest she has shown in our activities and her steady encouragement; her representatives Rudolf Burger and Sigurd Höllinger for their unselfish efforts in promoting a successful conference; Generaldirektor Heinz Kienzl, Rupert Gmoser, the Österreichische Länderbank, the Zentralsparkasse of the City of Vienna, the City of Vienna, and the Land Steiermark for their financial help. Above all, they want to thank Director Gerhard Schwödiauer and the members of the Institute for Advanced Studies for their generous support, be it intellectual, financial or administrative. Ms. Ilse Baumgartner should be given credit for her conscientious work in helping to prepare the conference and for her multifaceted participation in the activities of the conference. Finally, we owe much gratitude to Mrs. Carolyn Tucker for her aid in producing the manuscript.

Vienna,
January 15, 1975

KARIN D. KNORR
HERMANN STRASSER
HANS GEORG ZILIAN

LIST OF CONTRIBUTORS

Gernot Böhme studied mathematics, physics, and philosophy, and has received his Dr. phil. and Privat-Dozent. He is currently a staff member of the Max-Planck-Institut zur Erforschung der Lebendsbedingungen der wissenschaftlich-technischen Welt, Starnberg, and teaches sociology of science at the University of Munich. He has published work in classical philosophy, philosophy of science, and sociology of science in learned journals such as *Zeitschrift für Soziologie*. Since 1971 he has worked mainly in the area of the science of science.

Nicholas Capaldi was educated as an undergraduate at the University of Pennsylvania and received his Ph.D. in philosophy from Columbia University. He has taught at Columbia University, the State University of New York, the University of Rochester, and since 1967 he has been at Queens College in the City University of New York. His numerous publications include the books *The Philosophy of Science* (1967), *The Enlightenment* (1967), and *Human Knowledge* (1969).

Lewis A. Coser studied at the Sorbonne and Columbia University where he received his Ph.D.. He is the author of *The Functions of Social Conflict* (1956), *Men of Ideas* (1965), *Continuities in the Study of Social Conflict* (1967), *Masters of Sociological Thought* (1970), and *Greedy Organizations* (1974). He is currently Distinguished Professor of Sociology at the State University of New York at Stony Brook, and President of the American Sociological Association.

Jerry Gaston holds a Ph. D. in sociology from Yale University. He is the author of *Originality and Competition in Science* (1973) and has published articles in scholarly journals in his areas of interest: the sociology of science and of education. In progress are *The Sociology of Science in Europe* (ed. with R. K. Merton and A. Podgorecki) and *The Reward System in British and American Science*. He is currently Associate Professor and Chairman of the Department of Sociology at Southern Illinois University, Carbondale.

H. Rom Harré obtained his Ph.D. from Oxford. He has taught at a number of universities and is presently Lecturer in Philosophy of Science at Oxford University and Fellow of Linacre College, Oxford. Among his publications are: *The Anticipation of Nature* (1965), *Principles of Scientific Thinking* (1970), *The Philosophies of Science* (1972), *The Explanation of Social Behavior* (with P. Secord; 1973), and *Causal Powers* (with E. Madden) shortly to be published. He is working on a study developing the social and structural side of ethogenics as a sequel to the 1973 publication.

Walter Hirsch studied at Columbia University and Northwestern where he received his Ph.D.. His publications include *The Sociology of Science* (ed. with B. Barber; 1962), *Explorations in Social Change* (ed. with G. K. Zollschan; 1964), *Scientists in American Society* (1968), and *Social Change: Explorations, Diagnoses and Conjectures* (ed. with G. K. Zollschan; 1975). He is currently Professor of Sociology at Purdue University and involved in research dealing with the gatekeeping process in scientific communication.

Karin D. Knorr studied sociology, linguistics, philosophy, and anthropology at the University of Vienna where she received her doctoral degree. After postgraduate studies in sociology at the Institute for Advanced Studies in Vienna, she became head of its project group on the development and organization of science. She is also a Lecturer in sociology at the University of Vienna and participates in an UNESCO project on the organization of research in the natural sciences. She is currently working on a major publication on the state of the social sciences in Austria and the nature of argumentative processes.

Tibor Machan earned his Ph.D. in philosophy at the University of California, Santa Barbara. His articles have appeared in such scholarly journals as *Educational Theory, Inquiry, Theory and Decision*, and *Philosophy of Science*. His monographs include *The Pseudo-Science of B. F. Skinner* (1974) and *Human Rights and Human Liberties* (1975). He currently teaches philosophy at the branch of the State University of New York in Fredonia, N.Y.

David Morrison graduated from Hull University with a degree in sociology. After a few years as a teacher, he studied at the University of Leicester and received an M.A. in the Sociology of Education and Mass Communications. He has recently completed his doctoral thesis on a biographical study of Paul Lazarsfeld as institutional innovator. He is currently engaged on a study of *The Social Basis of Middle Class Pressure Groups* (with Michael Tracey).

Nicholas C. Mullins did his graduate studies in sociology at Cornell and Harvard Universities where he also received his Ph. D.. He taught at Harvard, Dartmouth College, and Vanderbilt University and is currently Associate Professor of Sociology at Indiana University, Bloomington (with an associate appointment in History and Philosophy of Science). He is the author of *The Art of Theory* (1971), *Science: Some Sociological Perspectives* (1973), and *Theories and Theory Groups in Contemporary American Sociology* (1973), and of numerous articles in *American Sociological Review, Minerva, Science Studies,* etc..

Maria Nowakowska is a graduate from the University of Poznan and holds a Ph. D. in psychology from the Jagiellonian University, Krakow, Poland. She is presently Assistant Professor and head of the Laboratory of Psychology and Action Theory at the Institute of Organization and Management of the Polish Academy of Sciences in Warsaw. Her publication records include the books *Language of Motivation and Language of Actions* (1973), *Action Theory: An Attempt of Formalization* (1973), and *Quantitative Psychology with Elements of Meta-Science: Selected Methodological Problems* (1975).

Hilary Rose has taught sociology and social policy at the London School of Economics and has been Visiting Fellow at Essex University. Recently she was appointed to the Chair of Applied Social Studies at the University of Bradford. She is the author of *The Housing Problem* (1968), *Science and Society* (with S. Rose; 1969), and *Ideology in the Natural Sciences* (ed. with S. Rose; in press). Her main interests are in the areas of sociology of community action, of inequality, and of scientific knowledge.

David Silverman completed undergraduate work in sociology at the London School of Economics, obtained an M.A. at U.C.L.A. and a Ph.D. at L.S.E. He has also taught at these institutions and is currently Senior Lecturer in sociology at the University of London, Goldsmiths' College. His books include *The Theory of Organizations* (1970), *Reading Castaneda* (1975) *and Organizational Work* (1975). He is currently working in the area of ethnomethodology and semiology.

Hermann Strasser obtained a doctoral degree in economics from the University of Innsbruck and a Ph.D. in sociology from Fordham University. He has taught sociology at Fordham University, the University of Oklahoma, and the University of Vienna. He is now staff member of the Institute for Advanced Studies in Vienna. His publications include *Introduction to Theories of Social Change* (with S. C. Randall; 1974), *Sozialstrukturelle Bedingungen der Schul- und Lehrerbildung* (with F. Kast; 1974), and *The Normative Structure of Sociology* (1975). In progress are *Explaining Social Change: A Theoretical Introduction* (ed. with S. C. Randall) and *Die Idee der Entfremdung in der soziologischen Forschung* (ed. with M. Brenner). His current interests are in sociological theory and social stratification.

Robert I. Watson received his Ph.D. from Columbia University. He is presently Professor of Psychology at the University of New Hampshire, Durham, where he directs the Graduate Training Program in the History of Psychology. From 1965–1975 he was editor of the *Journal of the History of the Behavioral Sciences.* Among his numerous publications are: *Psychology as a Profession* (1954), *The Great Psychologists* (1963) and *Eminent Contributors to Psychology: Primary and Secondary Bibliographic References* (1974) as well as several papers on prescriptive theory. He is also Charter President of the Division of the History of Psychology, American Psychological Association.

Richard D. Whitley holds degrees in sociology and communications. He has been Research Fellow and Lecturer at the Manchester Business School (University of Manchester) since 1968. Besides conducting several studies of communication organization in scientific laboratories, he edited *Social Processes of Scientific Development* (1974) and published a dozen papers in the area of the sociology of science. He is also Convenor of the British Sociological Association's Sociology of Science Study Group.

Hans Georg Zilian studied law, political science and philosophy of science at the University of Graz, Austria. He did postgraduate work in sociology at the Institute for Advanced Studies in Vienna, where he subsequently became an Assistant Professor. There he co-directed a study on the state and institutionalization of social science disciplines in Austria. He is presently on a fellowship at Oxford University where he ventures to prepare a book on the philosophy of the social sciences.

SECTION A

AFTER THE KUHNIAN REVOLUTION: NEW TRENDS IN METASCIENCE

INTRODUCTION

The following five papers deal more or less with issues in the philosophy of science, but a good deal of what is said is said about philosophy of science; to a subset of the articles there is a distinct 'sociological' flavour. This heterogeneity reflects some recent developments in metascience: we shall try to characterize them only very cursorily and shall point out in greater detail some problems arising out of these developments.

Thomas S. Kuhn's *The Structure of Scientific Revolutions* was very much intended as an attack on classical philosophy of science, which we can imagine to be exemplified by books like Ernest Nagel's *The Structure of Science* – incidentally, Nagel's book was first published only one year before Kuhn's, namely in 1961. The attacked tradition of logical empiricism thus received a memorable summary (some, no doubt, would use the term 'tombstone') at the same time when Kuhn's book proved to be a revolution itself. For the sociologist of science, it is also great fun to note that *The Structure of Scientific Revolution* first appeared in *The International Encyclopaedia of Unified Science* whose editors – the very cream of logical empiricism – therefore proved themselves to be quite an open-minded group. Furthermore, the discussion with Kuhn was taken up on a larger scale only by the Popperians; this was probably because they had always given much attention to the problem of scientific growth, with which Kuhn was centrally concerned. Another reason may be that the Popperians at that time may have felt somewhat like medieval knights who enjoy fighting, but are bored by the species of dragons there are to fight against. Anyway, there was a discussion, which resulted in some books and many articles, most of which are referred to in the papers below.

It is to be hoped, and indeed, it can be expected, that someone will someday undertake a careful analysis of the Kuhn-Popper-Controversy which could throw light on some interesting features of the game of philosophical argumentation; I think one could observe, when this game was played by men like Lakatos, Feyerabend, Kuhn, etc., many instructive

Knorr, Strasser, and Zilian (eds.), Determinants and Controls of Scientific Development, 3–10.

moves (some of them highly original). Be that as it may, here we can only state that the controversy has been of the utmost importance for thinkers interested in metascience; of course, especially for the younger among them. And as may easily happen when people with a more or less different scientific and/or philosophical socialization earnestly confront each other, there have been some serious misunderstandings, which luckily enough went along with a sharpened awareness for the puzzles which have given rise to the misunderstandings. The following papers, in one way or another, reflect this state of affairs. We shall try to sketch some of these issues.

(1) Initially, Kuhn gave primarily an account of how physics has changed in its paradigmatic phase by revolutions of thought: whole world views were overthrown and connected theories and methodical tools abandoned. Part of his claims were of a factual nature and were also treated as that, by producing historical evidence to corroborate or reject his account. This alone wouldn't have created a problem for traditional philosophy of science. More disquieting was the 'world view'-part of it, which provided a link between fundamental epistemological and metaphysical presuppositions of a scientific community and the way it coped with nature by constructing theories and doing experiments. Here, on the one hand, seemed to be a careful and learned account of the history of physics which produced evidence for things like conventionalism, 'cognitive relativism', various assumptions (or results – sometimes the difference is hard to tell) in the sociology of knowledge, etc. Not surprisingly, Kuhn's analysis was most heartily accepted by social scientists. One could assume that the main reason for that lies in the theoretical situation within the social sciences: which is not all that good. The theories tend to be empty or inapplicable or just nonsensical; so it's not surprising that predictions either cannot be made, or, if made, rarely work out. An account of science where not theories are tested, but the individual scientist's ability at puzzle-solving – whatever that may mean for the social sciences – is well suited to soothe frustrations generated by the unwieldiness of social reality. Apart from that, one still has to wonder how historical findings can become relevant for fundamental philosophical issues. A problem of circularity arises at once: because if it is held that we arrive at our historical and sociological findings in the light of a certain epistemology, and if this should somehow influence the assessment of their validity, then of course any argument proffered with the help of these

findings and designed to demonstrate that different epochs or cultures have different world views is contaminated from the outset: the outcome of our investigation falls back on our argument. We are confronted with puzzles like these when we ignore the distinction between philosophical statements and statements of fact. So the very first thing many students have asked – and we assume they will go on doing so – when confronted with Popper's falsificationism is whether the criterion itself is falsifiable. We are confident that the usual strategy here is to point out that Popper is talking about empirical science, not doing it, and that it would therefore be nonsense to subject his statements to his own criteria. But this strategy is not viable if the distinction between empirical science and philosophy is challenged.

We see that Kuhn's attack on traditional philosophy of science necessitates exercises in the philosophy of philosophy. It would be very easy to simply declare that we should prefer a 'science of science' which does not take the old borderlines between the various fields too seriously; but that would probably be dangerous too.

(2) We already stated that Kuhn's main interest was focused on scientific change; he held that within mature sciences incommensurable paradigms succeed each other. This, of course, is anathema to the Popperian, because the very idea of 'incommensurability' implies that there is no way of comparing different paradigms, which rules out the possibility to speak sensibly of 'scientific progress' – and the latter has always been very near to the Popperian's heart. The matter of cognitive incommensurability is taken up in the Introduction to Section B of this book; what interests here is the fact that after Kuhn many people have come to the conclusion that either there are no criteria to judge ways of doing science, or that those vindicated by, say, the Popperians have to be supplemented (or even substituted!) by other criteria, for example moral ones. One fondly remembers the late Imre Lakatos' belligerent astonishment, when at some conference a brash young man had the audacity to venture that Aristotelian physics was in some sense 'just as good' as the Newtonian one. The disagreement was naturally about which criteria we should employ when evaluating research programs. The proffering of some such criteria by the philosopher of science is now by no means an arbitrary act. We see two logically distinct constraints on this activity. The first would be the following: The philosopher of science has to start from an object that is given

to him in the sense that science constitutes a system which is distinguishable from other cultural systems, e.g. religion, and which, as a matter of historical fact, has been distinguished. If a person were to say 'Well, religion is just a brand of science' one surely would be entitled to reply that the speaker did not understand the meaning of the term 'science' as legitimately as if he had said 'Science is a black and white striped elephant'. It is simply impossible to try to ignore the characteristic features of what has counted as 'science' for centuries.

Something different happens if some criterion is postulated, which is supposed to enable us to call, for example, one scientific theory 'better' than another; 'better' means here that the choice of one theory over another is recommended. These recommendations are not arbitrary, because one always can ask 'Why?'; i.e. require reasons for the choice. A justification will in this context hardly point out what the term 'scientific theory' means, but refer to the purpose of scientific theories, and in the last analysis to one's conception of what the proper aims of the undertaking 'science' are. 'Scientific progress' is then, of course, defined relative to those aims.

Considerations like these lie at the basis of the claims that philosophy of science is 'normative'. These one hears and reads so often in just this simple form, that we think it justifiable to make a few comments on the phrase 'philosophy of science is normative'. One not very farfetched objection is that it is probably not very informative thus to predicate something of the philosophy of science as a whole; because we can safely assume that within the field we find activities like descriptions or explications, of which it would be awkward or even self-contradictory to say that they are normative. The second problem is, of course, the word 'normative', which is an extremely difficult predicate. It tempts people to confound the philosopher of science with a sovereign who lays down the law for his subjects, which leads to the misunderstanding that there is no need for rational justification or criticism for a given brand of philosophy of science, just as a sovereign legislator need not worry about justification and criticism. It also must be pointed out that there are important differences between prescribing, issuing norms and describing norms, which cannot be accounted for by the fuzzy word 'normative'. Still, we have seen that within philosophy of science there are certainly to be found many prescriptions or recommendations. This again refers back to the relation between philosophy and history (and sociology) of science: it is obvious that it is no argument against a recommendation that no one follows it or has followed it so far. Only when we inquire into the reasonableness of a recommendation factual considerations play a role; but even then, the question whether people follow, have followed or are likely to follow it, is not necessarily among those considerations. All this is further complicated by the fact that the is-ought-dichotomy – which we presupposed – can itself be questioned. But, understandably enough, we cannot go into this now. What we can state is that a prescriptive conception of the philosophy of science and the creed that history and sociology of science is relevant to the philosopher, if expounded together, generate grave difficulties.

(3) Be that as it may: Kuhn's work has not only contributed towards shaking traditional philosophy of science out of its complacency, it has also kindled a vigorous interest on the part of metascientists to look at what scientists actually do or have done, and to do this in an empirical way as well. His work of 1962 had been compelling reading, but we do not think it unfair to consider *The Structure of Scientific Revolutions* rather a fascinating sketch than a set of wholly convincing arguments. So there were all his assumptions about the behavior of scientific communities which could be elaborated or made explicit and then put to an empirical test (although one can assume that these 'tests' never have been or can be as severe as some people would like them to be). The Popperians had dealt quite interestingly with the 'context of discovery' and relegated it to the realm of psychology, sociology and so on. This sounded fine enough (and may, despite Kuhn, on the whole still be philosophically sound) but the only problem was that while the philosophers were gloating over their 'context of justification', there were practically no psychologists or sociologists to take advantage of the generous concession. This resulted in the state of affairs where – if we consider the system of science in the broadest sense as a whole – more people were concerned with justifying what scientists were doing (or supposed to be doing) than were interested in what was actually going on in science. This has changed, and much of the credit is Kuhn's. Not only have social scientists taken increased interest in metascience, but also many philosophers, encouraged by the scope and success of Kuhn's thinking, have broadened their approach towards their object. The five papers gathered in this section testify to that.

In 'Prescriptive Theory and the Social Sciences', Robert I. Watson takes as the starting point of his discussion Kuhn's contention that the social sciences lack a paradigm and are hence preparadigmatic. Watson agrees with this and also with the alleged usefulness of the notion of 'part'-paradigm, which helps us understand such sciences as psychology. Now it is the case that within a paradigmatic science, certain aspects of reality can acceptably be dealt with only with the help of certain definite methods, employing agreed-upon theories etc. Within preparadigmatic sciences this is different, because subject matters of the field can be explained from points of view at conflict with each other. Watson proceeds to argue that within preparadigmatic sciences the behavior-guiding role of the paradigm is taken over by so-called prescriptions, which are attitudinal in character. These prescriptions are characterized for psychology by some twenty pairs of concepts, e.g. determinism-indeterminism, molecularism-molarism, quantitativism-qualitativism etc. Elements from these pairs form patterns, which constitute psychological schools. Watson elaborates his point by giving numerous examples from the history of psychology and by reporting on empirical data which have been gathered on the perception of psychological schools, typical clusters of prescriptions and the change of the latter over time. We could add that the emphasis on the prevalently attitudinal character of prescriptions seems quite important compared with the cognitional account Kuhn gives of the fundamental assumptions of a paradigm; whereas the latter are hardly ever questioned (and within normal science even cannot be questioned), prescriptions can be criticized (or justified). This accounts for the bitter feuds between schools on the one hand, and the almost morbid preoccupation of some social scientists with demonstrating the methodological excellence of their work, on the other hand.

Richard D. Whitley's paper, 'Components of Scientific Activities, Their Characteristics and Institutionalization in Specialties and Research Areas: A Framework for the Comparative Analysis of Scientific Development', starts from the observation that it is quite futile to try to give a cogent account of the development of science as a whole: It is more modest, but also more reasonable to analyse different sciences at different times. He therefore proposes a conceptual framework which isolates five components or aspects of scientific activity, which he deems suitable for describing and understanding scientific development: research practices, techniques, explanatory models, specialty concerns, metaphysics. Whitley holds that within specialties or disciplines the various components may be developed and articulated to a varying degree; high development of one or two components does not necessarily imply that all the other components are highly developed as well. Still, there are systematical restrictions on the possible relationships between degrees of articulation of the components, and these are thoroughly discussed, along with an analysis of the integration between, and the institutionalization of, the components, and their role in scientific innovations: differently developed components in different disciplines are receptive to innovation to a differing degree.

We believe that Whitley's emphasis on the need to differentiate various components of scientific activities when looking at the latter's development is warranted. Still, there is a possibility that his components, although analytically distinguishable, belong into different logical categories, which could create some formidable complications as far as the applicability of his analytical scheme is concerned.

A third paper critically concerned with Kuhn is Tibor Machan's 'On the Possibility of Objectivity and Moral Determinants in Scientific Change'. He begins with a version of the self-referential argument sketched above, by pointing out that Kuhn uses evidence from psychology and sociology which is, according to his own theory, subject to frequent reinterpretation – which leaves us with the odd case of a theory that says something about its own evidence; with the still odder consequence – we use here for convenience the much-eschewed term 'true' – that if the theory is true, there is something wrong with the evidence, and if the evidence is unobjectionable, then it seems to follow that the theory is false. By pointing out this and other philosophical weaknesses of Kuhn, Machan comes to a rejection of the latter's relativism. Machan further argues for a reconsideration of moral determinants in the development of science, on which to fall back seems more sensible than simply to say that a scientist's choice is a matter of 'faith'. Machan also holds that moral criticism is especially important in periods of crisis, i.e. when intra-field standards of evaluation are in turmoil. Machan's paper exemplifies a type of reaction against Kuhn where one tries to establish a certain amount of respectability for objectivity and rationality within science, without returning to the tenets of Logical Positivism.

Something similar holds true of Nicholas Capaldi's contribution, 'The

Moral Limits of Scientific Research: An Evolutionary Approach', at the center of which is the concept of rationality. Capaldi presents four views of science by sketching the model-science pertinent to the respective view and the conception of rationality, explanation, values etc. espoused within that view of science. Capaldi criticises the *a priori*-(Platonic)-view, that he calls 'Structuralist' (Aristotelian) and the Newtonian version of the 'Copernican' conception of science. He rejects all of those on several grounds, and proposes an evolutionary variant of the latter view of science. The evolutionary approach takes evolutionary biology as its model science. In the same spirit as the Newtonian version, knowledge, science and rationality are conceived to be the product of the interaction of men and the world: hence a description of science and rationality as mere correspondence to a structure independent of men is ruled out. Capaldi infers that this view controverts any distinction between facts and values and theory and practice. The evolutionary approach links 'rationality' to the notion of falsifiability, and the latter to the idea of the revision of basic premisses over time. The idea of 'objectivity' is reinstated and has – in this view – also applicability for value-systems. Capaldi bases on this analysis a policy of rational science control where competition between different theoretical proposals is considered to be both healthy and useful and where learning by trial and error is not a necessary evil but a consciously held value.

Finally, David Silverman, in his paper 'Preserving Science: Virtuosity as Virtue', contributes an account of modern science where 'virtuosity' has taken the place of 'virtue' in the sense that mastery of technique is now the touchstone of scientific success. Silverman deplores this state of affairs where the form of life of commodity exchange is represented through the embeddedness of scientific discourse in the grammar of market speech. He also refers to the controversy between Kuhnians and Popperians, which seems to be a debate located within the 'mathematical project' with its attempt to dominate language through technique. Science has become forgetful of its own human roots, thereby presenting only a special instance of the capitalist crisis, where 'virtuosi' of science are mere functionaries of a system of commodity-exchange and therefore as alienated as any worker who loses himself through his being treated merely as a factor of production. Silverman exhorts the scientist to recollect his 'virtue' by taking responsibility for the grounds of his speech.

ROBERT I. WATSON, SR.

PRESCRIPTIVE THEORY AND THE SOCIAL SCIENCES

According to the prospectus for this Conference, one facet of our concern is a theory of scientific development in terms of its conceptual and methodological presuppositions. I propose to emphasize prescriptions – one form of presuppositional attitudes which I consider to be an extremely significant factor in the dynamics of the history of psychology. Some thoughts also about the extension of this view to the other social sciences will be offered.

I participate as a working historian in psychology, not as a philosopher of science. I follow May Brodbeck (1953) in considering the philosophy of science as talking about science and not as a part of science. Since I am "talking about" history rather than "doing" history and since I am relatively unversed in the philosophy of science, it is inevitable that I draw upon personal experiences – my reactions to what I know about the philosophy of science and the philosophy of history. I will not go as far as B. F. Skinner (1959) who, when asked to give his theoretical position following a detailed outline conformed to by all other participants in a cooperative, many-volume critique of the current state of the science of psychology, responded by writing his autobiography! Nevertheless, there are certain consequences of my professional career that influence my view of my historical task. One direct consequence of being a psychologist is adherence in historical endeavors to methodological individualism.

Indeed, the prescriptive approach, the major focus of this paper, is part of a more general emphasis on methodological individualism – a position that calls for the concepts used to be interpretable in terms of characteristics of individual human beings, a position peculiarly fitting to a psychologist who is concerned with experience and behavior of *individuals* whether they be in isolation or in groups. Now, all historians must come to terms with the inescapable fact that they are concerned with individuals; but here, historians divide into those who see this as a trivial statement and those who consider it a basic one. I take it seriously but not so one-

Knorr, Strasser, and Zilian (eds.), Determinants and Controls of Scientific Development, 11–35.

sidedly that I would argue, as did Watkins (1953), that *all* concepts in social science are to be based on this doctrine.

In part, prescriptive theory was a reaction to evidence presented by Kuhn (1962) that some sciences possess a paradigm, i.e., a universally-accepted contentual model that guides the activities of the practitioners of that science for an appreciable time. Although he has said relatively little about it, he did indicate the existence of pre-paradigmatic sciences, either as historically earlier stages of what are now paradigmatic sciences or as contemporary sciences which have yet to achieve paradigms. What little he said about contemporary pre-paradigmatic sciences showed that he had psychology and other social sciences in mind. Although what is to follow most often refers specifically to psychology, you would be correct in presuming that I am inclined to believe that other pre-paradigmatic social sciences face the same issues. As I proceed, you may wish to ask yourself whether or not this is the case.

In the original paper on prescriptive theory published in 1967, it was contended that psychology lacks a paradigm as demonstrated by our vociferous failure to agree about fundamentals, the continued presence of schools of psychology despite their weakened appeal, and the presence of national differences which make it possible to speak of German psychology, Russian psychology, and the like.

It would seem that American sociologists, too, share a conviction that their field does not have a unitary, over-arching theoretical posture; or as I would put it, it lacks a paradigm. Thirty eminent sociologists, including Kingsley Davis, Everett Hughes, Daniel Katz, P. F. Larzarsfeld, Seymour Lipset, Talcott Parsons, David Riesman and Neil Smelser were queried on this point by Popovich (1966). Twenty-four took the position that sociology had no general, inclusive theory. Six believed that sociology had a prevailing view as expressed in the functional-structural approach. Moreover, on the issue of whether there was a close relationship between general theory and empirical research, only two thought the relationship to be very close while twenty-two thought it to be very loose.

Turning to the European scene, I imagine, from the point of view of a European scholar, that this lack of a universal contentual model can be most expeditiously recognized in the provincialism, massive though it may be, that he attributes to American psychology. Moreover, European psychologists, when speaking of their countrymens' work, share my view about the lack of a contentual model. The French historian of psychology, Reuchlin, in an article published in 1972 and entitled 'Psychology or psychologies?', characterized psychological activities as exhibiting increasing divergencies and contradictory viewpoints; while Graumann (1972), speaking as Past President of the German Society of Psychology, contended that psychology lacks the coherence of the exact sciences and went on to support a pluralistic view of the field.

Psychology and sociology, and perhaps the other social sciences, tend to avoid grand or large-scale theories, and, instead, seek small-range or middle-range theories. But the brute fact is that many of the old controversies which were integral aspects of the larger theories of the past have not been resolved; we pass over them in silence or are even ignorant that they are merely latent, not non-existent. But, in renunciating all-inclusive theory almost to substitute an untidy eclecticism, are we not also saying that we make no claim to have adherence to a broad enough position to be acceptable as an all-embracing paradigm?

One expansive tendency does seem to contradict this interpretation. Since my original paper (1967), there have been several articles, also stimulated by Kuhn's work, which argue that psychology *does* indeed possess a paradigm. However, what seems to have happened is that these individuals are arguing that present adherence to a limited paradigm is the forerunner of psychology's paradigm-to-come.

Limited paradigms do have their place in non-paradigmatic sciences. Kuhn, himself, distinguishes between paradigms as universally-accepted contentual models and 'part' paradigms, accepted for a time by specific groups within a particular science. Acceptance of contentual models of less than universal character does occur in psychology and other social sciences. The Skinnerian model, adopted by an appreciable number of American psychologists, is illustrative of a perhaps even more inclusive Behavioristic model which embraces it. A functional paradigm might be seen as acceptable by a considerable number of American sociologists; but the study by Popovich, referred to earlier, would show it is by no means universally accepted; hence, it is a 'part' paradigm.

Because a paradigm as a contentual rallying-point is lacking, the pre-paradigmatic stages of a science are those in which different subject matters of the field are either to be explained from widely different points of view or one remains content to work with what is acknowledged to be but a segment of that field. To embrace the scope of the field calls for a considerable degree of complexity and a plurality of themes or trends. These themes I have called prescriptions.

As aspects of the dynamics of the history of a science, prescriptions are orientative or attitudinal: "telling how a psychologist-scientist must or should behave. .... They help to direct the psychologist in the way he selects a problem, formulates it, and the way he carries it out" (Watson, 1967:437). And, to avoid becoming entangled with ephermeral themes, it is also essential to have evidence that interest in these particular themes had been maintained for appreciable lengths of time, preferably from the beginnings of the modern period. Therefore, the second distinguishing feature of a prescription is the maintenance of a trend over a considerable temporal period, despite changes in how and why it is manifested.

Far removed from the formal programmatic research design that is occasionally espoused in the textbooks, my search for these themes more resembled oil-prospecting than anything else. My first sounding, before any formalization of the theory, was taken on the occasion of a paper delivered at the Seventeenth International Congress of Psychology in 1963 at a symposium which had been organized to discuss national trends in psychology in several European countries and in the United States. To quote from my final paragraph concerning the contemporary scene:

> It has been seen that national trends in modern American psychology follow certain dominant prescriptions. Determinism, naturalism, physicalism and monism, although very much operative, are judged to incite relatively little opposition. Functionalism, operationalism, quantification, hypothetico-deductivism, environmentalism, and nomotheticism are likewise dominant, but there are counter-prescriptions which tend to oppose them.... Serving as counter-prescriptions to those dominant in psychology are those calling for increased complexity in theorizing, for an increased attention to philosophical matters, for general acceptance of phenomenology, for increased attention to existential psychology and in a somewhat amorphous way almost all of the areas of personality theory calls for counterprescriptions of one sort or another. (Watson, 1965: 137.)

It might be added parenthetically that I later saw that some of these just quoted trends, operationalism and physicalism for example, were short-time attitudinal expression of changes in thinking and better incorporated in more general prescriptions of longer duration.

I then began another book on the history of psychology beginning with the seventeenth century and based on prescriptive theory. As I wrote, I found I could organize my thinking in prescriptive terms when dealing with the content of the thinking by Descartes, Locke, Leibniz, Hume and others. This book is still incomplete, and the only related publication is a published paper devoted to Descartes (Watson, 1971). In effect, prescriptive theory at this point consisted of finding the approach meaningful for two temporal periods: the seventeenth century and contemporary psychology. I found the same prescriptions appearing in relevant and significant fashion in the thinking of individuals separated by centuries. I was therefore emboldened to consider that prescriptions, in addition to a guidance function, could be characterized as persisting over long periods of time – despite many changes in detail and context – and that they thus fulfilled the dictionary connotations of the term: both something of a directive character and something hallowed by long usage.

In this search for a framework of themes for the history of psychology, I arrived at the list, given in Table I, which summarized prescriptions, that are at present isolated, in terms of contrasting parts. Although there is a heuristic value to presentation in terms of bipolar opposites since each extreme helps to explain the other – what one pole is, the other is not – it is necessary to immediately dispel any belief that this is the only or even the major way they are related one to another. All I hold on this point is that at some temporal period they were seen as opposed. Prescriptions also show many other relationships. As they change at some point in time, both of the 'opposing' pairs may be accepted according to circumstances, as would be inductivism and deductivism in contemporary research according to whether one was proceeding from observations-empirical generalizations to theories or from theories to hypotheses derived therefrom; synthesized, as is rationalism-empiricism in logical empiricism; ignored, as are supernaturalism or vitalism in large measure today; conceptualized as degrees of difference, as in quantitativism-qualitativism; or practiced in different settings, as are purism and utilitarianism. Other relationships also hold. Some are implicit, for example, as was the molecular constancy hypothesis which was unidentified until the Gestalt psychologists, the critics of the older structural view, brought it into being. Sometimes it will be possible to speak of one or another prescription as being dominant, or of others as being counter-dominant but viable, as I did in the quotation from the earlier paper. Other usages, especially of patterns of prescriptions, will become clearer as I proceed.

I find myself in agreement with Skinner about the primacy of attitude.

TABLE I

The prescriptions of psychology arranged in contrasting pairs

Conscious mentalism-Unconscious mentalism (emphasis on awareness of mental structure or activity – unawareness)

Contentual objectivism-Contentual subjectivism (psychological data viewed as behavior of individual – as mental structure or activity of individual)

Determinism-Indeterminism (human events completely explicable in terms of antecedents – not completely so explicable)

Empiricism-Rationalism (major, if not exclusive source of knowledge is experience – is reason)

Functionalism-Structuralism (psychological categories are activities – are contents)

Inductivism-Deductivism (investigations begun with facts or observations – with assumed established truths)

Mechanism-Vitalism (activities of living beings completely explicable by physiochemical constituents – not so explicable)

Methodological objectivism-Methodological subjectivism (use of methods open to verification by another competent observer – not so open)

Molecularism-Molarism (psychological data most aptly described in terms of relatively small units – relatively large units)

Monism-Dualism (fundamental principle or entity in universe is of one kind – is of two kinds, mind and matter)

Naturalism-Supernaturalism (nature requires for its operation and explanation only principles found within it – requires transcendent guidance as well)

Nomotheticism-Idiographicism (emphasis upon discovering general laws – upon explaining particular events or individuals)

Peripheralism-Centralism (stress upon psychological events taking place at periphery of body – within the body)

Purism-Utilitarianism (seeking of knowledge for its own sake – for its usefulness in other activities)

Quantitativism-Qualitativism (stress upon knowledge which is countable or measurable – upon that which is different in kind or essence)

Rationalism-Irrationalism (emphasis upon data supposed to follow dictates of good sense and intellect – intrusion or domination of emotive and conative factors upon intellectual processes)

Staticism-Developmentalism (emphasis upon cross-sectional view – upon changes with time)

Staticism-Dynamicism (emphasis upon enduring aspects – upon change and factors making for change)

To quote, "Science is first of all a set of attitudes. It is a disposition to deal with the facts" (Skinner, 1953:12). However, I lack his assurance as to what precisely are the 'facts' because I do not adhere to the admirably consistent and explicit part paradigm that guides him in his choice of truth values. While accepting attitudes as primary, to me they are prescriptive in nature. In terms of one of the themes of the Conference and central to this paper, prescriptive attitudes are to be presented both as a unit of change and as a cause of change in the history of psychological science.

Prescriptive attitudes are a cause of change because prescriptions are motivational in character. Conception of prescriptions as attitudes make it possible to characterize them as, in keeping with contemporary usage, favorable or unfavorable. Prescriptive allegiance or rejection is not often a bloodless, neutral affair. Emotion, feeling and wishful thinking enter. A behavioral scientist, for example, who has a favorable attitude toward quantification of data and the use of objective means, views unfavorably investigations that do not have these characteristics. His attitudes may extend to the persons and the materials which he judges in these terms: Professor Smith (a humanist) whom he views unfavorably; a computer (a tool of quantification) that he likes to use whenever possible; and Rorschach cards (dubiously quantifiable) which he views with mixed feelings.

Attitudes, however, are not confined to affective responses; they have a considerable cognitive component as well. A psychologist may not only prefer to use quantitative approaches or to consider his task as a psychologist to be an empirical one, but he also has a considerable cognitive component from which he may defend the essential rightness of his attitudes. As for latent attitudes, there at first may be no cognitive component. But when a cherished, heretofore unverbalized attitude is brought to his attention, this gap is soon filled with either 'wishful' or 'real' reasons for the belief.

Generally when one refers to theory, one has references to a principle supported by data from the content of the discipline under discussion. Disregarding the data content for a moment, I submit that some important and remaining aspects of the theories that are espoused are colored by prescriptive attitudes toward them. For example, having somewhat less a tendency in this direction, it seems to me that some colleagues, when responding to a problem, a behavior, or an experience that strikes them as worthy of scrutiny, immediately turn to the question of how it can be quantified. They do not, mind you, stop to ask if it is suitable for quantification or whether it might be premature or even impossible to do so. Adherence to a quantitative prescription is probably the basis for the tremendous appeal that operationalism has had for American psychologist: What you measure *is* it meaning, which you must admit is rather neat – if you can do it. When one holds this view, one *perceives* the measures used differently.

Negative attitudes find prescriptive expression through what individuals are against. The phenomenologist, Buytendijk (1957), nearly twenty years ago wrote a short paper, with a still-contemporary ring, in which he challenged the overwhelming concern in experimental psychology with objectivity and causality. He advocated as being more appropriate what he referred to as the manipulation of personality theory from models and schemata; to reach the objective of understanding not in the laboratory but in meaningful situations. What I have called methodological and contentual objectivity, determinism and molecularism, are being cirticized. Even from his short critique, one may hazard that Buytendijk supports at least a contentual subjectivity, some form of indeterminism, and a molaristic approach to psychological problems. What he affirmed may be said to involve some aspects of a phenomenological approach.

The relation of a pattern of prescriptive attitudes to a particular theory helps us to understand incongruous or non-salient elements in a theory and, thus, helps to distinguish between prescriptions and theories other than on the basis of content. Despite the possibility of their being confused, a pattern of prescriptions is not a theory. Within the history of psychology there has been more than one instance where a particular pattern of prescriptive allegiances are formed by an individual and, once brought together around contentual problems have become inseparable – despite any necessity for them to be related – and are seen in historical perspective as a 'theory'. A personal pattern of prescriptions does not a theory make.

Early behaviorism in the hands of John B. Watson and others, can be characterized as being militantly objective in its rejection of the mind as subject matter for psychology, empirical in espousing learning as the core problem, and violently non-hereditarian. This pattern, or some approximation of it, is often referred to as behavioristic *theory*. There is confusion here: they are not a logically coherent set of propositions; their juxtaposition is derived from Watson's pattern of prescriptive allegiances. They were simultaneously present in the thinking and writing of Watson; but if the objectivity he espoused is to be considered salient to behaviorism, it does not follow that behaviorism must also be anti-hereditarian. A behaviorism with an allegiance to a hereditarian prescription would still be behavioristic.

Since attitudes are means of expressing prescriptions and *vice versa*, it follows that one can appeal to role theory for their elucidation. In society, maintenance of groups depends to some degree upon the functioning of the majority of the participants toward shared goals, with only some individuals performing certain tasks (such as those performed by social scientists). There are implicit and explicit expectations held about other members of a particular group. In the setting of the present discussion, one or another of the social sciences is that group. Not only members of society in general but also fellow social scientists and the individual social scientist expect members of his and other social sciences to behave and think in certain ways. These expectancies are attitudes. It would seem that prescriptive-role theory could be used in demarcating science from non-science. Psychologists, sociologists and other social scientists are members of social groups; there are certain expectations held by each kind of social scientist, as well as by others, by virtue of their respective roles. For example, social scientists would reject as being foreign to their role expectancies that among their tasks was the writing of fiction about society – social-science fiction as it were. A behavior to be expected when one is involved in the role of a novelist, is rejected when one is performing in the role of a social scientist.

Analysis in terms of roles is commonplace in social psychology and sociology with, for example, Parsons and Shils (1951) referring to the social system, at least in one sense, as a variety of roles. Nor has it escaped attention that among the roles in the social system are those that are identified with the social sciences themselves. Indeed, much of the work in social psychology and sociology of science is cast in terms of roles shared by members of a community. Ben-David and Collins (1966) in their study of social factors in the origin of psychology in Germany, France and England, argue that a new role is necessary for a new scientific community to come into existence. Prescriptive theory would ask what were the prescriptive attitudes the members associated with their emerging roles. Tacitly, what I have just been saying answers one question raised by the organizing committee: sociology of science *is* relevant to the historian. I leave to others to define more precisely the nature of this relationship between sociology of science and history; but I would add that the bridge between the two, the mutual use of role concepts, suggests that it would be worthwhile to also look for cross relationships through those concepts from each which might be shared with the other.

Although it would be interesting to examine the literature dealing with psychologists' conceptions of their roles, I will content myself with asking you to picture what psychologists would say when asked to recount their attitudes toward themselves as psychologists. Is it not plausible that their accounts would touch upon at least some matters that might be subsumed under prescriptive rubrics? 'I work with people for their own sake, not to find generalized laws of human nature.' Is it not plausible to think that idiographic and utilitarian attitudes are being expressed?

The content of research must often be distinguished from the attitudes which it, relevantly or irrelevantly, reflects. A typical content of research may be the rate of verbal learning by timed practice trials, with the score the number of trials to learn to a specified level of proficiency. Prescriptive attitudes adhered to by the psychologist carrying out the study may include quantitativism, empiricism, contentual objectivism and methodological objectivism but with no direct reference being made to any of the terms or to any of the concepts. By the very nature of the study, there is dependence upon knowledge that is quantifiable; qualitative findings, if noted at all, are relegated to the debriefing session or to an occasional note. The perhaps implicit assumption is made that it is experience, i.e., practice, that is important, while nativistic possibilities of influence are tacitly ignored. (Nativism is considered to be a historical outgrowth of rationalism after an intermediate step of adherence to innate ideas espoused by rationalists.) Contentual objectivism, the behavior of the individual, is that which is being observed with no attempt being made to infer what is experienced. Methodological objectivism is expressed through using a method open to repetition by other competent research men.

Prescriptions may have been contentual issues in philosophy or in earlier psychology: say, a defence of empiricism against rationalism by Locke or the explicit advocacy of contentual objectivism by John B. Watson. There are also theoretical discussions by contemporary psychologists directly on these topics or on their modern counterparts; but this is but a small fraction of the total productivity of psychologists. This larger productivity does not deal with prescriptions as content but, instead, reflects them attitudinally.

Aside from the semantic shock that occurs when a contemporary psychologist is faced with the contention that his thinking is still influenced by such antiquarian entities as dualism or rationalism, it is not considered a disadvantage that many of the trends concern what is or was contentually aspects of philosophical study. On the contrary, a field that grew out of some of the contentual problems of philosophy should be expected to show its influence.

Kuhn (1962) has held that scientific revolutions depend upon a shift in the fundamental image scientists have of their *subject matter* paradigm.[1] A psychologist lacks this unified subject-matter contentual image; and therefore, he has to fall back in part upon his image of himself as a particular kind of scientific agent. Prescriptions then, concern roles.

Since prescriptive theory involves roles, it makes possible the utilization of the subconstructs usually discussed in connection with roles such as role expectation and role perception. It would not be too difficult to devise methods to study how a particular kind of social scientist holds certain prescriptive expectations about the nature of his role, to investigate what he pays attentions to, to what behaviors he emits qua social scientist.

Role perception of another person is also germane to the task of the historian-sociologist of science. Here he asks the question, what role or roles does a particular individual or a particular identifiable group take? This has been investigated for characteristic groupings of psychologists – the schools of psychology.

A study by Fuchs and Kawash (1974) inquired into patterns of prescriptions adhered to by the various schools of psychology. They took as their point of departure a definition I had suggested for schools of psychology as "a set of interlocking prescriptions espoused by a group of scientists generally with an acknowledged leader" (1967:441).

Seven-point scales to indicate ratings ranging from 'no' (1) to 'great' (7) significance, were prepared for each extreme of the bipolar arrangement resulting in a doubling of the prescriptive terms to 36. All members of the Division of the History of Psychology of the American Psychological Association who replied to their questionnaire by a certain date, numbering 68, rated the degree of significance they perceived the adherents of the five schools would attach to each prescription. Aided by a glossary of the meanings of the prescriptive terms, structuralism, functionalism, behaviorism, Gestalt and psychoanalysis were so rated. Despite the many obvious deficiencies of this exploratory study, the results have some interest.

In order to reduce the amount of data given in Table II, results are reported only on those twenty-five prescriptions which would allow each school to have at least three significant prescriptions available for presentation. To do this, it was found that the lowest mean rating selected for further discussion to be 5.50. Since even in this lowest case the standard deviation is about 1.10, practically all individuals rated adherents of the schools as considering the prescription as significant, i.e., above the midpoint of four on the seven-point scale. For higher mean ratings, say of 6.5, practically everyone had rated their estimation of adherence as 6 or 7 and hence of "great" significance to the particular school in question. Significant precriptions for behaviorism, the school offering the largest array of prescriptive allegiances, were found to be adherence to a contentual and a methodological objectivism, determinism, empiricism, inductivism, mechanism, molecularism, monism, naturalism, nomotheticism, peripheralism, and quantitativism. Psychoanalysis showed high and unique positive emphasis on developmentalism, dynamicism, idiographicism, irrationalism and unconscious mentalism; and shared with one or the other schools, emphasis on centralism, contentual subjectivism and determinism. Its idiosyncratic pattern of prescriptive adherences confirms the view that it is more isolated within the psychological field than are the other schools under discussion. Functionalism, as might be expected, showed high ratings on functionalism, along with the relatively common empiricism and naturalism, but also displayed a positive utilitarian attitude. Gestalt shares with psychoanalysis an emphasis on centralism, but most emphatically was seen as emphasizing molarism. It was the Gestalt psychology rating of 5.50 for nomotheticism that supplied the lowest rating singled out for attention in the decision to draw the line so that at least three prescriptions for this school were mentioned. It follows that it was the school that seemed to supply the least in the way of pronounced peaks of adherence. It was followed by functionalism which was considered to emphasize four prescriptions. Structuralism, in common with other

TABLE II
Average rater judgments on five schools of psychology as found by Fuchs and Kawash (1974)

Scale	Schools of psychology				
	Behav-iorism	Function-alism	Gestalt psychology	Psycho-analysis	Structur-alism
Centralism	3.07	4.35	5.75	5.82	5.25
Conscious mentalism	1.73	4.60	5.07	3.68	6.41
Contentual objectivism	6.43	4.82	3.43	2.78	2.65
Contentual subjectivism	1.40	4.13	5.10	5.93	5.90
Determinism	6.66	4.93	4.72	6.15	5.37
Developmentalism	4.75	5.19	3.91	5.87	2.01
Dynamicism	3.85	4.94	4.85	6.01	2.18
Empiricism	6.28	5.63	4.57	4.12	5.81
Functionalism	4.84	6.84	4.13	4.41	2.07
Idiographicism	2.37	3.40	3.56	5.97	2.19
Inductivism	5.91	5.41	4.68	4.18	5.43
Irrationalism	2.58	3.03	3.09	6.48	2.13
Mechanism	6.33	4.06	3.56	3.68	4.06
Methodological objectivism	6.85	5.01	4.50	2.50	3.87
Molarism	2.82	4.54	6.44	5.19	2.43
Molecularism	5.73	3.41	1.79	2.60	5.96
Monism	6.22	3.50	3.88	3.56	2.87
Naturalism	6.51	5.63	5.35	5.19	5.48
Nomotheticism	6.46	5.40	5.50	4.57	5.98
Peripheralism	5.58	3.93	2.68	1.98	3.24
Purism	4.46	3.18	4.94	2.91	5.81
Quantitativism	6.28	4.62	3.34	1.68	4.91
Structuralism	1.86	2.44	3.34	3.88	6.90
Unconscious mentalism	1.63	2.82	2.85	6.81	2.13
Utilitarianism	5.06	5.74	3.22	5.28	2.37

(Standard deviations of those singled out for discussion average 1.10.)

schools, they saw as showing emphasis on empiricism, naturalism and nomotheticism, but also shown was high acceptance for conscious mentalism, purism, contentual subjectivism, and highest of all, structuralism at 6.90. All of the latter, except contentual subjectivism, are unique to this particular school. Aside from psychoanalysis, it appeared to be the most unique school in terms of prescriptive allegiances.

The results of this study also provide something akin to divergent validity. If one considers the eleven prescriptions which have not yet appeared in discussion and were omitted from the Table because they did not qualify to appear, the evaluators are saying that less or non-relevant to the schools of psychology, at least in the immediate past in the United States, are prescriptive adherences to deductivism, dualism, indeterminism, methodological subjectivism, qualitativism, rationalism, staticism, supernaturalism and vitalism. The lesser relevance accorded then seems appropriate, although the relative irrelevance accorded deductivism may be somewhat harder to account for.

Is this a study of stereotypes or of role perception? There may be few surprises and our prejudices may seem to have been confirmed by the results, but I prefer to think of this as a study of role perception. It is *not* a monolithic characterization of the schools, and, hence, does not have this characteristic of a stereotype. Instead, considerable complexity of characterization emerges from the patterns. Moreover, this study is independent in the sense that it was not carried on by one of my students and it does confirm that prescriptive dimensions may be illuminating concerning views held about the various schools of psychology.

In hearing of such guiding concepts as rationalism, dualism, mechanism, and naturalism, many of you will be reminded that these and the other themes are the very fabric of the approach known loosely as the 'history of ideas.' Again, as a methodological individualist, I find the history of ideas, as usually expressed, to be an unnecessarily disembodied reflection of our intellectual heritage from Plato, Descartes and Kant. Ideas are too often treated as having a ghostlike life of their own with ideas propogating, contradicting, or supporting other ideas as if we had explained them when we had sorted out their chronology. Attitudes, however, have both a firmer personal reference and a more compelling character than do ideas. To be sure, there are ideas of reference as well as compulsive ideas; but it is precisely the point that to convey this, modifying terms must be employed. From the perspective of prescriptive theory, the history of ideas is a rich reservoir from which to draw materials for studying change in history.

Direct from the mythology of science is the notion that the scientist is a disinterested observer who, without emotion, allows the results of his findings or those of other scientific workers, to decide the truth. Enough has already been said to show that I do not subscribe to this view. Rather, in their search for truths, scientists *try* to be as impersonal and knowledgeable about their motives as human nature permits. In a Hobbesean vein, one could even say that our instruments, our statistical and experimental controls, are necessary because we are venal or stupid or both. We are so unable to grasp the meaning of raw data, so open to wishful thinking, that we introduce external controls and conventions of statistical significance to protect us – we hope – from errors of interpretation. Investigation of prescriptive attitudes enters as part of this pattern of controls. Dualism as an idea may be neutral; one's attitudes toward it are not!

It is appropriate at this juncture to remind ourselves that the historian as a social scientist is human too! He, too, needs the sobering effect of methodological controls. He, too, must allow his cherished beliefs, his fond anticipations, to be exposed to the possibility of contrary evidence. Thus, with this in mind, I am going to address myself to methodology pertinent to the study of prescriptions *and* particularly to the historian as social scientist.

My somewhat crude but workable distinction between functions of a psychologist *per se* and those of a psychologist-historian, is that in the latter role I am confined primarily to interacting with documents, not persons.

In all historical writing there is the inevitable selection that the historian makes among the available documents. History as a humanistic discipline with the narrative depicting not only the contents of the documents but the selection that the skills of the historian emphasize and otherwise make from them, their organization by the historian, his ability to penetrate into them, are crucial to the art of writing history. I value the artistic aspects of historical writing. I would not wittingly denigrate or destroy them. As I see it, my task is to walk that narrow line between falling victim to my enthusiastic intepretation of the historical material and committing the equally excessive error of distorting the material by overobjectification to the point that idiosyncratic light and shadow, depth, perspective and feeling have been squeezed from the historical material: 'The goal… is rigor, but not rigor that distorts'.

One of the tendencies that marks the historian as human is that he has to resist projecting into these documents what he would like to find in them. Some historians are recognized by their peers as more successful in resisting projection than others. We refer to them as 'fair', 'judicious', and 'well-balanced'. At a collective level, criticism of a particular brand of historical writing as unfair takes such the forms of referring to the 'Whig theory of history'. Aspiring to fairness, I would supplement, not supplant, this art with objective techniques.

Potentially at least, a document may have both an overt and latent content – what is actually said, and what is implied. To search for implicit meanings, a task on which all historians embark whether they have made this explicit to themselves or not, is not to be undertaken lightly. What I have just stated about documents should be placed in a larger context no matter how one 'does' history.

It would be naive for historians to rest content with what men of the past have had to say explicitly about a particular problem. One does not have to accept the existence of an unconscious of the dynamic psychologies to appreciate the possible correlative functioning of unverbalized, implicit assumptions in historical events. Whitehead (1925) has remarked that, when one is examining the philosophy of a period and, by a not unreasonable extension of this view, the history or a problem areas in a science, one should not direct attention primarily to those statements that have been made, but rather to search for the assumptions which have remained unstated. Unverbalized presuppositions may have seemed so obvious to those of the period as not to have required statement. Prescriptive theory in which the attitudes being expressed are not necessarily overt and consciously held may be helpful in this regard.

Prescriptions conceived as attitudes renders relatively easy and straightforward the transition from theoretical discussion to application of content analysis. Commonly-accepted definitions of this technique often refer to it as being a means of assessing the relative extent to which specified *attitudes* permeate a given set of documents.

One level of application is informal or qualitative. Quantification is not attempted. Instead, the historian examines the pertinent documents while continuously asking himself whether the material can be interpreted in prescriptive terms. An informal application of prescriptive theory was used by Ronald Mueller in his doctoral dissertation (1974) devoted to James Mark Baldwin and his era at the turn of the century in the United States. One phase of this study concerned Baldwin's three major books: *Mental Development, Social and Ethical Interpretations* and *Development and Evolution.* Familiar with prescriptive theory, Mueller examined the contents of these books and found developmentalism and functionalism to dominate. Developmentalism is manifested in Baldwin's genetic method, his dialectic of personal growth and his adherence to recapitulation theory. Functionalism was Mueller's means of integrating Baldwin's theory of imitation, his views on suggestion, habit and accommodation, organic selection and social heredity. As Mueller used prescriptive theory, quantitative findings were not sought; rather, the theory helped him to integrate the material meaningfully.

To serve as an introduction to quantitative research with more formal application of this approach, I turn to an equally-straightforward application of prescriptions. Prescriptions, as categories for content analysis to the seventy-five Presidential addresses of the American Psychological Association between 1892 and 1970, were chosen by Kenneth R. Gibson (1972) as indicators of change in the conceptual bases of American psychology. These addresses were believed to be important documents because of the eminence of those who had delivered them, their presumed sensitivity to the then current issues, and their continuity from 1892 to the current time. Three raters, using prescriptions as content categories, made a blind analysis of their contents. Although ratings were made for neutral stands and for rejections, 78 percent of the codings were for acceptance. These acceptances form the basis for what further discussion time considerations permit.

One of the characteristic struggles through which psychology passed during the years encompassed was the shift from psychology as the science of the mind to psychology as the science of behavior. This had most direct expression in contentual objectivism-contentual subjectivism. In the decades 1891 to 1900, 1901 to 1910, acceptance of contentual subjectivism dominated, appearing in 85 or 90 percent of the papers; contentual objectivism was accepted in less than 20 percent. In the decade 1911 to 1920, that of John B. Watson's famous manifesto of 1913, acceptance of *both* was characteristic with 70 to 80 percent of the papers paying explicit, positive attention to them. The decade 1921 to 1930 actually showed a puzzling drop in contentual objectivism to 40 percent, while contentual subjectivism almost held its own. Nevertheless, there was a steady drop in contentual subjectivism although it never fell below 40 percent right up to the last decade. This degree of acceptance seems to contradict the oft-repeated generalization that American psychologists had little use for subjective phenomena. Nevertheless, contentual objectivism did dominate from 1931 on.

The findings are even more dramatic when *methodological* objectivism and subjectivism are considered. In the first two decades, acceptance of both was about equal, ranging between 35 and 45 percent. In 1911 to 1920, methodological objectivism surged forward to 80-percent acceptance, while subjectivism dropped to 30 percent. This drop in adherence to methodological subjectivism continued until it hit a low point of 5 percent in 1931. For the immediate past three decades, there has been a slight resurgence of acceptance of methodological subjectivism, but never more than acceptance by 10 percent. From 1911 on, methodological objectivism has clearly dominated.

As an indication of contemporary American psychology, using the last two decades from 1951 to 1970, Gibson found empiricism to be the most generally-accepted prescription. While methodological objectivism, quantitativism and nomotheticism are very widely accepted, methodological subjectivism, qualitativism and idiographicism are sufficiently viable to be counter-dominant. More or less equally and widely accepted, are both deductivism and inductivism, purism and utilitarianism, and molarism and molecularism. Almost disappearing from expression on the current scene were supernaturalism, irrationalism, indeterminism, and monism and dualism. It would seem that prescriptive attitudes can be used as a unit of scientific change.

It is one of the myths of our history that psychology emerged from the matrix of philosophy as late as the nineteenth century with previous consideration of psychological matter being essentially philosophical or prescientific in character. Barbara Ross (1970) in her doctoral dissertation questioned this and turned for evidence to the pages of the pioneer scientific journal, *The Philosophical Transactions*, from its founding in 1665 to 1700, despite omission in its policy statement of any concern with psychological or social phenomena.

Distortion is inevitable when one tries to summarize in two or three pages a dissertation that totals 520 pages. What will be most conspicuous by its absence in what follows, will be the narrative itself which fleshes out the bare outlines of the findings to be mentioned here. Her method otherwise was content analysis, using the services of two raters other than herself, with the content categories being the prescriptive variables. The raters were given practice in understanding the directions through a manual and pilot study on materials from the *Philosophical Collections* which had been published for the years 1679 to 1682 when the *Transactions* had lapsed publication. Adequate reliability being established (summarized as over 0.80), she classified all the articles in the *Transactions* (from 1665–1700) and found that 367 of 1,924 papers, or 19 percent, were relevant to the study of man.

To arrive at these papers, she used twelve major content categories of indices of interest in man, each with more specific subcategories, which combined to give the 367 papers relevant to man. The six largest of these categories were, successively, physiology, sensation, social phenomena, physiology, perception, abnormal and comparative phenomena. About 30 percent of these papers made reference to social phenomena, more specifically identified by its subcategories: social habits, language, value systems, somatic characteristics of various peoples, their intellectual systems, aspects of their material culture, interpersonal behavior, sub- and inter-cultural behavior. The over-all results supported her first hypothesis that men in the latter half of the seventeenth century sought reliable scientific knowledge about man, his experiences, and his behavior.

These 367 relevant papers were those to which the prescriptive categories were then applied. In prescriptive terms, the second hypothesis was that the prevailing scientific attitude of the late seventeenth century reflected methodological objectivism, probablistic determinism, nomotheticism and quantitativism. The third hypothesis, more specific to psychology, was that a behavioral view of man predominated in the scientific literature. This involved these same prescriptions and, in addition, mechanism, *new* empiricism and *new* rationalism.

Instead of talking about the overwhelming support for these hypotheses that the data engendered, let me discuss two points which, at first glance, might seem to be either discrepancies between her hypotheses and her findings or a weakening of prescriptive theory itself. Prescriptions other than those hypothesized were found to be present to a statistically significant degree since the entire array of prescriptions were studied. Non-predicted emphases in over 70 percent of the papers were found in acceptance of contentual objectivism, an inductive approach, and a naturalistic emphasis. Ross might well have demonstrated the presence of the predicted emphases by the usual narrative, qualitative approach; but then she would not have been forced, as she was, to use the entire repertoire of prescriptions to consider non-hypothesized prescriptions that emerged in a significant way. This safeguard in historical document analysis reduces the subjectivity of interpretation that is a characteristic of historical study. The unpredicted findings became a challenge to broaden and modify her view which, without being able to go into further detail, as I see it, she did successfully.

What appeared to be novel prescriptions, i.e., 'probablistic' determinism and the 'new' rationalism and 'new' empiricism as she called them, emerged in her preliminary analysis of the factual data. In her opinion the data called for introducing them as prescriptive categories while still investigating 'old' rationalism and emprircism as defined in the table. To describe only one of these, new empiricism, while agreeing that the source of knowledge is experience, also was concerned with, as she put it, "what was really there, not only what seemed to be on the basis of everyday experience with careful observation and accurate recording" (1970:328).

Locke, for example, would in this sense be a rationalist despite his vaunted empiricism. Her results bore out her contentions. The raters coded only 19.3 and 11.7 of the papers in terms of acceptance of the old, but 65.1 and 66.8 in terms of the new empiricism and rationalism, respectively. It is still possible, however, that if the new categories had been omitted, the raters would have coded them in the original categories.

These results illustrate not only the kind of modification or expansion that prescriptive theory permits, but also that more subtle meanings can be sought with adequate, advanced planning. It is intriguing, for example, that there has been found within empiricism and rationalism in the seventeenth century what Ross referred to as "rational considerations of the empirical data" (1970:320). This can only be seen as a beginning of synthesis of rationalism and empiricism to which modern logical empiricism was to be a culmination.

Prescriptive theory, it is being said, is capable of modification and refinement when a particular combination of historical circumstances demand it, thus partaking of the advantage often attributed to models of discarding or modifying, if the findings seem to so indicate without thereby falsifying the theory.

The relation of prescriptive theory to the other social sciences demands consideration. Prescriptive theory, itself, is a cross-discipline venture since the constructs, attitude and role, are at a point of interaction of the disciplines of psychology and sociology. The relation, however, of prescriptive theory to the social sciences has not been explored heretofore, and what I have to say is no more than illustrative and very tentative.

To me, the most convincing line of evidence in support of the supposition that prescriptive theory is relevant to the other social sciences is to be found in our common heritage. Before the emergence of the specializations that are the social sciences, each of them was more or less discernable strands in the history of philosophy. Surely psychology and the other social sciences share something today from our forbears, and prescriptions are rooted in this heritage. It is conceivable that the particular patterns of prescriptions suggested have limited application in the other social sciences, but I hope I am forgiven for believing that there is at least somewhat more relevance than that. Even if *all* specified prescriptive attitudes that have been demonstrated to some extent to be relevant in psychology are rejected in the other social sciences, there would still be the formal phase of prescriptive theory – pluralistic attitudes held by practitioners within a given social science that shape to some extent the nature of the field.

I will now try to sketch some possible investigations from the social sciences. This is perhaps a foolhardy thing to do since I am stepping out of whatever area of expertise I can claim as my own. I ask only for the forbearance that might come from your remembering that details of my illustrations may be naively, poorly or even falsely expressed without thereby vitiating prescriptive theory itself.

Time permits but three vignettes of possible studies and one somewhat more detailed proposal. The culture-concept approach in anthropology is one in which molaristic contentions might be expected to predominate with the apparent decline in this approach being traceable to the intrusion of more and more molecular considerations. Jeremy Bentham's indeterministic rationalistic (as opposed to irrationalistic), utilitarian, dynamic, nomothetic approach to human nature would seem particularly appropriate for prescriptive study since it was an effort to supply a view of man appropriate for economics. The nature of the functionalistic approach in sociology might be investigated prescriptively, not for its functional character alone, but to ascertain what patterns of other prescriptions cluster within this view as distinguished from patterns of prescriptions held by individuals opposed to this view.

A detailed illustration may be chosen from the work and influence of Durkheim. That he refused to acknowledge the relevance of individual psychology to sociology lends a certain piquancy to this choice. The immediate setting for my illustration is the recent, stimulating discussion by Terry N. Clark (1973) on the rise of the French university.

Clark argued that the basic unit of academic organization in the French universities was the cluster or informal grouping of a dozen or so persons "who shared *a minimal core of beliefs about their work* [italics mine] and who were prepared to collaborate to advance research and instruction in a given area." Their professional tasks included working for the advancement of one another, placing men in strategic places, universities, committees, journals and the like. The cluster around Durkheim Clark considers typical, and he proceeded to document this contention in considerable detail. I would now focus on that "minimal core of beliefs" that was present only as incidental background for Clark's study.

Durkheim's quasi-positivism,[2] engendered in part by admiration for Wundt, would suggest that he would be found to stress empiricism. His strong belief in progress and his use of a comparative method and a general evolutionary hypothesis would lead to a prediction that his position would show stress on developmentalism. Durkheim's finding of a molar social level beyond individual minds at the molecular level, challenges us to answer his stricture that "Every time a social phenomenon is directly explained by a psychic phenomenon one may be sure that the explanation is false." That he is a functionalist in one sense or another, I believe I can, for the sake of brevity, take for granted.

One of the challenges that this study would face rests in Durkheim's stress on social solidarity. Does this call for a social preference-individual preference prescriptive attitude not yet predicted? The path of caution would be to postulate it and see whether it turns out to be essential or whether the findings can be more parsimoniously explained without it.

My intent to formulate the relationship of prescriptive theory to the study of the history of the social sciences in no sense implies rejection of the other aspects that make up a reasonably complete pluralistic approach to history. To place sole reliance upon a prescription approach would give only a segmented and therefore distorted picture of the study of the historical process.

In my original paper on prescriptive theory (Watson, 1967), in point of fact, it was indicated that historians must concern themselves with the contentual topics of psychology, such as motivation and learning; the hypotheses, laws and theories that have been stated, defended and criticized; the methods used, observations and experiments to state them in broadest terms; other personal characteristics of individuals with whom one is involved; and the social circumstances in which these events take place. Content, in the conventional sense, theory, methods, personal characteristics and social circumstances require attention. In a later paper (1971) a description was given of how prescriptive theory could be related to and utilized with the Great Man and *Zeitgeist* approaches. And, as I indicated earlier in this paper, I have since grown to appreciate the significance of contentual part paradigms.

Implicit in what I have been saying has been a stress on internal determinants of history considered central for the conference discussions on scientific development.

Sciences are what scientists do and think. Scientists are influenced by their scientific heritage *and* by external social influences beyond the boundaries of the individual or scientific community. It is an empirical problem to determine the extent to which they are influenced from beyond their own personally-held traditions. To go beyond my topic for a moment, it may even be a prescriptive pattern studies could be extended to questions of organized scientific research. To take a clue from research on marriage, I would find it intriguing to investigate whether successful research organization shows complementarity or congruence in acceptance or rejection of prescriptive attitudes.

One of the advantages of preparation for participation in a cross-discipline conference is the reminder it gives; that perspectives other than one's own have a legitimate and vital role, and one's own personally satisfying position is perhaps no more than an area of the ground with a different figural emphasis.

Speaking more generally, psychologists wearing methodologically individualistic spectacles, as it were, make the individual the figure, while sociologists and most other social scientists would see the social circumstances as the figure and the individual aspects as the background. Sectarian dangers arise when figures are perceived as a unity segregated from the ground. The contour, the boundary where figure and ground meet, becomes crucial if it is to become possible to shift figure and ground.

I accept as a methodological imperative to attempt to reconcile the social and the individual approaches by trying occasionally to become flexible, to see as figure what comes naturally as ground! Perhaps other participants may be stirred in a reciprocal fashion to consider using the prescriptive approach in our common search for the dynamics of history and social science.

University of New Hampshire,
Durham, N.H., U.S.A.

NOTES

[1] Although not of direct relevance to this paper, it is probable that prescriptions are operative in so-called paradigmatic sciences but that their influence is probably less crucial. The presence of a paradigm makes it less necessary and, with some individuals, even reprehensible to be concerned with attitudinal components since to do so is both subjective and, heaven save the mark, philosophical. Practitioners of a paradigmatic science, secure in its data base, can eschew philosophy more easily than pre-paradigmatic scientists. A case in point is the sheer inability of some practitioners of a natural science even to see that philosophical problems are involved.

[2] I speak of quasi-positivism because of his epistemological sophistication influenced by Kant and his categories and Durkheim's belief in the reality of the categories of understanding which, according to the rationalist, are inborn or come from divine reason and a position he could not accept fully either. Social origins in collective representations was his answer. Moreover, as Parsons has pointed out, Durkheim's stress on goals, values and sentiments as social phenomena of great importance, are not susceptible to sense observation as demanded by the positivist.

BIBLIOGRAPHY

Ben-David, J. and Collins, R.: 1966, 'Social Factors in the Orgins of a New Science: The Case of Psychology', *American Sociological Review* **31**, 451–465.

Brodbeck, M.: 1953, 'The Nature and Philosophy of Science', in H. Feigl and M. Brodbeck (eds.), *Readings in the Philosophy of Science*, Appleton-Century-Crofts, New York, pp. 3–7.

Buytendijk, F. J. J.: 1957, 'Femininity and Existential Psychology', in H. P. David and H. v. Bracken (eds.), *Perspectives in Personality Theory*, Basic Books, New York, pp. 197–211.

Clark, T. N.: 1973, *Prophets and Patrons: The French University and the Emergence of the Social Sciences*, Harvard University Press, Cambridge.

Fuchs, A. H. and Kawash, G.: 1974, 'Prescriptive Dimensions for Five Schools of Psychology', *Journal of the History of the Behavioral Sciences* **10**, 352–366.

Gibson, K. R.: 1972, 'The Conceptual Bases of American Psychology: A Content Analysis of the Presidential Addresses of the American Psychological Association, 1892–1970', unpublished doctoral dissertation, University of New Hamsphire.

Graumann, C.-F.: 1972, 'The State of Psychology, 1', *International Journal of Psychology* **7**, 123–134.

Kuhn, T. S.: 1962, *The Structure of Scientific Revolutions*, University of Chicago Press, Chicago.

Mueller, R.: 1974, 'The American Era of James Mark Baldwin (1893–1903)', unpublished doctoral dissertation, University of New Hampshire.

Parsons, T. and Shils, E. A.: 1951, *Toward a Theory of Action*, Harvard University Press, Cambridge.

Popovich, M.: 1966, 'What the American Sociologists Think About Their Science and Its Problems', *American Sociologist* **1**, 133–135.

Reuchlin, M.: 1972, 'Psychologie ou psychologies?', *Psychologie* **31**, 181–196.

Ross, B. C.: 1970, 'Psychological Thought within the Context of the Scientific Revolution, 1665–1700', unpublished doctoral dissertation, University of New Hampshire.

Skinner, B. F.: 1953, *Science and Human Behavior*, Macmillan, New York.

Skinner, B. F.: 1959, 'A Case History in Scientific Method', in S. Koch (ed.), *Psychology: A Study of a Science. Vol. 2. General Systematic Formulation, Learning and Special Processes*, McGraw-Hill, New York, pp. 359–379.

Watkins, J. W. N.: 1953, 'Ideal Types and Historical Explanation', in H. Feigl and M. Brodbeck (eds.), *Readings in the Philosophy of Science*, Appleton-Century-Crofts, pp. 723–743.

Watson, R. I.: 1965, 'The Historical Background for National Trends in Psychology: United States', *Journal of the History of the Behavioral Sciences* **1**, 130–138.

Watson, R. I.: 1967, 'Psychology: A Prescriptive Science', *American Psychologist* **32**, 435–443.

Watson, R. I.: 1971a, 'A Prescriptive Analysis of Descartes' Psychological Views', *Journal of the History of the Behavioral Sciences* **7**, 223–248.

Watson, R. I.: 1971b, 'Prescriptions as Operative in the History of Psychology', *Journal of the History of the Behavioral Sciences* **7**, 311–322.

Whitehead, A. N.: 1925, *Science and the Modern World*, Mentor, New York.

RICHARD WHITLEY

COMPONENTS OF SCIENTIFIC ACTIVITIES, THEIR CHARACTERISTICS AND INSTITUTIONALISATION IN SPECIALTIES AND RESEARCH AREAS: A FRAMEWORK FOR THE COMPARATIVE ANALYSIS OF SCIENTIFIC DEVELOPMENTS

1. Introduction

Recent discussions of scientific development have tended to emphasise two points about sociological accounts. First, a simple, unilinear, neo-evolutionary approach which encompasses all sciences is inadequate and future work must be based on the comparative analysis of developments in different sciences at different times (cf., e.g., Elias, 1974; Martins, 1972; Nowotny, 1973; Weingart, 1974; Whitley, 1974). The second point, linked to the first, is that in reconstructing scientists' accounts, and especially the scientific elite's, of actual developments, social scientists must take these accounts as problematic. It can no longer be assumed that participants' recollections of past events are non-indexical or context free and the manner in which such recollections are constructed is itself a matter for investigation (cf., e.g., Edge and Mulkay, 1972; McAlpine and Bitz, 1973). This does not imply, of course, that the social scientist can therefore gain access to an underlying reality and reconstruct what 'really' happened but that a further understanding can be gained by going beyond simple translation of famous scientists' memoirs (cf. Naess, 1972).

These points, and other similar critiques of an immanentist history and sociology of science, suggest that to begin to satisfactorily understand paths of scientific development we need a conceptual scheme which can encompass scientists' reconstructions in different fields and relate them to each other in a systematic manner. To understand, that is, processes of scientific development we need some means of describing developments in different sciences which enables us to conduct comparative analyses in a fairly coherent way. This is not to suggest that it is essential to formulate an elaborate abstract framework of categories and their interconnections

Knorr, Strasser, and Zilian (eds.), Determinants and Controls of Scientific Development, 37–73.

before conducting any empirical work but to point out that all such work involves, usually tacit, assumptions about appropriate modes of describing phenomena and therefore appropriate ways of accounting for them. If these assumptions are not clearly outlined and vary between different case studies then it is impossible to systematically relate one historical account to another except, implicitly, through some imputation of a common 'common sense' rationality which is held to encompass all such studies. At the same time, of course, any conceptual scheme which does attempt to formulate appropriate modes of description relatively explicitly, implies some appropriate, or perhaps more commonly rules out certain inappropriate, ways of accounting for phenomena so described and thus implies some model of science and scientific activity. Modes of comprehension and modes of description are inextricably linked. Any attempt at understanding paths of scientific development, then, requires a means of characterising them and so this paper is concerned primarily to outline aspects of scientific activity and their institutionalisation since scientific development is seen as involving changes in these.

The characterisation of scientific activity outlined in this paper is based on a view of science as an essentially productive or constructive (cf. Piaget, 1972: 85–93) practice. The different categories or components of research work which will be described are derived from a position that sees science as a system of theoretical production (cf. Althusser and Balibar, 1970: 41–43) and research as a process of creating and transforming cognitive objects or structures by certain procedures for various aims and in accordance with a general set of values and norms. In accounting for different modes of scientific development this approach offers a way of describing what it is that is changing, i.e. components of scientific activity or the process of knowledge production. By characterising science as a system of production the focus of attention is shifted from the problem of how knowledge of nature is possible to the processes by which different 'knowledges' are produced. Given that a naive realist position is no longer tenable in epistemology, nor, for that matter is simple adherence to the central tenets of logical positivism in the sociology of science (cf. Alexander, 1963; Harré, 1970; King, 1971; Kuhn, 1970; Martins, 1972; Popper, 1959; 1963; Whitley, 1972), this shift represents an attempt to grapple with the implications of a non empiricist but equally, non idealist, epistemology in the field of the sociology of scientific development (cf.

Schmidt, 1971). Without, that is, attempting to deal with what Althusser terms the problem of 'knowledge effects' (Althusser and Balibar, 1970: 61–69) or the mechanism relating cognitive constructions to natural material processes the present approach at least recognises the implications of the reformulation of traditional epistemological and ontological concerns. In viewing science as productive and amenable to sociological investigation we are affirming its social, man made nature, its plural modes of development and, of course, its institutionalisation as a system of production related to others in a determinate fashion. It does not seem necessary to adopt the thoroughgoing rationalism of Bachelard (1968, 1972) or Hindess (1973) to accept their strictures against empiricism and common sense epistemological obstacles; the main point being to formulate an approach that enables us to examine changes in scientific work occurring in determinate circumstances in different sciences. The broad aim is to situate developments in their context and further explicate processes of scientific change by reference to such aspects as competition or conflict, institutionalisation and organisation of work and control mechanisms, legitimation and centrality of specialties and disciplines and developments in other systems of production such as the cultural or ideological.

The first step in this endeavour is to distinguish different components or aspects of scientific work which vary in certain ways and changes in which constitute scientific development. These can be derived from the idea of science as a productive activity and will be sketched in the next section. In any one activity these components are interrelated in terms of certain characteristics, notably those of articulation, integration and permeability. Scientific activities, then, are constituted by these five components characterised by varying degrees of these features which are related to likely changes in individual components and their ramifications for others. In a sense a scientific activity so described represents a 'state of development' and alterations in its components and their relations constitute scientific development. Obviously it is crucial to this approach of be able to identify units of organisation and their boundaries if changes and development are to be clearly isolated and investigated, and here the notions of research areas, specialties and disciplines are particularly useful (cf. Whitley, 1974). Scientific activities are organised into these units, and others such as disciplines and sciences, and are related to each other and to the wider society through them. The distinct components or aspects are variously institutionalised and organised into research areas, specialties and disciplines and changes or developments imply some alteration of the particular mode of organisation of the component or components in question. Changes in scientific activities, that is, take place in institutionalised settings and so necessitate consideration of the nature of such settings and surrounding units of organisation. They also, of course, involve changes in commitments to cognitive structures and social groupings and so conflicting or competitive relations between practitioners of different activities need to be taken into account although a full consideration of this point cannot be undertaken here. Initially, then, distinctions between the five aspects will be sketched and their characteristics of articulation and integration outlined. Subsequently, processes of their institutionalisation are discussed and how these are related to the formation of research areas and specialties. The degree of permeability or receptivity of institutionalised components is then considered and some tentative suggestions made about likely developmental processes in particular situations.

2. Five components of scientific activity

From the general idea of scientific research as a productive activity it seems possible to distinguish at least five separate components which are implicated in any such activity. First there is the set of day to day operations which define and circumscribe the immediate object of concern. The production and delineation of the cognitive object or structure by means of specific and immediate, local procedures constitute this component. It represents the concrete actions of carrying out procedures for obtaining information about some phenomenon or material. Essentially we are here concerned with the construction of the focus of immediate concern, the delineation of current uncertainty. General procedures of coping with this object constitute the next component, these are not specific and restricted only to the area of immediate concern but may be applicable to a number of them. They are the tools available for working on the original structure. The third component deals with the evaluative framework which, implicitly or explicitly, organises the first two, it 'makes sense of' and orders the results obtained by application of the procedures to the object. It also, of course, is involved in the formulation of the immediate concern in as much as this is done with some framework in mind.

The fourth component constitutes the general purpose of conducting the activity, the basic objective underlying the interest in, and modification of, the particular cognitive structure. As well, that is, as some evaluative framework that is implicated in the activity a more general purpose and definition of reality is implied that incorporates a number of such activities. Fifthly, and finally, there is the overall system of values and beliefs which serves to justify and integrate the activity with other systems of production, it legitimizes the basic objective and provides a general world view. These five components can, of course, be further subdivided but they do, I suggest, refer to distinct aspects of scientific work and can be characterised in more detail in the following way.

The first component I term research practices because it deals with the immediate day to day concerns of the researcher. It includes the ideas, norms and guidelines involved in the formulation of a specific area of concern. To conduct research, the scientist necessarily has some means of defining and constituting a specific question even if it is only of the trial and error type where *X* is done to *Y* to see what happens. The research practice aspect of scientific activity is precisely this description and characterisation of the object under investigation, it is the process of isolating and specifying the particular focus of interest in such a way as to imply – with varying degrees of ambiguity – the particular actions and analyses required to resolve the uncertainty. The definition, that is, of the area of immediate uncertainty implies the use of certain procedures and the observation of certain rules.

The more articulated is a research practice the clearer and more precise are these rules and prescriptions. They are likely to be used as evaluative criteria in the assessment of research work and to be part of the folklore surrounding a given activity which will be transmitted to neophyte scientists. A research practice includes non experimental as well as experimental work for it refers to any minimally systematic way of characterising and working on an object. These objects may vary from mathematical symbols, computer programmes, botanical taxonomies to magnetic properties of amorphous materials and survival rates of mouse liver cells under certain radiation conditions. The essential feature here is that there is some means of constituting an area of uncertainty and of resolving it, or at least of carrying out appropriate actions which might be expected to resolve it.

The second component consists of the procedures and techniques used in the definition of an area of uncertainty and its attempted resolution. While the actual application of the technique to a specific concern is part of the research practice, its description and general rules of use which affect possible interpretations of results constitute a separate component termed techniques. Again, it should be borne in mind that this incorporates algebraic manipulations and guidelines for interviewing procedures as well as fractionation processes and neutron irradiation. The analytical distinction being made here is that between a general tool for working upon a series of objects and the particular rules governing its application to a specific system and problem which derive primarily from the way that problem is defined. The general procedure of centrifugal separation of biological tissues, for example, is common to many activities in biology, but the actual use of this technique for a particular problem and material may involve consideration of factors which are irrelevant for other uses. Also, of course, the decision to use it is dependent on the specific formulation of the problem. The definition of an area of uncertainty leads to particular techniques being applied and evaluated in ways specific to that area but the techniques themselves are applicable to a range of objects and their characteristics and uses can be described independently of any one application. Factor analysis, for example, can be taught and discussed without referring to any particular research problem and so can be identified as a distinct way of working on data which is general over a range of objects.

The designation of a research problem and the selection of appropriate ways of proceeding to deal with it implies some more general evaluative framework which justifies and orders this process. However implicit and vague, some idea of appropriate explanations or accounts is involved in conducting research and it is this which I suggest is a third component of scientific activity, explanatory models. Of course, this term has been extensively discussed and analysed by philosophers and others (cf., e.g., Achinstein, 1968; Harré, 1970; Hesse, 1966) and is subject to large variation in meaning (cf., e.g., Canguilhem, 1963; Clark, 1972; Jevons, 1972) but here it is intended to cover the wide range of cognitive products which lead to certain views and accounts being preferred and are often called metaphors and analogies. Basically, this component refers to the various ways of 'making sense' of phenomena and criteria for assessing accounts of them. It is the scheme of understanding which delineates the kind of explanation which is acceptable. It also defines which problems are more 'interesting' and hence worthy of attention. When highly elaborated and integrated with research practices, the model forms the basis of descriptions of phenomena and sets the criteria for evaluating problem solutions. Research, nevertheless, can continue without articulated explanatory models directing it, here criteria for evaluation will be more concerned with procedural adequacy than with theoretical utility. Also, of course, criteria from more than one model may be invoked and it may be unclear which particular model a piece of research is relevant to. Even when a model is highly developed it need not necessarily determine the selection of techniques and problems directly but is more likely to set the limits to research practices and, in particular, to rule out activities which are regarded as pointless, trivial etc. Unless, that is, all components of an activity are highly articulated and fully integrated with each other some measure of developmental differentiation is likely such that adherence to a given model will not unambiguously lead to uniform research practices and techniques. While research problems may be defined in terms of the prevailing approach there will be room for more than one interpretation and differences of opinion over the most 'interesting' area within the general framework of the model. This component, then, covers the evaluative approach which underlies the attempt to define and resolve uncertainties; it indicates the field of relevant problems and criteria for assessing results of research practices.

The fourth component of scientific activity, I suggest, is the overall field of concern and interest that defines and characterises some reality. While the research practice delineates the area of immediate concern, an object which is to be worked on, this component defines a general problem or focus of attention. It formulates a fairly large scale object which guides and orientates research but cannot itself be the object of an isolated research practice. It represents the underlying problem which requires explanatory models, techniques and research practices for any resolution. It constitutes a specialty concern and often serves as the focus of commitment for scientists though it is not always clearly articulated. When it is highly articulated it will limit possible explanatory approaches and problem formulations although is unlikely to uniquely determine them. Specialty reality concerns constitute descriptions of fields of uncertainty, they delineate and demarcate the scope of interests invoked in scientific activities. Examples are the liquid state in physics, carcinogenesis in biology and the operation of markets in economics.

The fifth and final component of scientific activity is the general set of values and beliefs that define it as 'scientific' and legitimise it as such. These "paramount values" and "disciplinary matrix" (Kuhn, 1970a: 21, 1970: 182–187) constitute the metaphysical aspect and includes such beliefs as the ultimate rationality of reality as well as more discipline specific views and constructions. Indeed Weingart (1974) distinguishes highly general values from more specific "meta paradigms". This component, then, incorporates commitments institutionalised and organised in sciences and disciplines and necessarily deals with the characterisation of scientific rationalities, the prestige ordering of the sciences and the centrality or marginality of disciplines. Although this aspect is not greatly in evidence in many activities, it does of course operate as a background filtering device for ideas and personnel and becomes prominent during periods of severe, crystallized conflict as Kuhn has so clearly shown (1970; cf. Hallam, 1973; Martins, 1972). The point here is that any activity involves metaphysical assumptions and views even if they are not articulated or appear mutually contradictory.

Before continuing with a discussion of how scientific activities and "states of development" can be differentiated by the articulation and integration of these five components it might be worthwhile to briefly indicate how they could be applied to activities in two different sciences. If an activity in the sociology of organisations is being considered, the research practice might be concerned with relations between organisational size and bureaucratization and how these could be investigated in a number of organisations. The techniques used could include interview and questionnaire procedures and various statistical devices for manipulating the information obtained. The explanatory model would presumably be derived from a reading of Weber and later writers together with some idea of how their views could be applied. The specialty concern might be organisational development or, the "organisational society" and the metaphysical component could well be derived from logical positivism and utilitarianism. In contrast we can also apply the terms to work in crystallography. Here, a research practice might deal with the electron distribution of Berillium, the techniques could be X-ray and neutron diffraction and methods of sample preparation, general crystallographic models of the solid state would constitute the third component and the specialty concern would be the elucidation of crystalline structures. Metaphysical assumptions presumably would be common to physics and involve a mixture of epistemological realism with adherence to the Copenhagen interpretation.

3. Articulation of Components of Scientific Activity

Having sketched five components of scientific activity, I will now discuss one feature of them which differentiates scientific activities and leads to a means of characterising relations between components and so activities. This feature is also relevant to processes of institutionalisation of components and their permeability and receptivity to potential innovations. The degree of articulation of a cognitive structure refers to the clarity and precision with which it is stated. Accordingly, a low degree of articulation exhibits vague outlines, ambiguous phrasing, and a multitude of possible interpretations. Standards for distinguishing correct uses of the component are incomplete and imprecise. This allows instances where it is unclear whether a given application is justified. On the other hand, a highly articulated structure is exemplified by clear boundaries, internally coherent and consistent rules, and elaborate standards for identifying correct work.

For research practices, articulation refers to the degree of formulation of guidelines and evaluation criteria. This ranges from the considerable folklore and propositions available to Ph.D. students in solid state physics to disciplines with little general assistance available and each researcher rather blindly finding 'successful' and 'unsuccessful' practices by trial and error. A lack of articulation of research practices, nevertheless, need not be accompanied by a complete absence of evaluative standards. There are simply few shared exemplars and no developed, conscious research tradition. Criteria for assessing research will be vague and perhaps contradictory. Indeed, in this situation, the process of accounting one piece of work as important and another trivial will be obscure. In so far as a cognitive structure exists, it will be almost entirely "tacit" (Polanyi, 1966) and take the form of *post facto* evaluation rather than being applicable at all stages of the research process. The degree of articulation of research practices, in short, refers to the extent there are shared exemplars of successful work, clear standards for evaluating the constitution and resolution of uncertainty and identifying errors, and an extensive folklore of research failures and dead-ends.

Similarly, the articulation of techniques indicates the precision and clarity with which available means of transforming objects are described and characterised. This may vary from vague references to common features of a technique in a variety of applications to a formalised list of instructions and characteristics which delineate what the technique is, where it can be used, how it is to be used, and what the results mean. As well as referring to the degree of clarity of description of each technique, articulation also indicates the clarity of relations between techniques. Their characteristics and mode of use will be distinguished and relatively unambiguous. In the study of amorphous materials in solid state physics, for example, it is fairly clear how a laser should be used and what problems it is appropriate for. Similarly, electron microscopy is highly developed as a physical technique. When, however, it is used to analyse biological tissues the outcomes are more ambiguous; because it is not always clear how to interpret the application of a device developed for the analysis of relatively simple systems to much more complex and theoretically distinct systems. Often before a technique can be meaningfully transferred from one specialty to another, a new observation theory is necessary – one at least articulated to some extent – to make sense of results. The transfer of techniques between areas of concern is by no means simply a process of adopting a novel method of treating phenomena since novel ways of constituting objects are frequently part of techniques. The development of radio and radar techniques in astronomy created new objects and lead to modifications in optical astronomy and cosmology (Edge and Mulkay, 1972; Mulkay and Edge, 1973).

The articulation of explanatory models or mechanisms refers to the degree of elaboration of modes of understanding phenomena and the extent to which they direct and define research problems and contain unambiguous standards for evaluating research results. An articulated model distinguishes sharply between interesting and trivial problems and outlines acceptable ways of solving problems. The more articulated an explanatory model, the more it constrains scientists' research activities, while integrating their results into a coherent, systematic understanding. Conversely, a weakly articulated model constrains the type of acceptable accounts, but the boundaries will be vague. It will not be clear how different ways of accounting for phenomena can be distinguished or ordered. Neoclassical economics, for example, stipulates acceptable explanations fairly clearly. Functionalist sociology is much less clearly elaborated. Several types of functionalist explanations exist. Some overlap with rival modes of understanding such as exchange theory. In many cases, functionalists might accept an account in terms of exchange theory as adequate or would claim that the latter could be subsumed under the former. Here, as in other areas of social science, there is a general lack of precision concerning research goals and standards for evaluating research efforts.

As mentioned earlier, the basic concern of a specialty contains a way of defining constitutive phenomena and the degree of articulation refers to the extent this is formulated and implications are clear. The study of the market in economics is an example of an articulated specialty concern. The basic problem is clearly outlined and there are implications for proceeding with research. The definition of the problem delineates possible modes of resolution. As for the other dimensions of cognitive structures, a high degree of articulation implies a high degree of internal coherence and consistency which in turn implies formulated, or formulatable, rules, definitions and guidelines. The market may be contrasted with the considerably less articulated object of organisations in sociology (cf. Krupp, 1961; Storing, 1962; Whitley, 1975a). In this case, although there are competing schools, it is not entirely clear what they are competing about since relatively precise delineations of the basic problem and how it would be tackled are frequently not available. Indeed, often recourse is made to common sense notions without formulating theoretical objectives (cf., e.g., March and Simon, 1958: 1–2). The more weakly articulated a basic concern or problem of a specialty is, the more its members may share a general tacit understanding that leads to considerable disagreement and argument when attempts are made at formulating it.

The articulation of metaphysics is rarely as systematic as that of the other cognitive structures. It is well known that some scientific schools specifically deny the existence or relevance of metaphysical assumptions for their work. Nonetheless, occasionally research traditions and disciplines outline their dominant assumptions and basic values. Frequently the formalisations occur as part of internecine warfare or in the establishment of a new specialty or discipline. The more articulated its metaphysics, the more clearly the discipline is demarcated and the more systematic and coherent are its dominant assumptions and values. An articulated metaphysics clearly describes the world views of the discipline and its understanding of science. It provides criteria for distinguishing science from non-science and a general view of the reality it is concerned with. On the other hand, relatively weakly specified metaphysics offers little indication of the nature of science and the reality studied. Relations between vaguely defined values and views of the 'furniture of the earth' are unclear and contradictory opinions may coexist since they are not formulated as statements. Metaphysics require considerable care in imputation and may well exhibit, in any one case, an easy eclecticism as when, for example, a positivist theory of explanation and understanding is combined with an anti-positivist ontology (cf. Bershady, 1973).

4. Relations between Articulated Dimensions of Cognitive Structures

Having outlined how articulation can be applied to the different components of scientific activity, relations between them in terms of their articulation and, subsequently, their integration can now be discussed. What implications follow from the degree of articulation of one component for the others and how interrelated are they? Although the different dimensions might be seen as constituting hierarchical levels, this does not mean that the 'higher' logically incorporates the 'lower'. While there are some determinate relations between dimensions, they are intended to be used as hierarchies of generality as suggested by Weingart (1974). The types of connections remain to be ascertained and need not necessarily all be implicative.

A high degree of articulation at the research practice level implies some minimal articulation of techniques since their use and interpretation requires fairly detailed knowledge of their implications and limitations. For research practices to be systematic, and errors of procedure and interpretation clearly identifiable, techniques need to be at least moderately articulated so that mistakes of practice can be distinguished from purely technical difficulties. Similarly some, however vague, articulation at the explanatory model level is implied since the correct definitions of problems, use of techniques and meaning of results suggests that some system of understanding is available to indicate the adequacy of research albeit not clearly formulated and open to diverse interpretations. For it to be clear at all stages of the research process when errors of judgment and application have occurred some expectations are, of course, essential but these need not be fully integrated and formalized. *A fortiori* the basic concern of the specialty and the dominant metaphysical assumptions may be even less developed than explanatory models. While strong assumptions may be implicit in research activity, they need not be explicitly formulated or elaborated. Thus, an articulated set of research practices may coexist with weakly articulated concerns and metaphysics, although this is by no means necessarily the case. They may also, of course, be connected to a number of concerns rather than restricted to one specialty.

If, on the other hand, research practices are weakly articulated it is unlikely that techniques or explanatory models will be highly articulated and developed although it is possible that the specialty concern and metaphysics may be, but no means necessarily so. In this case, there would be considerable debate about fundamental presuppositions and objects of analysis, but little development of ways of understanding, tools for manipulating problem situations, or exemplary research practices. Insofar as there are cognitive structures being produced and worked on, these are likely to be highly permeable to 'common sense' rationalities and the general dominant culture. Little coherent and consistent development is probable here until some set of minimally autonomous practices have arisen.

Highly elaborated techniques need not, on the other hand, imply highly developed research practices, although it is unlikely that their development will have occurred without some articulation of research practices and ability to formulate immediate problems in the same specialty. Whilst it is possible for techniques to be developed in the absence of clear applications and uses, this seems unlikely in specialties where there are some articulated explanatory mechanisms. It may occur in fields where a strongly articulated metaphysics justifies concentration of techniques and where research practices and explanatory models are not developed, but presumably the techniques will then be used, albeit somewhat incoherently, and gradually create a system of research practices. However, it is possible for articulated techniques to exist without systematically elaborated specialty concerns, models or metaphysics and in this case research will consist of a series of technical applications and refinements which are not at all clearly connected. This situation may arise when a specialty concern has become the focus of general societal attention but it is not articulated and a number of techniques are used in some vague and tenuous manner to tackle it. In other words, an organisation may be established by the scientific and political elites to investigate a 'problem' which is not connected to current techniques and models; scientific work there consists largely of the application of imported techniques and models to objects which are only vaguely relevant to the organisational label.

A low degree of articulation of techniques, though, does have some additional implicative relations. Research practices are unlikely to be highly developed if available techniques are ambiguously described and assumptions unclear. If ways of manipulating objects are vague and unclear, then it is difficult for any coherent understanding of research activity to be possible. Research will consist of disconnected operations which are open to multiple interpretations. Little coherent interpretation of the results is likely even if there is an articulated model since there is no systematic understanding of techniques, hence, no clear view of the results of their use. Lack of developed techniques, thus is consistent with either a moderately or weakly articulated explanatory model. It is unlikely to occur, however, where there is a highly articulated model since the development of a detailed way of describing and understanding phenomena in a specialty will probably take place with at least a minimal elaboration of techniques for dealing with research problems indicated by that model. In other words, the formulation of an interpretative scheme, which occurs over time, requires discussion of techniques to proceed. This need not be the case for the specialty concern or metaphysics: any degree of articulation is consistent with a low extent of articulation of techniques.

As already indicated, a high degree of articulation of an explanatory model implies at least a moderate development of techniques and of research practices, although they need not be highly articulated. It is not likely to occur without at least some understanding of the basic concern of the specialty since the object of investigation limits the type of explanation sought and an elaborate model or mechanism assumes a relatively clearly defined objective which is being explicated. It, also, implies certain definitions of science and world-views, but these need not be formalised. While only a limited number of metaphysical positions will be consistent within a given articulated model, it does not follow that these will be elaborated or clarified.

A weakly articulated model is consistent with varying degrees of elaboration of techniques, nevertheless highly developed techniques are unlikely except in limited circumstances such as the importation of techniques into embryonic fields. Research practices in this situation may not be highly articulated but could be if techniques are and some general scheme is available for assessment, however vague. A thorough, detailed model, in other words, is not essential for coherent research practices as long as some referent, perhaps in another specialty or discipline, is thought relevant. Some way of coping with and ordering data is necessary but this need not be fully elaborated as an explanatory model. Undeveloped models are also consistent with different degrees of articulation of specialty concerns and metaphysics, but a highly articulated concern is unlikely without some model being proposed, however tentative, for approaching it.

A high degree of articulation of the specialty concern suggests a minimal articulation of explanatory models for the detailed definition, delineation, and formulation of the basic problem is unlikely without some development of preferred ways of tackling it. The systematic description of a concern in a specialty implies some understanding of how it is to be explained: the more detailed and clear the formulation the easier it will be to formulate "explanation ideals" (Toulmin, 1961: 57). Similarly, a minimal development of techniques and research practices would be expected since along with a clear articulation of the basic concern and a moderately clear set of explanation ideas, ways of defining and examining research problems are likely to have developed. Nevertheless this does not imply that the formulation of a detailed specialty concern necessarily, as a direct consequence, produces research techniques and practices, but that it is empirically unlikely that it should be highly developed without some procedures for outlining and tackling research problems. This is because the explicit and systematic formulation of a specialty problem takes time and requires some experience of dealing with the sort of problematics which might become part of such a formulation. The development of an elaborated, coherent specialty concern, as is the case for an explanatory model, requires the results of research experience for clarification of boundaries and scope of definition. This need not be the case for a concern which is only moderately articulated. It is possible to have a definition of a specialty which is fairly clear insofar as it can be distinguished from others, without much development or understanding of techniques and research practices. This is particularly likely when specialties are defined predominantly in terms of some highly articulated metaphysics, or some 'practical' problem, the development of neighbouring fields implies the formulation of a new object or concern, but it is unclear how available techniques and practices can be applied to the problems which this concern would incorporate. Regarding metaphysics, a highly articulated specialty concern need not be accompanied by an elaborate scheme at this level. As for explanatory models, there are likely to be fairly strong assumptions and values implicit in the formulation of a specialty concern but need not be explicit, although it will be fairly clear what is regarded as a *scientific* specialty.

A weakly articulated specialty concern may occur both with highly articulated techniques and research practices or with relatively undeveloped ones. A vague and ambiguous outline of the defining problem of a specialty is consonant either with well understood and consistently applied standards of research practice or with disconnected and unstructured research activities. As mentioned earlier, some development of the specialty concern is likely for highly articulated models to exist so a low degree of elaboration of the concern implies less than fully articulated explanatory mechanisms. When techniques and models are applied to novel phenomena and problems, i.e. when they are used in another specialty, they often redefine the object of that specialty to make it consonant with such use, but this redefinition need not be particularly strongly formulated. In many cases all that is required is a general orientation which allows employment of particular explanation ideals. Relatively developed models, that is, can be applied without being coterminous with elaborated specialty concerns if those concerns permit loose interpretations which incorporate these modes of understanding. As with other dimensions of cognitive structure, weakly articulated concerns are consonant with highly or vaguely formulated metaphysical positions.

Generally, the degree of articulation of metaphysics does not imply the extent of development of other dimensions. This is for two reasons. First, it applies over a range of specialties and so is compatible with a variety of specialty characteristics. Second, although a thoroughly formulated metaphysics undoubtedly constrains the sort of scientific work conducted in specialties, it need not imply a consistent, coherent, highly integrated 'research programme' (Lakatos, 1970: 132–138). A developed set of scientific values and 'way of seeing' limits explanatory approaches without suggesting how such approaches could be articulated, and without implying that techniques and research practices are particularly well understood and applied. Equally, although highly developed models and concern undoubtedly rest on some metaphysical basis there is no reason to suggest that it is clearly, unambiguously and consistently formulated or that when someone attempts to articulate it, scientists will form a consensus. Obviously, scientific practice has metaphysical connotations, but this does not mean that vaguely formulated and ambiguous metaphysical positions are impossible with highly developed research practices or models.

These implicative relationships between degrees of articulation of different components are summarised in Figure 1 (p. 54).

There are several instances where no particular inference can be drawn which means that the circumstances of other components need outlining before even weak relations can be suggested. It is also obvious that by the use of the category 'medium' many relationships appear similar. This is partially because of the difficulties of inferring the articulation of any one dimension of structure from that of another without taking the state of the remaining three into account, and partially because the number of actual combinations which are possible and which seem to occur is large and which one is empirically observable at a time depends upon other circumstances such as the stratification system of the sciences, the development of neighbouring structures and the type and degree of autonomy.

5. Integration of Components of Scientific Activities

As well as the logical relations between differentially articulated components just outlined there are, of course, empirical relations between them. While articulated models, specialty concerns and metaphysics limit and constrain the development of techniques and research practices, they do not necessarily determine the actual nature of these components and so the extent of connectedness and coherence of components in a scientific activity needs to be considered. That is, the degree of mutual influence, the immediacy and range of repercussions for all components

	Degree of articulation	*Research Practices*	*Techniques*	*Explanatory Models*	*Specialty Concerns*	*Metaphysics*
Research Practices	Hi		Medium/High	Low/Medium/High	Low/Medium/High	Low/Medium/High
	Lo		Low/Medium	Low/Medium	Low/Medium/High	Low/Medium/High
Techniques	Hi	Medium/High		Low/Medium/High	Low/Medium/High	Low/Medium/High
	Lo	Low/Medium		Low/Medium	Low/Medium/High	Low/Medium/High
Explanatory Models	Hi	Medium/High	Medium/High		Medium/High	Low/Medium/High
	Lo	Low/Medium/High	Low/Medium/High		Low/Medium	Low/Medium/High
Specialty Concerns	Hi	Medium/High	Medium/High	Medium/High		Low/Medium/High
	Lo	Low/Medium/High	Low/Medium/High	Low/Medium		Low/Medium/High
Metaphysics	Hi	Low/Medium/High	Low/Medium/High	Low/Medium/High	Low/Medium/High	
	Lo	Low/Medium/High	Low/Medium/High	Low/Medium/High	Low/Medium/High	

Fig. 1. Relations between variously articulated components of scientific activity.

resulting from a change in one, is relevant to understanding processes of institutionalisation of activities and their receptivity to potential innovations. The degree of integration, then, refers to the extent one component has implications for another. For example, an articulated model may be highly connected to certain techniques such that the types of explanation sought by that model necessitate the use of these particular techniques. Similarly, a highly developed metaphysical position may not only restrict the types of explanation admissable as scientific, but may also delineate the desired approach. Highly integrated cognitive structures – which are also highly articulated – are similar to the "paradigms" of Kuhn (1970) and "research programmes" of Lakatos (1970), but it is important to note that many of the sciences, unless they are defined in such a restrictive manner, are not highly integrated articulated structures. This is why the degree of integration – and of articulation – of distinct dimensions of cognitive structure is significant. The circumstances in which differing degrees occur and alter require investigation for an understanding of scientific development.

The degree of integration of different dimensions is not independent of the degree of articulation of those structures. For implicative relations to hold between models and techniques, for example, a fairly clear understanding of both must exist. If research results are to be connected to an explanatory model to have direct implications for the success or adequacy of that model, then they must be fairly unambiguous; the application of research techniques must be clearly correct which implies some elaborated criteria for that evaluation and the model has to be formulated so as to permit conclusions to be drawn which are not hopelessly vague and possibly mutually inconsistent. In other words, a low degree of articulation of two structures implies a low degree of integration of them, but the reverse is not necessarily the case. Likewise, a high degree of integration implies a moderate to high extent of articulation, but the reverse does not hold. It is possible to have a situation where highly articulated models and techniques exist which are not highly connected, because a range of techniques are consonant with the 'explanation ideas' and no particular ones are necessarily implied.

Alternatively, a number of techniques may be applicable to many types of understanding in that competing explanatory mechanisms may concur about admissable characteristics of research techniques, but need not specify the details. Similarly, research practices may be fairly well articulated, as may be competing explanatory models, but connections between these levels may still be weak and unclear. Competent research, that is, can easily be distinguished, but the precise model the results are relevant for, and the way they are relevant, remain undetermined.

A further point about the integration of components of scientific activities is that two activities utilising the same model or specialty concern need not be identically integrated. For example, the use of the crystalline model of the solid state in fracture mechanics and amorphous materials does not mean that developments in each area are equally important for 'refuting' or 'corroborating' it. Activities, that is, are not equally 'central' to the specialty concern or associated models; their marginality and their integration varies.

6. The Institutionalisation of Components of Scientific Activities

So far scientific activities have been discussed in isolation and their features only considered within activities. In examining relations between activities and processes of scientific development it is necessary to look at how patterns of activity are established and organised. Given, that is, that scientific work is characterised by these five components, their articulation and integration, how is it organised and how does this organisation change? In particular, how do novel activities develop and become institutionalised? Furthermore, how do institutionalised components become foci of commitment and the basis of social groupings? In an attempt to begin to deal with these questions, I will discuss process of institutionalisation of components in specialties and research areas in this section.

The articulation and development of a component of scientific activity leads to the formulation of norms regarding its correct description, use and interpretation. These norms are organised into units of cognitive and social structure that I have elsewhere (1974) termed specialties and research areas but also incorporate disciplines and, perhaps, sciences. These units form the framework for carrying out scientific research, they order activities and these, in turn, constitute paths of development of such structures. Scientific activities exemplify, or manifest variously organised norms and also develop them in so doing. Particular activities are defined by particular components patterned in cognitive structures and, at the same time, alter those patterns and structures by the particular constructions and activities undertaken. Components are developed and institutionalised to different degrees and they are also variously organised in research areas, specialties and disciplines. It should be noted that not all activities necessarily 'belong' to a single unit of organisation. Novel techniques can, for instance, be developed without forming a distinct organisation unit and activities may be undertaken which combine components from different structures.

Research practices have been defined as processes of describing and working on areas of uncertainty. They include means of evaluating procedures and actions taken to resolve this uncertainty. When a set of research practices focussing upon a particular concern become organised about that problem, i.e. they form a unit of cognitive and social organisation based on some area of immediate uncertainty, they form a research area. Techniques may also be institutionalised as a set of rules and procedures as they become articulated and formalised and often the distinction between a means of working on an object and the constitution of that object is not easy to make in a research area but in principle the norms governing the description and use of a technique are separate from those governing the formulation of a problem and the selection of appropriate ways of dealing with it. However, both are involved in the formation of a research area. This term, then, refers to the institutionalisation of an area of concern and rules for appropriate actions, it implies a minimal degree of articulation and consensus but it should be noted that research areas vary in degree of institutionalisation and it does not follow that they are always the primary focus of commitment and social solidarity for scientists. That depends on characteristics of specialties and disciplines and the articulation and integration of components.

Highly solidaristic research areas are most likely to occur in situations where there is no clear specialty organisation – or the immediate concern is very marginal to any existing specialty – and there is a distinct metaphysical commitment which is opposed to current rationalities and values or other trends in the discipline. Where there is some specialty organisation, and research areas are not distinguished by opposed metaphysics, it is unlikely that clearly identifiable, sect-like social groups will emerge around research area problems. Where the problem is not linked to a more general programme which is opposed to others, a lower degree of social solidarity would be expected than in situations where competition between developed approaches exists in the specialty. If it is part of a generally accepted programme in a monistic specialty, then, distinct social groupings are unlikely to emerge. In specialties characterised by weak consensus and articulation social groups may well be based on technical expertise rather than on particular problems. Where there is little overall cognitive order and scientists from a number of specialties and disciplines have transferred their expertise and concepts, only a low degree of group formation may be observed and that based on membership of other structures and concerns (cf. McAlpine *et al.*, 1974).

Weakly institutionalised research areas are characterised by little agreement on the definition of the problem or immediate concern and combine a variety of techniques in an eclectic mélange. They are unlikely to occur in specialties with articulated concerns and models and definitely not in monistic, integrated specialties but may well emerge in conditions of technical competition (cf. Whitley, 1975) or where specialty organisation as a whole is weak. It is possible for these research areas to occur in situations where there are moderately articulated techniques and research practices, but which are not linked to any clear research programme or elaborated object. Here, scientific work is characterised by a number of technical approaches and fairly systematic practices being applied to a wide variety of phenomena without any clear strategy underlying the choice of experiments or connecting results. In this situation, the formation of research areas focussing on particular problems for some time in some depth is unlikely. Similarly, social groups with a high degree of solidarity will not emerge around current research problems.

Specialties refer to the organisation of research around a common purpose or concern. They are the prime focus of orientation or reference for scientists in that they define – with varying degrees of clarity and consensus – the basic object under investigation. Generally, it is in terms of this object that scientists characterise their organisational identity and field of research. There are instances of disciplines fulfilling this role and specialties not apparent as distinct units of organisation, especially in the social sciences. In some fields, scientists may be unable to outline their major specialty as a separate entity (cf. McAlpine and Bitz, 1973). However, the reality constructed and organised in specialties does not deal with the focus of actual, current activities in that the general concern is too broad and general to delineate the area of immediate uncertainty facing the scientist.

The formation of a specialty as a unit of social and cognitive organisation, which circumscribes many activities sharing a common core of uncertainty, is dependent on both the development of reality definitions and boundaries and explanatory approaches and, indeed, highly articulated models are unlikely to occur without developed specialty concerns and vice versa. This raises a more general point in that the constitution of distinct units of organisation, such as research areas or specialties, which combine sets of scientific activities is distinct from the institutionalisation or patterning of components of activities and usually subsequent to it. In other words, not all developed models, techniques etc. are specifically contained within any one unit of organisation and any one activity need not be fully incorporated in any single discipline or specialty.

The distinction between specialties and research areas as discussed in an earlier paper (Whitley, 1974) is partly one of level of generality and partly one of type. Research areas deal with the organisation of the resolution of immediate uncertainty and technical means for accomplishing this, specialties are more concerned with underlying scientific purposes and interests and theoretical objectives. The difference may briefly be described as that between research tactics and strategies. Research areas organise the actual technical processes, the turning of 'data' into 'information' (cf. Ravetz, 1971: 76–88) while specialties organise patterns of reality and ways of making sense of it. They incorporate modes of understanding and systems of explanation. In developing a characterisation of scientific activity here, processes of institutionalisation of these units can be described as the outcome of various activities which combine components drawn from a number of units. The formation of new specialties and research areas, that is, is dependent upon the development of novel components of activity and their institutionalisation and these do not necessarily occur concomitantly. However, it does not follow that whenever novel components are developed, new units of organisation inevitably arise. This process is dependent on the general organisation of the discipline and science.

Before briefly examining the likelihood of novel components becoming institutionalised in variously institutionalised specialties the link between metaphysics as a component of scientific activity and the discipline as a unit of organisation should be clarified. Essentially the concept of a discipline incorporates, and in many cases is the institutional embodiment of, metaphysical beliefs and basic values. Again it should be pointed out that not all current metaphysical positions are expressed through such forms of organisation but that, in general, it is at this level that arguments about the scientificity of work and views of the world occur and identities as a scientist and xologist are established and maintained. Of course, disciplines are also the dominant form of societal institutionalisation of scientific activity in that they form the boundaries of education experiences and university resource allocation mechanisms. Indeed, it could be said that this is their primary function and they reflect more a combination of reified 'official histories' of earlier science and organisational needs than current controversy and activities. However, competition for resources and general scientific prestige between disciplines may involve conflicts of world views and the most 'interesting' scientific work so reflecting divergent metaphysical approaches to science or at least to 'important' science. As Hallam (1973: 111), following Harré (1970), has suggested these assumptions and predelictions governing criteria of intelligibility and plausibility are usually inexplicit until novel approaches become so developed and lead to open conflict as in the continental drift debate.

In dealing with the development and institutionalisation of novel components, it seems more likely that innovations in research practices and techniques will occur than models and specialty concerns on account of their lesser scope and hence field of potential influence. This should especially be the case in specialties which are highly institutionalised with one dominant approach and view of the basic problem – a monistic specialty – such as X-ray crystallography where development seems to mostly consist of the improvement and extension of coverage of the common technique (Law, 1973). As Law points out, the development of alternative foci of concern such as fibrous proteins which involved different techniques and explanatory approaches was regarded as deviant or at least not serious crystallography. However, not all disciplines are organised into monistic specialties and there may be cases when the articulation of novel models and specialty concerns is less likely to be resisted. In broad terms this seems possible in situations where no clear specialty structure exists and such articulated approaches that are available are not integrated with techniques and research practices, i.e., no clear 'research programmes' (Lakatos, 1970) exist. A low degree of institutionalisation, that is, will be more conductive to novel accounts and concerns being explored but this does not include specialties with oppositional competition (cf. Whitley, 1975) since here there are already a number of articulated and integrated scientific activities which form the basis of identity and commitment and are conflicting, e.g., the stabilist and the mobilist views of the earth (Hallam, 1973). The point here is that where little articulation of specialty concerns and explanatory models exists and the specialty as a distinct unit of organisation is not clearly demarcated then a variety of these components will be suggested, perhaps derived from analogies and metaphors imputed to more prestigious sciences. This is particularly likely in marginal disciplines struggling to establish themselves and here the various alternative approaches will often be associated with conflicting metaphysical positions and views of the discipline as has been observed *ad nauseam* of sociology. In other words in weakly institutionalised disciplines, novel metaphysical positions are likely to be formulated, together with specialty concerns and explanatory approaches although these may not be especially highly articulated. Frequently these positions will be conflicting leading perhaps to a form of metaphysical competition. The development of specialty concerns and models in this situation is unlikely to be rapid unless techniques and research practices are articulated which can be linked to the competing views. A similar condition is likely to be relevant when novel approaches are propounded in specialties which are apparently experiencing inconsistencies and difficulties. In particular, the high degree of openness of such disciplines to the dominant culture and 'common sense' rationalities is unlikely to lead to high institutionalisation of scientific activities since this requires a measure of autonomy and internal coherence.

Where specialties are formed by societal agencies designating a 'problem' and constructing organisations for dealing with it, as distinct from their formation on the basis of an institutionalised specialty concern – and of course this is not meant to be an all inclusive dichotomy but merely one distinction – and no clearly identifiable scientific activity with a similar specialty concern exists, then it is likely that a similar degree of receptivity will be manifested and activities from other specialties will be imported. Especially we could expect the transfer of techniques and research practices, perhaps with some modifications, although in many cases it will not be clear how these are connected to the societally defined problem and fully integrated activities including specialty concerns are unlikely; at least in the short term. This permeability of organisational forms and degree of cultural autonomy will be briefly discussed in the next section after considering the development of new research areas through the institutionalisation of techniques and research practices.

In many recent case studies the role of personal mobility into new areas in proceeding to the formation of new fields has been emphasised (cf. Edge and Mulkay, 1972; Law, 1973; Mullins, 1972; Watson, 1970). Also referred to is the importance of transferring new technical approaches to old problems and it is noticeable that these tend to be physical approaches transmitted by physicists. The institutionalisation of novel research practices and techniques seems to be combined in these cases with continuing identification with one's previous specialty or discipline until such manifest 'success' is obtained that there is sufficient basis for clearly separate models and specialty concerns to be formulated in terms which are more consonant with the 'host' specialty and lead to the development of an alternative identity. In the development of radio astronomy, for instance, the physicists at Cambridge and Manchester spent much time working on technical improvements and refinements before distinct 'objects' were discerned and it was even later that the optical and radio universes were related (Mulkay and Edge, 1973: 55). The formulation of distinct astronomical problems by radio astronomers as opposed to the articulation of physical techniques occurred quite late in the institutionalisation of the research area. This work did not seriously affect optical astronomy for some time and there is still little formal collaboration. The point here is that while the transfer of expertise may result in the formation of a new research area and transformation of existing specialties as, it may be argued, happened to biochemistry with the advent of molecular biology (cf. Mullins, 1972; Watson, 1970), it is by no means inevitable and the criteria according to which novel research practices and techniques are institutionalised and subsequently incorporated into a specialty vary and the receptivity also varies according to, among other factors, its institutionalisation, centrality to the discipline and location in the scientific stratification system. Astronomy, for example, was already well established and contained institutionalised components of scientific activities and so, in spite of the prestige of physics, as Mulkay and Edge (1973) point out, the criteria of technical adequacy of the radio astronomers' work were largely set by the existing specialty. The establishment of the research area in the first place followed the available expertise in radar work and its application to novel concerns such as meteors and radio emission. Essentially the institutionalisation of techniques lead to further refinement in dealing with the technical problems raised by their extension to new areas of uncertainty. This development led to the formation of a distinct area of research largely through its concentration at two major centres in Britain, the availability of resources for relatively expensive equipment (cf. Griffith and Miller, 1970) and clear organisational leadership as Mulkay and Edge (1973) fully document. The basis for the research area, then, was a common source of expertise and a common interest in its application to a distinct set of problems which could be easily differentiated from other work in physics and astronomy.

As Law (1973) shows, research areas can also be formed around a common core of concern using various technical approaches eclectically although he does point out the dominance of traditional crystallographic techniques in the development of X-ray protein crystallography. In general, for a separate unit of organisation to become institutionalised it does seem that distinct techniques and research practices are required although if the technical apparatus is very large, expensive and complex a concern with clearly different phenomena may not be essential as in the case of using nuclear reactors in neutron beam crystallography. In examining the development of novel components of scientific activity and the likely conditions of their institutionalisation it is obviously necessary to discuss the receptivity of existing institutionalised activities and related factors. As already indicated this is likely to vary among specialties and disciplines and so it is useful to have some means of characterising this aspect since different patterns of development can then be described and explicated. Although the notion of permeability incorporates the concepts of articulation, integration and institutionalisation it also refers to the degree of consensus and commitment to components and existing definitions of specialties and research areas.

7. Permeability and Receptivity of Components of Scientific Activity

The permeability of components of scientific activity deals with their degree of openness and closure. It is, obviously, connected to the degree of articulation for that implies boundary definition and maintenance. However, permeability implies more than simply the clarity and firmness with which components are differentiated, it also incorporates the degree of resistance or susceptibility to innovations from others. This in turn has implications for the relative states of development of activities. The degree of permeability of a particular component refers to its receptability to innovatory ideas and techniques from others which includes its articulation, but also deals with the 'meaningfulness' of possible innovations at a particular state of development. A weakly articulated component is obviously highly permeable. It is unclear what it includes so new ideas and techniques are frequently considered and applied. However, a moderate degree of articulation may exist with varying degrees of permeability. If available innovations are not applicable, in no way are relevant at a particular time, they will not be adopted and permeability is low. Alternatively, receptivity may be relatively high because of difficulties with current approaches and alternatives are sufficiently pronounced for innovations to be meaningul in the context of the current problems. For example, the application of electron microscopes to the analysis of biological tissues required a developed understanding of how they functioned in purely physical terms and the articulation of some 'craft' knowledge (Ravetz, 1971) in their use in physics combined with sufficient technical expertise in handling biological tissues. It also required the articulation of some way of making sense of the results obtained in biological terms. In other words some concept of the nature of biological tissues which could attribute meaning to them was necessary if there was to be any point in adopting this technique. Even then, it is by no means obvious that in fact this innovation would have been developed. There has to be an interest in doing so and, moreover, one which is not strongly opposed by other concerns or priorities. The transfer of innovations from one research area or specialty to another requires considerable development and extension of them if they are to be at all useful, which in turn implies the allocation of resources and positive commitment by some scientists. The permeability of a component, then, refers to the clarity of boundary definition, its state of development in terms of receptivity to innovation and the perceived need for ideas and techniques and, obviously, incorporates the ease of personal mobility into the dominant mode of social organisation. The more permeable it is: the less it is institutionalised, the more possible innovations are available, and the more likely a group of researchers, either from within the field or from other fields, will commit themselves to developing innovations so they can be usefully applied in the specialty or research area.

Permeability, also, involves the relations between different dimensions of cognitive structures. In a highly articulated and integrated set of components, commitment to new models, for example, would have major repercussions on techniques used and research practices adopted. Thus, the latter would be highly permeable to developments within the set, but relatively impermeable to those outside. In contrast, in a specialty containing components which were weakly articulated and integrated, they would be permeable since they did not exhibit clear boundaries and firm definitions, but the implications of developments at one level for those at another would, also, be hard to determine. There is a low degree of clarity and so connections might be weak and each dimension develop relatively independently. Where, for example, researchers use a multitude of techniques in dealing with rather vaguely defined problems and there are no clear rules about the 'correct' use of these techniques for any given research practice, the addition of a new technique or radical change in an existing one will probably not make a great deal of difference to what scientists actually do. Similarly, the introduction of a novel approach to explanation or accounting for phenomena may have little or no impact on current research practices because it is not clear how to apply it or use in day to day research activity.

8. Permeability and Relevance Criteria

The concept of permeability deals with relations between components of scientific activity in terms of the incorporation or consideration of developments in surrounding areas. The perception of a development as relevant and worthy of consideration to one's own work is obviously primarily in terms of whether it offers fruitful new approaches for current problems or general concerns which are considered important. The ways in which the fruitfulness of a development is evaluated are functions of existing activities, the potential innovation and the organisation of the specialty. The more systematic and integrated are a set of scientific activities, the more they are going to be impermeable to developments which are not directly translatable into the current problems and concerns. Where research is highly formulated, esoteric and 'technical' fruitfulness will be assessed in terms of direct connection to the current objects under investigation, e.g. as in optical astronomy in relation to radio astronomy (Mulkay and Edge, 1973). If a development is not easily expressable in the conventions currently being used – which are highly specific and formulated – then it will generally be disregarded except where, by virtue of the formalisation, it is seen that major inconsistencies exist.

Contrarily, the converse situation admits of broader relevance criteria which are not so clearly formulated and are more likely to be based on 'cultural' considerations than on current objects. These cultural factors consist of general metaphysical considerations of the 'scientific' nature of the development and so refer to, *inter alia*, the prestige hierarchy of the sciences and current disciplinary perceptions of the 'scientific method' as well as more societally based rationalities and values. Developments which are perceived to run counter to the dominant culture and rationality – ways of describing and understanding reality – are not likely to be seriously considered unless there are strong reasons for believing that they do offer substantial gains for current purposes as appears to be the case of the continental drift hypothesis in the earth sciences (cf. Hallam, 1973). But, as already indicated, grounds for such belief are much stronger in highly articulated dimensions where clear rules exist for assessing developments and – also most important – it is probably most likely that specialties with such activities are relatively well established and so moderately autonomous with respect to commonsense rationalities. In short, in the weakly articulated dimensions of scientific activity in embryonic specialties, the cultural aspect of evaluation will assume a much greater importance than in established specialties with articulated structures although, of course, it is still relevant.

The prestige hierarchy of the sciences – not necessarily a coherent unidimensional one – is, also, relevant when considering the characteristics of the development which might be institutionalised as part of a scientific activity in a specialty. The more it is consonant with current notions of 'scientific' and has been shown to be fruitful in other fields, the more likely it will be taken seriously. Here again, it must be sufficiently clearly formulated for it to be possible to see how it could be transferred and utilised.

In other words, the development must be sufficiently articulated to be distinguished and to suggest how it could be applied but, given this, it is more 'relevant' if it has been successful in a more prestigious specialty or discipline. This, likewise, applies to the innovator. If it is transferred by someone already prestigious in the specialty, or by distinguished scientists in established specialties, it is likely to be considered relevant. Again, the mobility of personnel into a specialty and set of activities are an aspect of permeability which focusses on the social organisation of research. Highly articulated components of activity are likely, particularly in established specialties, to have institutionalised their rules in the dominant mechanism of social control – professional journals and so form the basis of mechanical solidarity as Law indicates was the case for X-ray crystallography (1973). Scientists from other specialties will not be able to publish in these journals making alternative developments available unless they imbibe the prevalent system of rules and practices and are assimilated to the current institutionalised components. While social control may be even more vigourously exercised in specialties containing relatively weakly articulated components, if they are characterised by technical competition linked to opposing metaphysical positions and journals and organisations are based on these oppositions, their lack of clearly articulated rules and relevance criteria based on a coherent 'research programme' makes it more likely that scientists from more prestigious fields will be able to introduce developments without major modifications because more general 'science wide' or societal criteria for personal and scientific adequacy will be dominant. This dimension of permeability is linked to the scope and intensity of consensus among scientists in a given unit of organisation about institutionalised components.

The degree of consensus, here, simply refers to the extent of agreement among members of a specialty that a given component is significant and correct. It, also, covers agreement on its interpretation and application, i.e. its articulation. In specialties and disciplines where components are weakly institutionalised and it is difficult for scientists to identify any unit of organisation, let alone their major one, allegiance tends to be given to aspects of scientific research based outside the specialty and, indeed, their research is often defined in terms of these. In the field of cancer research, for example, scientists find it very difficult to identify their major specialty. They often refer to biochemistry or radiation biology which obviously cover much more than carcinogenesis. Indeed when pressed to do 'cancer relevant' research by sponsors, some say that it is unclear what the term signifies (cf. McAlpine and Carr, 1974). Most scientists are willing to admit the validity of different approaches in the sense that there do not applear to be any particularly strong grounds for rejecting virologically or immunologically based research in favour of purely biochemically based. The extension of recognised scientific fields into cancer studies, in other words, is accepted as legitimate by most workers. This is largely because the central problem is variously defined and weakly connected to research practices. The consensus on theoretical and technical eclecticism is based on a vaguely articulated concern which admits the extension of virtually any 'successful' approach to be correct. In one sense, consensus is high because there is little to disagree about. In another, it is high because articulated techniques and ideas are imported from proven fields and so legitimated.

There are, then, no necessary relations between the degree of articulation of a particular component of scientific activity and the degree of consensus. Similarly, the degree of integration of cognitive structures can be associated with varying degrees of consensus. In a monistic specialty with one set of highly articulated and integrated components, it would be expected that a high degree of consensus exists. In one with oppositional competition, the extent of agreement on the validity of any one technique or model would be low throughout the specialty, although, obviously, high within each opposed group. The degree of consensus on a particular component is, though, obviously related to that on others if they are highly integrated and so while the extent of articulation of any one structure may not be necessarily connected to the degree of consensus on it, when all aspects are taken into account empirical relations between articulation, integration, and consensus for one or two components have implications for others.

Since permeability includes the willingness of scientists to commit themselves to explore and develop innovations, it is inversely related to consensus. The more agreement there is concerning the validity and utility of a given component, the less likely scientists are to pursue possible alternatives. The greater the dissensus about a technique or idea the more permeable, in the sense of active interest in innovations, it will be. It does not follow, of course, that the transfer of innovation will actually occur in a situation of dissensus since the states of development of surrounding components may severely restrict the availability of possible alternatives. A relative lack of agreement about the validity of any one component need not result in an agreement on the importation of an alternative. Those who accept the current one will presumably continue to use, modify and develop it. The degree of consensus may thus continue low.

9. Conclusions

As indicated in the introduction the main purpose of outlining a number of concepts such as components of scientific activity, their articulation, integration and permeability and organisation into specialties and research areas was to provide a way of describing and accounting for diverse ways of scientific development. These terms are based on a view of science as a productive enterprise and emphasise the varied nature of knowledge production processes and their dynamics. The analytical separation of individual scientific activities and their components from units of social and cognitive organisation which follows this approach to describing science enables the investigator to consider changing patterns of activities without necessarily having to assume corresponding and equivalent alterations in modes of organisation. Obviously some implications do follow but the point here is that scientists may act in relatively novel ways without necessarily constituting new social formations or creating major cognitive innovations and so scientific development need not be always characterised as a process of sudden, earth shaking 'revolutions' (Martins, 1972) in modes of cognitive and social organisation but can be analysed in terms of processes of change of varying scope and intensity. This approach seems more adequate to recent accounts of particular developments and to the views and actions of scientists in a number of fields (cf. McAlpine and Carr, 1974).

A further point is that it is not necessary in this characterisation to take the individual scientist as the basic unit of analysis. Sociologists have always tended to accept the irreducible reality of the 'common sense' individual just as historians of science have, perhaps, treated current concepts of 'significant' ideas as their building blocks and so a gap has appeared between a sort of Popperian "third world" (1973: 106–190) account of scientific development and one based on 'invisible colleges' and solidaristic groups. While not suggesting that the approach sketched above closes or transcends this division it does at least offer the opportunity of taking the concrete individual as only one possible unit of analysis and focusses attention on activities rather than on some common sense idea of the atomistic individual scientist. In so doing, it opens the way to an understanding of science which distinguishes between individual actions, their institutionalised components, individual producers and their selection of activities and units of organisation and identity which constitute a background and framework for these activities and yet themselves are subject to the outcomes of them. This might be a start towards an account of how cognitive structures become articulated and change in relation to patterns of commitment and foci of identity such that scientific development can be analysed as a social process.

There are three further aspects which require consideration but could not be taken into account in the present discussion. First, the formation of different types of relations between sets of scientific activities within and between given units of organisation has not been seriously developed here. In an earlier paper (1975) I suggested four modes of competition or conflict which characterised relations between different views and approaches within specialties and discussed how these were related to particular features of specialties and research areas. These four types of competition can be seen as incorporating different scientific activities and so leading to a fuller characterisation of specialties which should enable a more adequate account to be given of processes of institutionalisation. This account would examine, in particular, processes of selection of components and the ways in which groups are formed around them and their relations with other, competing groups. Competition between alternative components and their combinations then, is seen as a major aspect of scientific work within specialties but there is also at least one other form of competition in science and this is connected with the second aspect requiring elaboration. Disciplines are seen here as performing largely a mediatory role between the exigencies of resource allocation procedures and university structures, developments in the dominant culture and in societal institutions and scientific activities. Within disciplines, specialties may be seen as competing for resources, just as disciplines compete within sciences, but they also compete for the allocation of prestige or relative scientific importance of 'their' reality and its problems. Specialties, that is, compete for being the most 'central' to a discipline and disciplines similarly conflict over their importance and so prestige and influence. This aspect of how specialties and organisational units containing scientific activities are themselves institutionalised into societal forms of legitimation and dominant structures obviously requires development and here notions of marginality, autonomy and the stratification system of sciences may be helpful.

The third aspect which needs consideration is the way of organising and directing scientific work. The importance of contrasting research styles and strategies in alternative developments has recently been emphasised by Mulkay and Edge (1973) in their account of radio astronomy and several other authors have pointed to the key role that certain personalities have played in establishing scientific areas (cf., e.g., Dolby, 1973; Mullins, 1972). The actual way work is organised on a day to day basis and priorities decided is of obvious importance in accounting for scientific development and is especially relevant in the comparative short term (McAlpine and Carr, 1974; McAlpine *et al.*, 1974). This has rarely been systematically studied in relation to scientific developments, organisationally oriented writers have tended to ignore the actual work of scientists and its changes while historians often appear to treat scientific research as occurring in some celestial location where nothing ever changed and all work was identically arranged. While it would be premature to suggest that this particular lacuna is anywhere near being successfully tackled at least an awareness of it may lead to future consideration (cf. Lemaine *et al.*, 1973).

University of Manchester

BIBLIOGRAPHY

Achinstein, P.: 1968, *Concepts of Science*, Johns Hopkins University Press, Baltimore.
Alexander, P.: 1963, *Sensationalism and Scientific Explanation*, Routledge and Kegan Paul, London.
Althusser, L. and Balibar, E.: 1970, *Reading Capital*, New Left Books, London.
Bachelard, G.: 1968, *The Philosophy of No*, Orion Press, New York.
Bachelard, G.: 1972, *La Formation de l'Esprit Scientifique*, Vrin, Paris.
Bershady, H.: 1973, *Ideology and Social Knowledge*, Blackwell, Oxford.
Canguilhem, G.: 1963, 'The Role of Analogies and Models in Biological Discovery', in A. C. Crombie (ed.), *Scientific Change*, Heinemann, London.
Clark, P. M.: 1972, 'Image and Symbol: the Role of Models in Modern Physics', in T. Shanin (ed.), *The Rules of the Game*, Tavistock, London.
Dolby, R. G. A.: 1973, 'Social Factors in the Origin of a New Science: the Case of Physical Chemistry', Leeds University, unpublished paper.

Edge, D. and Mulkay, M.: 1972, *A Preliminary Report on the Emergence of Radio Astronomy in Britain*, University of Cambridge, Department of Engineering Report TR7 and TR8 2 vols.
Elias, N.: 1974, 'The Sciences: Towards a Theory', in R. Whitley (ed.), *Social Processes of Scientific Development*, Routledge and Kegan Paul, London.
Griffith, B. and Miller, A. J.: 1970, 'Networks of Informal Communication among Scientifically Productive Scientists', in C. E. Nelson and D. K. Pollack (eds.), *Communication among Scientists and Engineers*, D. C. Heath, Lexington, Mass.
Hallam, A.: 1973, *A Revolution in the Earth Sciences*, Oxford University Press.
Harré, R.: 1970, *Principles of Scientific Thinking*, Macmillan, London.
Hesse, M.: 1966, *Models and Analogies in Science*, University of Notre Dame Press.
Hindess, B.: 1973, *The Use of Official Statistics in Sociology*, Macmillan, London.
Jevons, F.: 1972, 'Analogy versus Analysis: Biochemistry as a Mode of Biological Explanation', in T. Shanin (ed.), *The Rules of the Game*, Tavistock, London.
King, M.: 1971, 'Reason, Tradition and Progressiveness in Science', *History and Theory* **10**, 3–32.
Krupp, S.: 1961, *Patterns in Organisation Analysis*, Holt, Rinehart and Winston, New York.
Kuhn, T. S.: 1970, *The Structure of Scientific Revolutions*, University of Chicago Press.
Kuhn, T. S.: 1970a, 'Logic of Discovery or Psychology of Research', in I. Lakatos and A. Musgrave (eds.), *Criticism and the Growth of Knowledge*, Cambridge University Press.
Lakatos, I.: 1970, 'Falsification and the Methodology of Scientific Research Programmes', in I. Lakatos and A. Musgrave (eds.), *Criticism and the Growth of Knowledge*, Cambridge University Press.
Law, J.: 1973, 'The Development of Specialties in Science: The Case of X-ray Protein Crystallography', *Science Studies* **3**, 275–303.
Lemaine, G., Lecuyer, B. P., Gromis, A., and Barthelemy, C.: 1973, *Les Voies du Succès*, CNRS, Paris.
March, J. G. and Simon, H. A.: 1958, *Organisations*, Wiley, New York.
Martins, H.: 1972, 'The Kuhnian 'Revolution' and Its Implications for Sociology', in A. H. Hanson, T. Nossiter, and S. Rokkan (eds.), *Imagination and Precision in Political Analysis*, Faber and Faber, London.
McAlpine, A. and Bitz, A.: 1973, 'Some Methodological Problems in the Comparative Sociology of Science', A paper presented to an OSTI Workshop on 'The Comparative Study of Scientific Specialties', Civil Service College, London, December, 1973.
McAlpine, A. and Carr, I.: 1974, 'The Information Process in a Cancer Research Laboratory', Manchester University, unpublished paper.
McAlpine, A., Bitz, A., and Carr, I.: 1974, 'The Investigation of Patterns of Social and Intellectual Organisation in the Sciences', Manchester University, unpublished paper.
Mulkay, M. J. and Edge, D. O. :1973, 'Cognitive, Technical and Social Factors in the Growth of Radio Astronomy', *Social Science Information* **12**, 25–61.
Mullins, N. C.: 1972, 'A Model for the Development of a Scientific Specialty; The Phage Group and the Origins of Molecular Biology', *Minerva* **10**, 51–82.
Naess, A.: 1972, *The Pluralist and Possibilist Aspect of the Scientific Enterprise*, Allen and Unwin, London.
Nowotny, H.: 1973, 'On the Feasibility of a Cognitive Approach to the Study of Science', *Zeitschrift für Soziologie* **2**, 282–296.

Piaget, J.: 1972, *The Principles of Genetic Epistemology*, Routledge and Kegan Paul, London.
Popper, K. R.: 1959, *The Logic of Scientific Discovery*, Hutchinson, London.
Popper, K. R.: 1963, *Conjectures and Refutations*, Routledge and Kegan Paul, London.
Popper, K. R.: 1972, *Objective Knowledge*, Oxford University Press.
Polanyi, M.: 1966, *The Tacit Dimension*, Routledge and Kegan Paul, London.
Ravetz, J.: 1971, *Scientific Knowledge and Its Social Problems*, Oxford University Press.
Schmidt, A.: 1971, *The Concept of Nature in Marx*, New Left Books, London.
Storing, H. J.: 1966, 'The Science of Administration: Herbert A. Simon', in H. J. Storing (ed.), *The Scientific Study of Politics*, Holt, Rinehart and Winston, New York.
Toulmin, S.: 1961, *Foresight and Understanding*, Hutchinson, London.
Watson, J. D.: 1970, *The Double Helix*, Penguin, London.
Weingart, P.: 1974, 'On a Sociological Theory of Scientific Change', in R. Whitley (ed.), *Social Processes of Scientific Development*, Routledge and Kegan Paul, London.
Whitley, R. D.: 1972, 'Black Boxism and the Sociology of Science', in P. Halmos (ed.), *The Sociology of Science*, Sociological Review Monograph 18, Keele University.
Whitley, R. D.: 1974, 'Cognitive and Social Institutionalisation of Scientific Specialties and Research Areas', in R. Whitley (ed.), *Social Processes of Scientific Development*, Routledge and Kegan Paul, London.
Whitley, R. D.: 1975, 'Competition and Autonomy in Scientific Development', translated into German and published in *Kölner Zeitschrift für Soziologie und Sozialpsychologie*, special issue *Wissenschaftssoziologie*, Westdeutscher Verlag, Opladen.
Whitley, R. D.: 1975a, 'Management Research: The Study and Improvement of Forms of Co-operation in Changing Socio-Economic Structures', in N. Roberts (ed.) *Information Sources in the Social Sciences*, Butterworth, London.

TIBOR R. MACHAN

ON THE POSSIBILITY OF OBJECTIVITY AND MORAL DETERMINANTS IN SCIENTIFIC CHANGE*

> ... to make a communication to me to which I am absolutely prevented from imparting my own sense in returning it is to make one which I am in fact forbidden to make any communication of my own in terms of the message I have just received ....
>
> J. G. A. Pocock, *Politics, Language and Time*

1. Introduction

I begin with a short note on current efforts to develop a sociology of science. With the metaphysical (ontological) and epistemological underpinnings that most sociologists have accepted, a sociology of anything, including science is biassed in favor of determinism and reductionism.[1] Because of confusions about the nature of science – and the frequent explicit rejection of the very idea of 'the nature of science' – a sociology of science has problems about being scientific in its study of science. Several other self-referential problems arise in the field but these will be touched upon in my criticism of Thomas Kuhn's ideas (1970).

My paper will proceed along the following broad outline: first I will criticize Kuhn's approach to the discussion and characterization of science. This will involve an examination of the philosophical underpinning of Kuhn's book and his means for supporting his theory. I will focus on Kuhn's use of Wittgenstein's 'family resemblance' argument and his rejection of the notion of 'the nature of *X*' (or definitions).[2] Once Kuhn's problems have been identified I will offer an alternative approach to the study of history, specifically the history of science – i.e., people's scientific activities. I aim here to emphasize a different conception of human action from what Kuhn accepts and thus give rise to a form of explanation not available to Kuhn. I will borrow from the work done by Madden (1969), Madden and Harré (1973) and Louch (1966) and others. (Elsewhere I have made a beginning for this alternative approach (1974b).) My general

Knorr, Strasser, and Zilian (eds.), Determinants and Controls of Scientific Development, 75–111.

purpose is to lend support to what I consider a neo-Aristotelian, objectivist conception of science and its philosophical discussion. Thus some of what I will propose will appear quite orthodox – which, by the present approach, need not be a flaw.

2. Kuhn's theory in focus

Let me start with a consideration of Kuhn's preparatory remarks. He observes first of all that scientists "can agree in their identification of a paradigm without agreeing on, or even attempting to produce, a full interpretation or rationalization of it...," (1970: 44) He justifies this sort of procedure in science by citing Wittgenstein:

> ... For Wittgenstein ... games, and chairs, and leaves are natural families, each constituted by a network of overlapping and crisscross resemblances. The existence of such a network sufficiently accounts for our success in identifying the corresponding object or activity.... (1970: 44)

By this Kuhn is preparing us for envisioning that "Something of the same sort may very well hold for the various research problems and techniques that arise within a single normal-scientific tradition." (1970: 45). These remarks characterize his own approach in the book at hand. Kuhn's "aim is a sketch of the quite different concept of science [from that of 'a constellation of facts, theories, and methods'] that can emerge from the historical record of the research activity itself [within the history of science]." (1970: 1). So the method of Wittgenstein is evidently thought by Kuhn to be appropriate to what he himself is doing. He does not, for example, define the concept 'science' but talks of the 'image of science'. Instead of discussing the *nature* of paradigms he simply tells us what he 'takes' these to be. Both of these cases clearly accommodate his interpretation of Wittgenstein's idea – i.e., that talk of 'the nature of *X*' in any familiar way is inappropriate. Kuhn never considers any alternative to the idea of natures and definitions Wittgenstein criticized. This, too, indicates his full rejection of objectivism in formulating definitions.

Kuhn's problems start here. From my own investigations of the import of his work (Wittgenstein, 1953; Stroud, 1966; Machan, 1973) it seems unwise to extrapolate Wittgenstein's conclusions about the nature of identification of objects like chairs; we must take care not to misinterpret the idea that when we try to identify the essential features (components, properties, aspects, etc.) of such objects (and anything else we may want to), all we are entitled to expect is a statement of "a network of overlapping and crisscross resemblances." I am sure we can and ought to do better, at least once we have understood the mistake of approaches Wittgenstein focused upon his work. Wittgenstein seemed himself to be clearing a path to such better solutions to the problem of definitions, criteria, standards of judgment, etc. (Stroud, 1966; Wittgenstein, 1969).

Kuhn, however, takes Wittgenstein to be offering a clear alternative to the empiricist approach with the idea of family resemblance. Kuhn talks of the implications of this idea for the problem of identifying a paradigm. He says that the Wittgensteinian idea shows that no justification for such an identification could be provided – instead we have 'interpretation' and 'rationalization' (!) (1970: 44). In the 'Preface' to his work, Kuhn follows faithfully this conception of what a scientist needs (and does not need) so as to produce "the particular conclusions he does arrive at..." (1970: 4). He says that these "are probably determined by his prior experiences in other fields, by the accidents of his investigation, and by his own individual makeup." He tells of "that element of arbitrariness" that is "compounded of personal and historical accident... (which) ... is always a formative ingredient of the beliefs espoused by a given scientific community [and, he says elsewhere, by the individual] at a given time "(1970: 4). Accordingly, Kuhn gives us a plethora of data about himself as he launches into his own project – sources of insight, influence, guidance, "random exploration," and other bits and pieces. These, by his own account, should help us to appreciate (though not show the justification for) Kuhn's "objective to urge a change in the perception and evaluation of familiar data," as well as a "new image of science" – i.e., to help us witness identification without justification!

Am I fair to think that Kuhn's approach isn't just self-referential mutatis mutandis, but is, instead, explicit and by self-conscious meaning? Clearly he considers himself to be doing history as well as sociology of science based in part on 'historical evidence' and resulting in something very close to a law of the history of science – e.g., when he says that "if I am right that each scientific revolution alters the historical perspective of the community that experiences it, then that change of perspective should affect the structure of postrevolutionary textbooks and research publications" (1970: ix). (I would not be surprised if Kuhn conceived the recent work in the history, sociology, and even philosophy of science as undergoing just the kind of change he is talking about in consequence of the revolution his work – and that of others suggesting his ideas in greater and greater numbers throughout the last few decades – produced. Although, this would pose a problem for Kuhn when talking about being 'right',[3] in light of some of his epistemological ideas – as I have argued elsewhere (Machan, 1974b) and will come to later in this paper.)

The above considerations set the stage for my criticism of a sociology (and history) of science *à la* Thomas Kuhn. I will not deal with Kuhn's specific renditions of his historical data, although I believe that for him to come even close to providing a meaningful (!)[4] justification for his conclusions about what actually went on in science throughout history, he would have to do more than *cite* widely discussed cases. This because Kuhn would have to show that nothing else but what *he takes to have happened* (in terms of his own paradigm) did happen – scientists *actually* did what they and/or their biographers say they did; i.e., whether things in fact occurred as they would have to have occurred for Kuhn's conclusions to be right.[5] Obviously I cannot make any headway *here* on such a project. Instead I am concerned to challenge the soundness of Kuhn's approach by pointing out its methodological and epistemological faults.

In trying to criticize Kuhn, one faces a recurring problem. This is that by Kuhn's own account, any criticism *must* begin with arbitrary standards.[6] Although Kuhn directs this point at the dynamics of scientific thinking and research, it is evident from what I indicated above that Kuhn's own approach falls under the kind of thinking and research scientists carry out. Before anything even mildly hopeful can be said in criticism of Kuhn, we need to see if the philosophical underpinning of his approach (to tackling issues in the history of science) is correct. This will take us to Kuhn's epistemology right off, as well as to the metaphysical implications of his approach. For while Kuhn eschews all epistemological argumentation in his book, the underlying epistemology we find can make it clear what goes on at later stages of his discussion. Moreover, it is not as if Kuhn were oblivious to such issues, only that he introduces them without sufficient detail and self-criticism. By this introduction, however brief it may be, Kuhn opens the door to what we might call cross examination on the topic.[7]

Kuhn brings in Wittgenstein's point about family resemblance very early, and quite understandably so. He aims to rest a good deal on it. But there is reason to believe that Kuhn reads too much into this Wittgensteinian point. Kuhn thinks Wittgenstein presented us with an impossibility proof about definitions of concepts. What Wittgenstein did, in fact, was to show the error in assuming a particular model or 'picture' for what a definition must be. He showed that taking a definition to be a fixed, immutable, unchanging standard is a mistake. He objected to conceiving of criteria in such a way that once they are correctly understood and identified, no more could possibly be learned about them. Wittgenstein realized that this view held it impossible that we *could* find it necessary to later revise our criteria for identifying what something is. He clearly seems to me to have been after dislodging a variety of Platonism[8] where the model of what knowledge and truth must be restricted us to accepting only certain kinds of things as capable of being known. This, Wittgenstein seemed to have realized, amounted to a prejudice – a case of prejudgment about *what* human beings can know and about the nature of knowledge itself, what *could* constitute knowledge of reality. (Incidentally the Platonist *model* need not be restricted to idealism; so Wittgenstein's objections could hold in all cases where the model guided investigation, independent of the particular account of knowledge by which it was to be satisfied.)

Kuhn is not unjustified in his invocation of Wittgenstein in connection with his criticism of the positivist ideals of knowledge, truth, and in turn, science. Wittgenstein's argument against an attempt to find what "all games and only games have in common," namely certain "features" and "elements," (1953: 66) is placed within the context of a certain idea many philosophers – including Wittgenstein himself at one time – had about *what* these features *must* be. He does not conclude that under any interpretation of 'common' it would be impossible to require common features of things, etc., to be identified as one *kind* of thing, etc. He says that "we see a complicated network of similarities, etc.," but this would appear to apply to efforts that aim at *perceiving* common features.[9] In Wittgenstein's context the conclusion about coming up only with family resemblances need not imply that we *could not* do better by producing a different account of the nature of criteria for identifying the essence of games. (Indeed, Wittgenstein goes to some lengths to provide clues for the formulation of an alternative approach, one I think expressed most cogently by Barry Stroud (1966).)

Kuhn drops the context of Wittgenstein's discussion. He then extrapolates (what he wrongly takes to be) Wittgenstein's point to the subject matter of the justification of a paradigm. So Kuhn begins on the wrong foot epistemologically. Thereafter several other points are used by Kuhn to support this beginning – e.g., Piaget's experiments, the difficulties associated with making correct distinctions in the cases of cards that have their colors switched and the duck/rabbit drawing (which he introduces at one point but appears to drop at a later one) (1970: 63, 113). This approach to establishing the foundation for what he will conclude (about criteria for making justified selections from among competing paradigms) allows Kuhn to speak of the ultimate incommensurability (1970: 44) between what he maintains are equally adequate *scientific* paradigms. For Kuhn does end with an *impossibility* proof. He warns that his position does not "suggest that new paradigms triumph ultimately through some mystical aesthetic." Nevertheless he proceeds as follows:

> ... But if a paradigm is ever to triumph it must gain some first supporters, men who will develop it to the point where hardheaded arguments can be produced and multiplied. And even those arguments, when they come, are not individually decisive. Because scientists are reasonable men, one or another argument will ultimately persuade many of them. But *there is no single argument that can* or should persuade them all.... (1970:158, italics added)

Because there is no single argument of that sort, "A decision between alternate ways of practicing science is called for, and in the circumstances that decision must be based less on past achievement than on future promise.... A decision of that kind *can only* be made on faith." (1970: 158, italics added) (This last does not follow from *his* premises!)

It is with this element of impossibility that I want to take issue. Kuhn does not *show* that "there is no single argument that can or should persuade them all," or that "A decision of that kind can only be made on faith," albeit he reads the historical material to be *indicating* such an impossibility. But, first, Wittgenstein's philosophical investigations do not offer an impossibility proof. Even if the family resemblance argument goes through – and there are grounds to claim that it does not[10] – the context of that argument is misread by Kuhn. Second, Kuhn is talking about what *persuades* scientists, and persuasion is not what is at issue in arguments, only in seeking converts. What is at issue is whether some (set of) arguments might be conclusive in the context of solving some problem or settling some dispute. Perhaps many instances of paradigm introduction *have* not been accompanied by conclusively sound or unsound justifications. But this does not prove that they *could* not be so accompanied. There are instances in most areas of science where even from within the existing context more care and attention could have precipitated better selections. Jean Rostand (1960) tells that the history of *N*-rays involved both honest error *and* wishful thinking. He also chronicles the story of radiesthesia and metapsychics, the first generally avoided by reputable scientists but the second still going quite strong. In these and other cases, a paradigm gained a significant degree of success unjustified by the relevant considerations. What is crucial, of course, is that the 'relevant' considerations be identified, and Kuhn does not help us understand in the least when scientists have *good* reasons for following through on paradigms, what with his consignment of decisions or choices to the area of faith. This Kuhn seems to do without researching the kind of considerations that might have to be appealed to at the crucial stages of drastic developments within some field of science.[11] Are we to understand that these stages involve no epistemological issues? Nothing of metaphysics could have a bearing on what will be taken to be good reasons for making selections between competing paradigms? Or if we should consider those factors, does it not follow from his views that in *all* areas mentioned, and others that could be relevant, we can only make a decision on faith? Granting that in periods of normal science there is a kind of 'objectivity' to scientific reasoning – based on the accepted paradigm and what counts as good reason within it – must we accept Kuhn's broader, or what I want to call transcendental, subjectivism because at the critical stages no preestablished basis for selection is readily available, that is, since none are agreed to or identified just then?

This is no meager point. Kuhn and I may both agree that scientists themselves might not *actually* bother with justifying their paradigm choices. Especially when such choices accord with some widely accepted doctrine about the nature of knowledge, argumentation, evidence, or other extra-scientific and more properly philosophical beliefs. But in Kuhn's scheme these too *must* end in arbitrariness. Kuhn is actually proposing a variety of (Peirce's) pragmatist epistemology[12] and implicitly, at least, affirms the impossibility of identifying some metaphysical fact that can function as the ground of justification. Yet he has argued none of this, leaving us, nevertheless, with a very powerful polemic. I intend to point out how crucial, paradoxical and, indeed, dangerous the consequences of this view are.

But for now let's consider another feature of Kuhn's method of argumentation. Kuhn makes several of his points on the basis of evidence drawn from history, psychology, and sociology. Evidence from history means case histories in Kuhn's context. But this is not so with evidence from psychology and sociology. Kuhn himself tells us that the weight, significance and even the factual content of the claims of scientists are paradigm-determined. He tells us that in the role of "a vehicle for scientific theory... [a paradigm] ...functions by telling the scientist about the entities that nature does and does not contain and about the ways in which those entities behave" (1970: 109), Kuhn believes that scientists who work within a paradigm are limited by it as to what they may admit as evidence, as to what they are entitled to call a fact, and regarding what they may take to be an entity of the sort involved in their investigations.

Before turning to the consequences of Kuhn's ideas for his own method and results, it should be mentioned that Kuhn assures us that "Whatever he may then see, the scientist after a revolution is still looking at the same world" (1970: 129). But this is very difficult to reconcile with the main trends in Kuhn's theory, since he also says that during revolutions scientists "see new and different things when looking with familiar instruments in places they have looked before" (1970: 111). This leads to difficulties. By not entering into the philosophical issues involved in these remarks, Kuhn renders it impossible for us to understand him clearly. What is it to be looking at the *same* world? More specifically, how could one tell *that* if, necessarily, *we all may* see the world differently?

When we turn these puzzling ideas upon his own "scientific theory," Kuhn immediately faces difficulties. The evidence for his theory is no evidence at all. Instead it must be construed as a case of the world (being looked at) "seen differently from the way it had been seen before." We have to ask whether the material cited by Kuhn could ever be invoked to support him – be that material facts about the Priestley/Lavoisier affair, Piaget's discoveries about children's perceptual development, or problems about telling the difference between cards ordinarily colored one way when the colors are switched. Kuhn could reply that he is merely trying to make sense of the world by his interpretation of some important events of concern to us. But this will do for data from the history of science only, not for Piaget's data (1970: vi, 112, 113), for in the latter case Kuhn is not trying to *make sense* of what Piaget discovered but is *using* the discovery to make sense of the history of science. What he says about seeing the world *via* a paradigm must apply to Piaget, also. Once a paradigm switch *à la* Kuhn occurs in perceptual psychology, and Piaget's facts take on a different significance (as determined by it), not much is left of Kuhn's evidence. The same can be said about the card problem and the duck/rabbit case. These too would have to be reinterpreted if a paradigm switch obtains within epistemology and psychology. And when we push the self-referential point further, we are left with no materials that can, in principle, give support for Kuhn's own position – not even in some kind of network of overlapping and crisscross resemblances.

Let us now turn to a philosophically central idea in Kuhn's system, namely the model of 'being right', or truth, knowledge of nature, that he appears to invoke as he drives home his conclusion that decisions about which paradigm ought to be selected "can only be made on faith." This last issue is most crucial for purposes of assessing Kuhn's metascientific theory, especially its impact upon a concern with how best to conceive of science as a human activity. Kuhn here is unfortunately giving support to the absurd and damaging contention that the human mind, in this case that of any scientist, is unable to make reliable, justifiable, conclusive, and correct identification of what there is in reality; he is affirming the impotence of the human mind in the matter of learning whether we have identified reality correctly.[13]

Kuhn says, late in his book, that he rejects the view that science evolves toward anything (1970: 160, *passim*). He concludes this from a criticism of the (so-called) positivist notion that science began with ignorance and is moving toward some conclusive, completed truth about nature. His account rejects this idea in favor of "a process of evolution *from* primitive beginnings... whose successive stages are characterized by an increasingly detailed and refined understanding of nature" (1970: 170). How he is able to say this much is unclear, but for him the question of whether this understanding is valid cannot be asked except from the viewpoint of a paradigm – one that is selected, in the end, on faith.

A word here about what Kuhn might mean by 'faith.' He does not discuss it. The idea is generally used to mean an approach to reaching conclusions or accepting beliefs without reliance on evidence and proof.[14] Existentialists have used the idea to mean an act performed with nothing to justify it, without a possible rationale – an irrational or arational choice each of us faces in human life, constantly and essentially. We find in Kuhn a combination of these two senses of faith.

Clearly Kuhn is after something important, quite likely true, in these allusions to faith, something positivists and mechanistically bent thinkers have simply ignored or actually denied. This is that people, including scientists, need to face some problems all alone, with nothing outside of themselves to fall back upon. Kuhn as well as Toulmin (1953) point to something like this in the lives of professional scientists. I shall say a good deal more about this in my positive remarks.

But as with many Existentialists, Kuhn also omits a thorough examination of those types of circumstances – although Sartre hints at its nature when he locates their source in human consciousness. When we identify the nature of this "act or leap of faith," this initial step that has been so important to the Existentialist philosophies and to anyone concerned with human creativity, it could turn out that some very definite traits are required to produce satisfactory results (Koestler, 1964). It need not turn out that such an act necessarily excludes objectivity. But for now I am concerned with the conclusions Kuhn draws from his "leap of faith" doctrine for purposes of understanding the concept of scientific truth, including whether we might not, after all, be able to speak of a correct or even true paradigm. These conclusions may now be turned upon Kuhn's own ideas.

I noted before that Kuhn uses the concept 'right' in connection with what he himself has identified (1970: ix). In that context Kuhn is offering an objective conditional statement, one he believes could be true: if his theory in the book is sound, then his statement is true. So in this case, at least, Kuhn believes one *can* be right. And since he also says he is working here as a historian and sociologist of science (1970: 81), he allows objective knowledge to enter the realm of science in at least one instance.

Yet by its own meaning that conditional proposition denies to him the prospect of such objective truth, i.e., knowledge of history. Consider what he says about what scientists can see and know:

... during revolutions scientists see new and different things when looking with familiar instruments in places they have looked before. It is rather as if the professional community had been suddenly transported to another planet.... (1970: 111)

Unless we confine the meaning of "scientists see" to the strict sense of "visually perceive" (which is not what Kuhn's usage indicates), Kuhn renders it impossible from his own framework to speculate on his being *right* about anything in the above objective sense of the term.

Kuhn's problem, I submit, lies in the rejection of the possibility of objective truth in science – the idea, that is, that scientists can know reality. This rejection follows from an invalid conception of the nature of truth offered within the positivist/inductivist framework Kuhn is challenging. As that framework prescribes it, the idea is indefensible. But that does not prove the impossibility of scientific truth. The idea of what truth is must be supplemented by a clarification of the nature of meaning. J. L. Austin's advice must be heeded in this connection when he tells us that something is quite wrong if "we suppose that language (or most language, language about real things) is 'predictive' in such a way that the future can always prove it wrong. What the future can always do, is to make us revise our ideas..." (1961: 56–57). At any given stage of human knowledge one could well say what is true – in or outside science. A true statement would be the best expression of (a judgment about) what is the case, some fact (state of affairs, principle, law, etc.) – i.e., one could possess the best,[16] most rational, idea about the fact in question.[17] What is true is a statement, and the criterion for ascertaining this could only specify something that is possible for people to do in terms of what is available to them to invoke as evidence, argument, sense, meaning, etc. Facts themselves may be reported, explained, described, etc., better later when meanings had been revised; those facts remain forever what they had been when stated in a true statement. Thus scientist could and often did know reality. It is philosophical ideas about truth that need to be called into question, not the capacities of scientists (Madden, 1973: 129). Our concepts of truth, knowledge, right, etc. have, for the most part, suffered utopian characterization by philosophers – although exceptions can be found. Yet this does not mean that *people did and do not actually know reality*, that *true statements have not been uttered* by scientists in virtually any period of human history (even if within the corpus of a scientist's work there will most likely appear a mixture of true, partly true, and false claims – sometimes undistinguished, other times *mistakenly* distinguished by the scientist and his contemporaries). The idea that truth is impossible is absurd – it amounts to holding that all of science rests on (a chain of) claims that include none that are true; nor any that are false!

But philosophers can make mistakes; they can also be closer and further from the truth and from what is in fact the best characterization of the requirements for truth itself. Kuhn seems to be right enough in indicting the "traditional" (positivist) philosophers of science for their characterization of truth (1970: *passim*). But instead of setting out to identify what amounts to a revised and better characterization, he abolishes the very possibility of objective truth.[18] (And by "objective" I do not have in mind the narrow/impossible sense of that term used by absolute idealists, nor what is meant it by positivists, behaviorists and other empiricist oriented thinkers, namely, roughly, 'capable of being supported by references to sensory experience'. I am using the idea to mean something more closely akin to 'capable of being provided with a justification in terms of standards appropriate to the subject matter.'). In his efforts to show the errors and misleading implications of positivism and empiricism, Kuhn goes on to deny something that is not at all strictly tied to the views he criticises, namely the possibility of objectivity in any way, shape or form.

Interestingly, however, Kuhn runs into difficulties beyond simply excluding his own position from the realm of (the set of) objectively true statements and conclusions. He may try to extricate himself from the self-referential problem with the admission that 'right' (or 'true') means for him what *his* paradigm forces upon him, no more. And who are we to challenge his choice about how he *will* use familiar concepts – everyone else is playing havoc with language, why can't Kuhn! A more severe problem arises for Kuhn at a point where he adamantly distinguishes between being *unreasonable* and being *illogical* – in the case of Priestley, for instance (1970: 158). The latter was unreasonable but not illogical! This distinction is made even more explicit by Morton L. Schagrin when he tells us, in what amounts to a defense of at least some of Kuhn's main points, that "Priestley was always rational... he was merely 'unreasonble', not irrational." (1973: 9) The meaning of 'rational' and 'logical' in these passages is clearly "loyal to the canons of sound reasoning, valid argumentation." Kuhn, as well as Schagrin – who is actually closer in his views to Feyerabend than to Kuhn – would admit that these canons remain, throughout the history of *bona fide* science, entirely indispensible. Indeed, to deny this much to science, at least, would clearly invite admission of the widest variety of kookism as bona fide scientific activity.[19]

Yet with the insistence on logic it becomes impossible that objective truth could be dispensed with. The ideas of 'rational' and 'logical', in the above sense, involve an explication in terms of the concept of validity. But validity is, in turn, defined as the attribute or aspect of arguments found just in case an argument is such that if the premises are *true*, the conclusion must also be *true*. But this definition presupposes and requires the applicability (meaningfulness, intelligibility) of the concept 'truth'. The idea of *proving* some conclusion assumes that there exist premises that can have truth value, that can provide the material in support of the conclusion. If it is impossible to obtain such premises, if we cannot know whether premises are true, and, thus, cannot know reality, the source of premises, then it is impossible to conceive of logic, proof, rationality and the host of related enterprises as *meaningful*. For a method such as logic to exist as a proper method for some purpose, it must be applicable; i.e., it must be capable of having a subject matter. There can be no point in insisting that scientists or philosophers or any other group of people proceed *rationally* unless such a procedure *can* secure for them some measure of contact with reality. Much else may be required for particular purposes – e.g., studying the nature of the physical aspects of reality – but rationality as a minimum is indispensible. Kuhn seems to recognize this (Aristotelian) doctrine, as does Schagrin. But because of a commitment they have to a doctrine about the relationship between logic and reality – e.g., a form of the analytic/synthetic dichotomy – they do not accept the presupposition of rationality, namely that human beings can know[20] that they have identified reality correctly.

I should make clear that the suggestion that false premises are sufficient for logic invites the reply that falsity is inseparable from truth. But the objection that to understand the meaning of the concept 'truth' does not require any knowledge whatever cannot be treated here – a full theory of concept formation and concept validation would be required (Rand, 1966). There are traces in Kuhn, but even more in Feyerabend (1963), of a full blown conceptualism. The above points must suffice for now as crucial objections to this view, even if an adequate alternative cannot be offered here. If anything entitles someone to reject another's views, then the fact that those views deprive him of the possibility of even ascertaining the subject matter of his discourse (ideas, theories, discoveries, etc.) must suffice for that purpose. In the end, despite the many very valuable and correct conclusions that comprise much of Kuhn's work, one is led to unacceptable conclusions. Kuhn's position disallows the possibility of anyone's knowing whether what he has concluded about some topic is true, whether one's beliefs are correct – in science or in any other field of concern to human beings.

It must be evident by now that my criticism of Kuhn has a relatively special focus. I have been concentrating on those features of his position that have significant consequences for the theory of knowledge and the philosophy of science. Kuhn says of Descartes' philosophy that it is a paradigm (1970: 121) .Thus he must accept that his own position qualifies as part of an alternative paradigm. By showing that, when turned upon itself, his theory of paradigms leads to the impossibility of our ever learning if it is correct, that theory is seriously, if not fatally, undercut. Kuhn seems to me to have set out to present us with several notions that are valuable but with a philosophical position that, like Kant's attempt to save rationality by conceding the impossibility of knowing reality, leaves us in a quandary as to its truth. Kuhn may be well aware of this, and the aptness of his term 'revolution' to characterize what must constitute the development of science needs to be fully appreciated. For Marx violent revolutions are (at least in some of his works) indispensible for the dissolution – not resolution – of inevitable class conflicts. These conflicts *cannot* be resolved (in the 'middle' Marx at least) because the consciousnesses of the members of the classes are set by factors not within human control. (Both Marx and Kuhn soften their theses at later stages (1970: 38).)

In Kuhn, also, certain features of the human makeup serve as the ultimate sources for the selections scientists will make from among different, incommensurable (but in some sense equally plausible) paradigms. Other problematic features of this aspect of Kuhn's position are treated in another paper in which I discuss his ideas (1974). One point I make there bears reiteration. That is that when Kuhn ends with his (self-mitigated) subjectivism, where determinants such as "personal and inarticulate aesthetic considerations" produce the choice (?) of paradigm candidates, he is proposing that scientists do not act freely at all but are moved by factors outside their own control (1970: 56, 75, *passim*) – at least in the crucial stages of their professional lives. (How they might still act freely at other stages is not clear from Kuhn.) But here we arrive at a problem Kuhn does not face up to, one that seems to be far more general than Kuhn's announced topic poses for him. This and other more general problems are the province of philosophy as such, not of science, since to understand what it is to be acting as a scientist requires, first, an understanding of what it is to be acting as a human being (Chein, 1972; Louch, 1966).

To shed some light on how the issues Kuhn discusses could be approached without meeting with the difficulties I have identified above, it will be helpful now to consider an alternative in terms of which these difficulties may be avoided. It will be clear from the outset that I have no intention of reintroducing the main and explicit target of Kuhn's criticisms, namely positivism. Whether my alternative is novel or ancient (or some other historically characterizable approach) is unimportant – except to someone who finds it axiomatic that philosophical controversies follow some inevitable dialectic of their own. It could well turn out that, fully developed and then identified, the viewpoint I will outline takes us back to earlier ideas.[21] I do not find that to be a defect, not especially in terms of my own implicit thesis that both philosophical and scientific *knowledge* has existed in virtually any stage of human history, even if the *statements* of what philosophers and scientists knew *then* no longer require exactly the same wording, nor the justification of these statements the same (albeit contextually suitable) procedures.

3. Towards a Theory of Scientific Conduct

There are some points of my discussion I will barely touch on, with the hope that, while my statements may not meet with full approval by those whose acceptance of them I wish for, their meaning will be uncontroversial enough to let us move on. I begin with the point hinted at earlier.

Relying on sociology to inform us about *what science is* raises problems that emerge because of what sociology, as a field of science, has been taken to be (by many involved in a sociological approach to science). Most social scientists, including sociologists, believe themselves to be responsible to engage in work within their field by living up to standards laid down outside of the sciences as such. This is natural, since scientists are most likely to know more about what they study than about what at minimum is required to study anything at all. Epistemology is not a branch of any of the sciences, yet epistemologists do the most work on problems related to answering the question of what knowledge is.

Epistemology, however, is not the first or most basic branch of the field of philosophy. To study knowledge, for there to exist a subject matter for epistemology, *something must be*. If nothing is, then there is no knowledge either, and epistemologists might as well quit at the start. Before epistemologists make sense as people inquiring about something, some statements about reality – about the realm where whatever we may know might be found – must be in order. The traditional field of metaphysics comes to mind to fill the role of the pre-epistemologist's domain.[22]

For reasons speculated upon by most historians of philosophy, metaphysics has not fared well for some time, except in less than dominant circles. Suffice it to note that those in philosohy, and, with their guidance, many others in different but not unconnected fields, chose to turn their attention to epistemology, especially since the time of Descartes. So most philosophical investigations have produced theories in terms of which questions of metaphysics took a secondary position. Fields such as sociology, taking their clues from epistemologists mainly, had plenty of guidance as to what knowledge might be but hardly any about what reality is a such – what anything that is, *must* be, at least. Method took precedence over content or subject matter. The latter had to be constructed out of the requirements and limitations of the former. Kuhn is to be credited with illustrating just how elusive the world seems after generations of living under the influence of this philosophical 'ordering of priorities' – and he ends in a philosophical position that is a result of turning that influence upon itself. (Dialectical conceptualism!?)

Positivistic/behavioristic sociology – pre-Kuhnean, pre-phenomological – tended, in the main, toward an implicit, indefinite adoption of the (anti-) metaphysics that could be salvaged for most of the sciences. For members of a growing but still somewhat disorderly field it would appear to make some sense, albeit not good sense (Louch, 1966: 7, *passim*), to suppose that getting to be scientific must involve adopting methods previously found fruitful elsewhere, methods well suited to the study of different areas of reality.

I said earlier that this resulted – i.e., that making these choices resulted – in the adoption of questionable methods and a very restricted ontology. The significance of this is that in social sciences we find, in the main, an insistence upon the narrow sorts of observations, interpretations and on results culminating in causal laws, involving *relations of events*. But events cannot do anything – they occur and contribute to the occurrence of other events, *ad infinitum*, in an everlasting chain (Chein, 1972; Taylor, 1966; Madden, 1969; Fisk, 1974). Even the exceptions, like Marx, built on this idea instead of challenging it at the roots. (Marx did homage to the empiricist methodology but gave it a supposedly empirically derivable dialectical twist, avoiding thereby as does Kuhn, the image of history as a smooth, upward moving, curve.)

In discussing science in general, as well as in the history, sociology and philosophy of science, we find now the consequences of the above choices. Granting some exceptions – e.g., Louch (1966), Polanyi (1969) and Agassi (1963) – the bulk of those thinking in these areas talk about scientists as if they were basically ineffectual, incapable of initiating actions, mere processes in the progress of science. I mention these points for I want to suggest that the ontology spoken of above is wrong: it could be ontologically correct to suppose that different causal laws might be found when we study the behavior of entities with physical properties, chemical attributes, biological features *and* human characteristics. I am suggesting that by turning to this alternative ontology, a non-reductionist conception of scientific understanding is made at least possible – i.e., suggestions of free will are no longer open to *a priori* dismissal from the point of view of the sciences that study human affairs. Human beings could be causes of some events.

I am not at all denying the possibility of a science of human affairs – or many sciences of those affairs. Economics, political science, psychology, sociology could all be scientific in the sense of seeking as thorough and precise an understanding of aspects of human life as the tools of the fields will allow – always acknowledging, however, that an approach would be dogmatic, not scientific, if it *presupposed* the ontological type of its subject matter, pushing man into a framework instead of developing the framework through a better and better understanding of human affairs by its inquiries. When sociology, for example, imposes the ontology of events upon human affairs, then the sociology of science will tend to do the same, and room for genuine choice – a creative (mental) act – is missing from the start. Rational, natural explanations of the events of the history of science are forced into the world view of an ever-evolving chain of events. Kuhn, antipositivist, falls prey to this but for his slight variation, similar in type to Marx's revision of mechanistic materialism (1968).

Thus far I have only made some admittedly sketchy attempts to clear the field for some constructive suggestions. What I want to argue is that once the room for choice, i.e., self-determined initiated action, is available, we can ask *whether* human beings can choose at least some of what they do, some of their actions. I will not enter here the dispute about freedom versus determinism because I could not do it justice. It does seem to me absurd to deny that people can choose (however much this is circumscribed by nature) some of what they do, most of all when we turn to their thinking (Boyle *et al.*, 1972; Jordan, 1969; Machan, 1974a). Without that fact secure, so much else we do would amount to nothing meaningful – knowledge, understanding, valuation, judgment, inquiry, research criticism, improvement, modification, revision, etc., to mention just some of the most pertinent areas – that I should suppose that writing papers and arguing about the validity of Kuhn's or Kuhn's critics' ideas would also lose their meaning. Be that as it may, let me for the moment simply explore what follows from the acknowledgment of human freedom of choice.

Let's admit that people can initiate some of their actions. These can include their scientific activities (e.g., research, reasoning, selection of explanatory schemas, etc.). If so, then a way to approach the history of science is to ask about what scientists chose to do and why, as well as about whether their choices were justified, reliable, objective, biased, irrational, hasty, stubborn, etc. One might suppose that Kuhn would consider this approach to be comparable to importing a *preconceived* view of human nature upon the study of history. Nonetheless, Kuhn would be the first to accept some of this – he himself imposes his own paradigm upon the affairs of science in the past. We are left with having to learn which is the *better* way of doing it. I am *not* suggesting that we judge what scientists did on the basis of what Agassi calls the most up-to-date textbook version of some field of study (1963). All I am suggesting is that we must have a clear conception of what human beings are, what they can and cannot do as such, as well as what they ought to do in general (and derivatively, in the context of their vocation or profession), before we can hope to provide a correct rendition of any part of human history of science. (Again, by 'correct' I do not mean 'final' but what is the only sensible rendition of that idea, namely 'rationally justified'.) If I am right that people can choose and that they are responsible to carry on objectively, especially in science (but no less so when they need to resist their bad habits or biases in, e.g., personnel hiring, evaluating or otherwise dealing with important matters, including people, in their lives), then the identification of the circumstances people have faced and the fullest possible knowledge of those who have faced these circumstances can, although will not *necessarily*, lead to an understanding of what they did, how well or badly they acted, and what results their actions met (Louch, 1966).

While such suggestions may appear to be rather mundane, they offer an avenue of approach in our attempt to understand human history not found by way of Kuhn's theory. I am particularly interested in the possible force of a certain kind of *ad hominem* argument that might be invoked in the explanation of scientific conduct (Gerber, 1974). We often do not have sufficient knowledge of the details of someone's thinking, the evidence he had available, the precise context of knowledge enjoyed within his field, etc. If we had to rely on such information alone we would be at a loss to explain his actions. (It seems that Kuhn chose to invoke "elements of arbitrariness" when he came to coping with such cases.)

I am suggesting that instead of settling just for 'social pressures' or supposedly subjective aesthetic preferences to make sense of conduct in such cases, we ought to try for a consideration of people's character. Elsewhere we find character reference quite important, we can – and often ought to – use it here as well. Scientific conduct is not immune from moral considerations – nor science itself uninfluenced by human virtue.

Of course the resistance to such an approach is overwhelming. While pundits, visionaries or laymen discuss the affairs and conduct of scientists and other professionals – e.g., military personnel, political leaders, bank tellers, businessmen, oil industrialists, and others – by reference to moral character, those close to science itself shy away from this (Merton, 1962; Polanyi, 1969). Several factors, some related and others distinct, account for this reluctance. One is that science is supposed to deal with facts, whereas most commentators and serious scholars consign matters of moral character to opinion or superstition. But there is very much wrong with this dichotomy – and Kuhn is among those who have helped us to realize that. Another is that many who dominate in these areas believe that explanations *must* be subsumed under general laws. Here again recent discussions in the philosophy of science point to this as either too simple or actually quite compatible with the idea that people are free and responsible (Louch, 1966; Madden, 1969; Scheffler, 1967; Machan, 1974a). There are some other ideas the acceptance of which bar moral considerations from entering where an understanding of human history is at issue. Among these we find the belief in the enormous difficulty of agreeing on what moral standards to apply to people in general, as well as to scientists as such. While Scheffler and some others have suggested an answer to what is morally central to being a working scientist (Polanyi, 1969; Scheffler, 1967; Skinner, 1967) the issue of what is central to morality as such is not easily approached – not to mention the fact that few agree that anything *could* be central in the first place (Beardsmore, 1969; Wheatley, 1968; Winch, 1968).

But lack of agreement is not a good reason for purposes of discovering whether there might be truth about some issue. That should be clear enough now when the methodologies of the various sciences are themselves at issue. So I believe my suggestions have a footing untouched by absurdity.

Having covered so much ground so hastily, to introduce additional tangents and elaborations would invite criticism on just that count alone. I will instead focus the rest of my comments and suggestions on the issue of the place of moral considerations in the history and sociology of science. These remarks will have to assume the possibility of the identification of objective moral values for human beings at any given epoch, with only the sort of changes possible in these that do not contradict values established in the context of different circumstances and epochs, with different possibilities available to people to choose from.[23] My suggestions – for these really cannot amount to more than that – assume that the very possibility of a sociology of science requires an understanding of at least some of these values. (Merton (1962), for example, consigns morality to the role of institutional norms). This is not an adequate approach in terms of the present hypothesis.) For the present purpose all that is necessary is that we allow for the possibility that wherever explanation of human action is at issue the matter be considered without prejudice as to whether some conduct can be chosen. There should, in short, be no *a priori* denial of free will and moral responsibility within the field of human activity we construe to be science. If my suggestions are fruitful, these possibilities may turn out to be important realities within the history of science, realities the acknowledgment of which could be indispensible to understanding and explaining human behavior on all fronts.

4. Applying a Theory of Scientific Conduct

In this final section of my paper I want to make an attempt to take up some of Kuhn's topic from the point of view of the present approach. I will consider some objections, also. With what I will do I hope to provide at least some initial grounds for invoking a more individualist – *not* atomistic or Hobbesean – and ethically oriented stance *vis-a-vis* the problems that puzzle us most in relationship to the history of human scientific conduct, judgment, development, even possibly progress.

One of the consequences of the present approach would be that events during periods of crisis in science would emerge as possible moral challenges to those who face them. In considering certain controversies and disputes we might have to ask ourselves what those involved should have done given the situation they faced. For this purpose we would have to reconstruct the situation of interest to us with considerable detail. We would have to invoke certain standards of conduct by which we could then evaluate what those facing the situation did. Of utmost importance would be the realization that those we are talking about might have done otherwise – that Priestley, for example, despite his protestations, might have approached his problems with greater receptivity, perhaps by going beyond the issues that lay at the surface of the dispute with Lavoisier, beyond the actual scientific problems all the way to a consideration of the metaphysical underpinnings that his approach took for granted. For this purpose it would not be enough to examine Priestley's own letters and papers. The 'raw data' of history would not be sufficient to yield an understanding and explanation of what happened. When Kuhn thinks that a "different concept of science... can emerge from the historical record of the research activity" (1970: 1), he indicates his opposition to the approach I am suggesting. Neither for the general understanding of science nor for the more specific understanding of scientific episodes is it sufficient, I think, to expect answers to simply "emerge from the historical record."

A general skeptical objection to these suggestions is worth considering,[24] namely that we cannot really expect to determine the value of scientific conduct, especially as regards its fruitfulness in discoveries and explanations, by paying attention to the character of the scientists' behavior. This is because we can find people who make what turn out to be very good decisions while conducting themselves quite irresponsibly in the process of arriving at them. Such individuals have been known to exist in the history of science. The conclusion emerges that unjustified, even irrational behavior could produce valuable, one might say entirely correct results. This, in turn, while other, entirely conscientious individuals and groups can be seen to have made only minimal progress in their endeavors. Surely in the light of such cases it is fruitless for purposes of understanding and explaining the nature and development of science to invoke such putative determinants as the scientist's moral character.

Clearly, the *method* I am suggesting simply cannot *guarantee* that one who always lives up to standards of good personal scientific conduct *will* produce great scientific results. In all areas of human life evidence exists that some of the worst people manage to succeed – those who abuse their health may yet live a very long life, those who drive carelessly may yet make it to their destination, and people who drop out of second story windows may yet survive the fall and even write a best selling book about the lucky event. Luck is clearly a factor in many a human endeavor, science not excluded.

In addition, neither is it possible to know always just what steps *must* lead to a discovery, a desired explanation or principle. There are surprises, even anomalies, in most areas of human conduct. The most vicious killer can carry out his vice with commendable results – as when he accidentally guns down an even more vicious fellon.

But to construe this as a failing of an explanatory scheme or approach requires that we entertain an impossible model as to what would be suitable. Kuhn is right that "methodological directives, by themselves... [are insufficient]... to dictate a unique substantive conclusion to many sorts of scientific questions" (1970: 3), including to those raised in the history of science.[25] Only to those who choose to entertain the false alternative of ideal certainty or skepticism will this be a failing on the part of a theory.

Having taken note of what the suggested approach cannot and does not promise, let me however indicate why the objection is not quite as forceful as even I have appeared to allow. A careful investigation of the Priestley/Lavoisier affair along lines suggested here will not rest with the observation that Priestley's obstinance managed, after all, to produce a great deal of useful research, some of which led to important findings. In such cases historians tend to focus on the beneficial consequences of someone's admittedly 'unreasonable' conduct (Schagrin, 1973). We could, however, approach the matter differently. What of the effort that *could have* been spared? If we admit that, despite protests, the stubborn men of science could have yielded what amounted to sound arguments in their context, that they could have gone a bit further in their inquiries to check their and the tradition's premises beyond what was conventionally appropriate, and so forth, then this line of inquiry will no longer appear hopeless. Kuhn's own protests about 'what might have happened' sort of claims notwithstanding (1970: 127), it is appropriate in our effort to make sense of human history to consider alternatives to the actual development of events. What this points out is that even if we admit that recklessness, general impropriety, unreasonableness in the light of high professional and moral standards, and the like can at times produce valuable results, we can still ask whether it might not have been better to abide by the proper canons. By this insistence upon the issue of what else could have happened, had the people involved made the effort to live by such standards as are (contextually) appropriate to their activity, the force of the objection involving anomalies is weakened.

Let me now return to some of the possible benefits of taking the approach I have been proposing. One area in which we may find it helpful is in the consideration of biographical data. The issues dealt with here can have a crucial bearing on what we will admit as determinants of scientific change, even of progress. A historian who takes the appropriate moral point of view will, first of all, be able to draw on such data – including the autbiographical presentation of it, the very nature of its exposition, the evidence concerning self-representation and how the person's actions, accomplishments, failures, etc. are characterized by him – so as to develop a general idea about the *moral* aspects of the scientist's life. I do *not* mean his sexual habits or how often he wrote to his parents, of course. The sense of 'moral' I am using here goes to the more basic aspects of a person's way of living his life *qua* person – whether he has *integrity*, whether he has cultivated an *honest* approach to his own professional competence, how he treats the material upon which he speculates, how he compares his work to those of others who may be working in opposite directions, and considerations such as these.

Of course to gain this kind of understanding of his subject matter, a historian *must* at the same time have a clear grasp as to what would constitute the scientist's actual and available professional universe. Just having a moral framework won't do to understand the actions of a scientists and to explain the consequences of those actions. But the moral point of view is crucial for understanding *the choices* of scientists in certain circumstances. (Incidentally, these remarks should not be taken to endorse moral supervision by governments. As I see it a decision having to do with science, research, funding, and the many related matters really should not be a political one in the first place.) Given that there are better and worse things that people can do with their lives, and that some people devote a good deal of their lifetime to science, I take it as given that there are better and worse things that people can do with their time in their profession also. The purposeless pursuit of bits and pieces of truth does not seem to me to be so commendable as I take it to be for some admirers of so called 'pure' science.

These points simply give a clue to what I would do in a broad examination of the history of science. But they assume a certain moral perspective already, one for which I have offered no justification. The present thesis does not defend a particular moral code to be cultivated in science, only that we consider the issue seriously. That is because general philosophical considerations touching on the issue of human nature and conduct indicate that people can be free to choose some of what they do, and that some of what they do is better than alternatives. And this would seem to be true of all areas of conduct, science not excluded. The historian of science would then be well advised to approach his studies in the light in which human conduct in general needs to be considered. While ordinarily relying on moral considerations for understanding the history of science may not be necessary, understanding and explaining the choices and actions of scientists in times of crisis – i.e., when 'normal' guidelines are missing – may require them.

Let me now turn to such 'revolutionary' scientific changes – Kuhn's central focus of interest. First we will have to consider just what Kuhn takes these eras to consist of and what makes them revolutionary. What exactly happens, what is it that is *in* revolution? Kuhn tells us that "during revolutions scientists see new and different things when looking with familiar instruments in places they have looked before" (1970: 111). Kuhn invokes the card-color-change example to illustrate his point. But ultimately he does not insist on pressing the example itself – and wisely, since people do come to understand that change as they learn of actual color *switching*. The analogy breaks down unless reality itself changes when the "scientists see new and different things." Still Kuhn does argue that, e.g., "Pendulums [not the idea of 'Pendulum'] were brought into existence by something very like a paradigm-induced gestalt switch" (1970: 120). Kuhn might well mean this literally, given his position and its implications.

The alternative I want to suggest, will help us in understanding whether Kuhn's points could be advanced in a light that does not result in what I have shown as Kuhn's insurmountable problems. No doubt the *concept* of 'pendulum' may have been introduced to identify a phenomenon that had thus been rendered more sensible by Oresme and, later, by Galileo. But is it correct to talk of the 'invention' of a paradigm? And thus of the invention of the pendulum? It is philosophically more accurate to speak of the discovery or identification or acknowledgement of a more meaningful, more sensible, more coherent, more consistent, more parsimonious conception of reality by the formulation of a concept for which the material and understanding gave ample support. Even if those who introduced it had only a so-called 'hunch' of the need for this new concept – does it not make a shambles out of science to talk of inventions, to extrapolate from the language of technology, of design, of engineering in this way? Not that Kuhn has no insight here. But akin to certain philosophically eager scientists (e.g., B. F. Skinner, W. Heisenberg, A. Eddington) Kuhn extrapolates from an area where the concepts of creation and invention apply, e.g., in art and technology, into one where they have only questionable functions (i.e., science). Kuhn admits that he "cannot yet specify in any detail the consequences of [his] alternative view of scientific advance" (1970: 171). Despite this and other cautionary remarks, however, Kuhn's general proposals are far too firm, considering the shaky foundations of his approach.

Among many valuable features of Kuhn's book, he points to an important feature of concept formation in the development of human knowledge, and he is making an important departure from the dominant view of science when he proposes that "If we can learn to substitute evolution-from-what-we-do-know for evolution-toward-what-we-wish-to-know, a number of vexing problems may vanish in the process" (1970: 171). Yet he undercuts this suggestion with his relativist metaphysics in which the very possibility of scientific knowledge fails to find a place, and in terms of which not just knowledge but the facts, entities, relations – i.e., existence itself – change during revolutions. We have seen Kuhn defend this metaphysics by way of some extrapolations from psychology and the history of science (as he interprets it). Another aid Kuhn invokes is the idea of political revolution (1970: 92, 160).

Concerning this last piece of ammunition Kuhn fails to consider the crucial point about political knowledge. Such knowledge purports to be about what people *ought to* do for purposes of organizing a good and viable human community. One need not embrace the is-ought dichotomy to acknowledge that the 'is' of political theories does not function in saying what people actually do in relation to politics; the "is" in such areas functions to point out what is right and good.

Thus the reference to political revolutions can be very misleading. The subject matter of political theories purports to be the political principles people *should* accept and practice. Only revolutions in political *science* could serve to satisfy Kuhn's illustrative purposes! Do political revolutions actually change *what should* be pursued within the context of organized human communities – i.e., *is* constitutional democracy the best, most sensible account of how a human community should be organized *only after the revolution* whose victors advocate it? Is socialism how human communities ought to be organized only after the Bolshevik revolution? – Only if that is the case will the example of political revolutionary change give support to Kuhn's thesis. But it is far from well established that such a Nietzschean approach to values is valid. Kuhn's analogy begs the question of his own ideas.

And this invites possible alternatives. First of all Kuhn's schema fails to account for the good possibility that we might be able to identify alternative choices scientists could have made – what they might have done, what might have developed, what advances might have been made, were it not for their actual choices. Kuhn uses the language of hard determinism throughout his work; for him scientists *had to* advance their findings as they did, they could not see what they did not see, they had no other alternative than to do what they in fact did (1970: 56, *passim*).

But there is reason to believe, based on a general philosophical theory of human action (Chein, 1972; Louch, 1966; Yolton, 1973), that scientists could have done otherwise than they did. And this suggests that some of them probably permitted themselves to advance theories they had not fully checked (by reference to a broader framework of understanding reality); some might have been stubborn, harbored irrational fears of controversy, lacked the courage to speak up when (objectively speaking) they should have, yielded to fads, failed to check on reports given to them by eager researchers, concocted research projects not warranted by what they knew, acquiesced (avoidably) to those not tolerant of new developments in their field, etc. All these and other possible issues can arise, despite Kuhn's protestation about the difficulties with 'what might have happened' statements (1970: 127).

Secondly, Kuhn's account allows that in science those on the upswing call themselves normal, established, respected and trustworthy scientists, while those at the outskirts – even outright irrationalists – characterize themselves as the wave of the righteous future. Only social pressures, which can clearly be wrongheaded, even irrational, operate to maintain the usefulness of critical concepts, of a semblance of orderliness. By adopting the more cumbersome, difficult, not at present very promising road of focusing on what *might* have been done by scientists, these paradoxical consequences could be avoided – without falling prey to the pitfalls of a positivistic analysis. Kuhn's historicist (conventionalist, dialectical) account isn't the only alternative.

Thirdly, in human affairs individuals must at times be trusted or condemned not on grounds of sufficient independent reason gathered carefully in support of their actions or against what they have decided to do, but on grounds of what kind of human beings they are. *Ad hominem* arguments have been abused in logic texts – yet character references are used in courts of law, where drastic measures are taken with their aid. Gerber puts the point well:

> By using some abusive, neutral, or laudative terminology one can say something which implies that a person fails to tell the truth. One can even say something which

implies of a man the chronic failure to tell the truth. One can call someone a liar, a cheat, a fraud, a swindler, a humbug, a quack, a charalatan, and a faker, using strictly abusive language, as well as a storyteller, a perjurer, and a pretender, using language which may on particular occasions be abusive, and which on others may be neutral or even laudative. Each term on either list in its own distinctive way sometimes can be used to indicate that the person spoken of does not tell the truth at all, or does not tell the truth with regard to some special matter. In calling Jones a faker (a man who *always* fakes) it is implied that when Jones complains of a throbbing migraine he is lying, from which it follows that what Jones says in this case is false (Gerber, 1974: 27).

Similar points can be made about arguments from authority – a form of argument clearly useful and capable of yielding not only valid but sound conclusions in our dealings with experts in many fields (e.g., medicine, education, engineering, auto-mechanics and science). I am not recommending, of course, that we invoke such arguments indiscriminately. But trusting others' judgments in cases where it is warranted can be a means to a rational account of events – *even* in science. In view of the enormous difficulty of gathering sufficient data in the history of science – as in trials concerned with establishing what must or might have happened – these sorts of arguments can be useful for purposes of yielding explanations of scientific conduct, even progress or regress!

For example, historians face so-called revolutionary periods in science and find that the knowledge they have of the state and dynamics of the field does not suffice to yield an explanation of what scientists have done, of the choices they did in fact make; or they may find that even detailed autobiographical reports indicate confusion, lack of sufficient support for one or another alternative approach to the handling of some problem. In short, the historian is at times unable to unearth the (possible) reasons that have led to *success*, i.e., to the eventual rational justification of the alternative chosen. (Consensus is *not* the valid standard for evaluating success in fields where *knowledge* is the goal).[27] Then and only then can such arguments as I have mentioned above serve the goal of explaining human action. It is at these points we can induce Louch's idea that "In appealing to reasons for acting, motives, purposes, intentions, desires and their cognates, which occur in both ordinary and technical discussions of human doings, we exhibit an action in the light of circumstances that are taken to entitle or warrant a person to act as he does" (1966: 4). (In the less than technical but not quite ordinary area of detective work, a hunch, for instance, may be a good reason for someone's action, just in case we know enough of the person to testify to his 'feeling' for the relevant issues, even when he himself either will not or cannot give adequate support for that hunch!) Kuhn might find these sorts of considerations inadequate for purposes of explanation because, as we have already noted, he complains of the "insufficiency of methodological directives, by themselves, to dictate a unique substantive conclusion to many sorts of scientific questions" (1962: 3). This is to set an unrealizable, thus incoherent, standard for what is to count as rational and reasoned scientific conduct and judgment. The 'God's eye view' of reason, rationality, and objectivity needs to be abandoned, not the view that reason should guide judgment and conduct.

In this connection consider Toulmin's point about Einstein's idea of "physical theories as 'free products' of the human imagination." Toulmin admits that scientific discoveries are not inductive inferences, nor conclusions based on strict deduction; but he says that "we must not be tempted to go too far. This is not the work for untutored imagination. It may be an art, but it is one whose exercise requires a stiff training" (1953: 29). Yet in studying the methods of the history of science, we are entitled not to focus on such cases. Kuhn fits the case here well captured by Louch: "The extreme case, employed to make the contrasting point, is taken as the standard or universal case, and thus the condition for any inquiry whatever" (1966: 7).

I cannot claim that Toulmin's reference to "stiff training" and "untutored imagination" mean for him morally significant aspects of a person within a field. I do want to urge that we consider such references in the light of man's moral nature. (The exceptions to so approaching someone would be when he acts under coercion or physical limitations.) Is it not appropriate to require that someone in a field of science should undertake such training? (Quite wrongly, many think that morality pertains only to cases of human *interaction*, that is rests primarily on *duties to others*, which must subsume other moral notions such as responsibility, integrity, honor, and similar *virtues*. We are in an era where philosophers tend to opt for *rule* moralities.)

An understanding of the development of science can utilize the knowledge we have of a scientist's moral character. Such knowledge can aid us in our efforts to make sense of what someone (and some group) did in some particular circumstance. It may help one to understand also why others interacted with him as they did when faced with his judgment and conduct. It is unnecessary to conceive of this approach as imposing on scientists of the past, virtues that *could only be developed* and practiced in eras other than theirs, e.g., those of the historian. Rational reconstruction must take heed of moral context!

I agree with Joseph Agassi about the advisability of a contextual approach to understanding the history of science, in preference to both the inductivist and the conventionalist alternatives. Kuhn's position is a variety of the latter. Yet even Agassi says something in this connection that lends support to conventionalists. He tells us that "The rise of an idea is the outcome of a work of genius, an unaccountable development" (Agassi, 1963:55). Joining Koestler (1964), Agassi (1963:26–7) is here invoking a notion of 'accountable' that he has found wanting in connection with the Marxist view of scientific interest-generation. 'Accountable' need not mean 'caused by factors'. Not at least when the factors *must* be external. We can account for the rise of an idea by reference to the moral character of an individual, by his lifelong devotion to his task, by his commitment to solving problems, seeking the truth. This is surely one way to talk about Socrates – why is it barred out of court when talking about Galileo or Einstein?

Nor need the rise of an idea be unaccountable in the sense that we preclude the prospect of understanding the subtle rationale for it which the scientist himself may not be aware of. Why should a scientist be worried about *how* the idea emerged – he is concerned with the problem it may help him to solve.

It has been my aim here to counter and offer an alternative to Kuhn's approach to the understanding of the history of science. I have suggested that a clear understanding of ethics would aid us in an attempt to understand the dynamics of scientific development in both the individual scientist and, through that, in the history of science – without abandoning prospects for objectivity. I have also hinted at a conception of truth and knowledge that is not utopian. I suggested that a contextual view of knowledge will support Agassi's rejection of being wise after the event and his acceptance of different kinds of error – valuable, justifiable, irresponsible, negligent, etc.

In connection with the revolutionary stages of science I have urged that we recognize the possible extra-scientific features of revolutions as well as the responsibility of scientists to make the best effort to achieve a rational approach concerning these features of (or near) their respective fields. I would simply add that to do so might turn out to be the responsibility of those scientists who have recognized that serious changes need now to be made in their fields.

Finally, I want to explain to those interested that I find Kuhn's ideas – at least those I have focused upon here – objectionable on yet another ground, one mainly implicit in what I have said thus far. Let me put the matter this way: When I consider some topic that I have investigated thoroughly, and when I do so to the best of my ability, then I can be right – and often am. Now so can Richard Nixon, Albert Einstein, Henry Kissinger, Thomas Kuhn or anyone who is without relevant brain damage or very serious incapacities of other sorts. My arguments against Kuhn have not aimed at showing that he (inherently) cannot be right but at showing that in the case considered he is not right. He teaches a false and dangerous epistemology.

Scientists who are not concerned with being right in what they think and say professionally are negligent and irresponsible. They can, of course, be so concerned and yet err, and so can we all be closer to and further from the truth of the matter. But Kuhn bars us from it on principle. True enough, physics, chemistry or biology are not professions concerned with learning whether people in general and professional scientists in particular can be right. When a scientist concerns himself about this matter he is stepping outside his profession. Still, many *decisions* about what is right need to be made by scientists even though the topic falls outside the profession as such, and so called revolutionary periods can call for most of these decisions – e.g., 'Will I acquiesce to pressure, compromise on principle, will I resist sound arguments, etc.?' But in the main what scientists conceive themselves to be capable of is not expert testimony about the nature of knowledge, but about their own self-conception![28]

At any rate, my concern here has been to discuss whether scientists *can* know reality, whether they *can* be right in their judgments, whether they *can* tell us the truth about some of what they study.

Kuhn's thesis render's human beings impotent to achieve one of their essential goals – if we are to believe Aristotle. That applies to Kuhn, also – he cannot know.[20] But in fact Kuhn knows a great deal, he is not wrong on all counts – so his thesis isn't even valid as a confession – naturally, otherwise why would any serious student of his subject matter pay attention to him anyway?

But about what I have been concerned with here, namely whether scientists can be right – be it about their choice of paradigms, their theories, what they conclude about aspects of their field, etc. – and whether they can tell us the truth about various aspects of reality, about that issue Kuhn is wrong. You will permit me to say, I am sure, that this is fortunate for human beings everywhere who love truth – and life. Assuming, as I do, that seeking and obtaining the former has a great deal to do with coping successfully in the latter.

State University College,
Fredonia, New York

NOTES

* Preparation of this paper was assisted by generous grants from the Institute for Humane Studies (Menlo Park, California) and the Institute for Advanced Studies (Vienna, Austria). I wish to express my sincere thanks to Morton L. Schagrin, Carl R. Kordig, David Silverman, and M. L. Zupan for helpful comments offered on an earlier version of this paper. Clearly I am, however, responsible for the results and these may not at all reflect the full value of the assistance.

[1] By 'deterministic' I am not referring to types of self-determination – as when we correctly say of someone that "He was determined to make his point." Given the theory of causality and agency proposed by Davidson (1971) and Madden (1969) and the validity of the argument for the human capacity to make free choices offered by Jordan (1969) and Boyle *et al.* (1972) it is possible to include a person among the determinants producing his conduct.

[2] In several places Kuhn attack's the notion of a definition of 'science'. He does this, in one crucial place, by posing rhetorical questions such as "Can a definition tell a man whether he is a scientist or not? If so, why do not natural scientists or artists worry about the definition of the term?" (1970:160). These are inadequate objections, to say the least; I can name a dozen scientists and artists who do worry about what science and art are, what it is to be a scientist or artist – and almost every textbook in the infant sciences remarks on this matter, as do most artists whose works are thought by many to be somewhat odd. Just because when nothing much hinges on the matter people are unself-conscious about the necessary and sufficient conditions of being something one cannot preclude the usefulness of such concerns – e.g., when abortion or organ transplants are being considered the definition of 'human being' and 'life' does require looking into. For how to avoid the pitfalls of Platonist or ideal and fixed definitions see Stroud (1966) and Rand (1966).

[3] Kuhn notes that his own extrapolations "cannot be attributed to" Wittgenstein (1970:45). Yet in basing so much on the philosophical ideas of Wittgenstein concerning the possibility of criteria and justification for 'naming', Kuhn could be held responsible

to develop his points at length. His impossibility proof should be laid fully open for investigation, given its very drastic consequences.

[4] By 'meaningful' I do not mean only that we could *imagine* well enough what it would be for those things to have transpired; I have in mind here rendering it possible to *conceive* of the actual events as having transpired as he says they have, i.e., actually learning of what occurred. In history, also, we are not after imaginable episodes but after *knowledge* of what happened. Yet paradigm induced images of the past cannot provide that, as Kuhn must admit. (Observe how he speaks of what "Priestley was, to the end of his long life, unable to see..." (1970:56). This is not a statement of historical fact but a Kuhnean paradigm-colored rendition of history by Kuhn's own doctrine. Are we justified in using it as evidence, then, in support of Kuhn's theory of paradigm determination? Not without obtaining a circular argument.)

[5] I am aware of the controversy over the validity of circular arguments and I also appreciate that Kuhn cannot be asked to settle that controversy before proceeding with his case. But again I want to point up how dangerous it is to provide impossibility proofs when the bases are not sufficiently covered.

[6] This follows from Kuhn's view that (a) choice of paradigm rests, in part, on an "element of arbitrariness" and (b) philosophical positions, too, are paradigms or paradigm governed (1970:4–5, 121). It should be noted, however, that to make out this case one would have to argue first that all (including metaphysical and epistemological) truths or principles *must* be arbitrary. Yet the concept 'arbitrary' cannot be rendered intelligible if all of our knowledge rests on arbitrary standards – since that concept is meaningful only in contrast to the possibility (i.e., actual prospect) of some knowledge that is demonstrably or axiomatically (i.e., undeniably) established.

[7] I am thinking here of the rationale for the courtroom procedure involvirg the standards for cross examination – only points raised in direct testimony can be brought into the latter.

[8] Whether scholarship bears out the charge that Plato himself advanced this model, in view of his admiration for mathematical knowledge, is not important here. The point is that that model has been thought to guide efforts to provide an account of human knowledge. Rationalist as well as empiricist, up until recently, have found it necessary to characterize all knowledge in such a way that it would have to satisfy the requirements of that model. Only recently has it been emphasized that extrapolation from mathematical knowledge (or, alternatively, from the immediacy of sensory feelings) might be an unjustified restriction on what could count as knowledge of other aspects of reality.

[9] If we allow that Wittgenstein was focusing here in part on his own earlier commitment to logical atomism, with its affinities to empiricism, the emphasis on perception can be appreciated. But this should also warn against extrapolating the results of this emphasis to other forms of awareness of reality.

[10] E.g., if we revise the notion of 'common features' to allow for other than observable or sensible aspects of what we want to classify, and if we allow for contextually determined modifications of the applicability of our concepts.

[11] E.g., unconsciously exercised, possibly habituated, rational inferences. But this approach requires a reconsideration of our widely used concept of 'rational'. See in this connection Toulmin (1953).

[12] I.e., with its emphasis on the impossibility of finding ultimate truths – something pragmatism shares now with several other philosophical schools, e.g., existentialism.

[13] I.e., in the possibility of *philosophical* knowledge!

[14] In religion, for instance, one has faith not because nothing else is available to account for some features of reality and human life – that, as some theologians claim, would be too easy. Only when faith takes the effort to overcome even contrary evidence can it have its proper significance. This is not, however, typical in theism.

[15] Kuhn admits to this in his references to Wittgenstein and others who have had an influence on his thinking.

[16] This, of course, is very tricky and requires much work to make for an adequate point about the issue. Anti-reductionists – who have, nevertheless, stood fast by the absolute requirement of rationality in all fields of knowledge – have been making the case for this idea all along, for better or worse. Thus 'best' would here allude to rationally justified, including consistent with philosophically sound ideas of first principles and human knowledge.

[17] Kuhn does not seem to think that an approach such as the one I am sketching can work (1970:101–109, 136–143, *passim*). I suggest that Kuhn is emphasizing anomalies altogether too much and does not consider that they might or could have been avoided in at least some instances – while those that remain would not, then, justify drawing Kuhn's conclusions about the *nature* of science. It could turn out that we just *do not* know enough to account for those instance.

[18] The expression 'objective truth' is used to mean the theoretically distinctive characterization of what 'truth' must mean in terms of the objectivist approach. Outside a philosophical context 'truth' would mean just what such an approach argues it must mean, assuming the argument is itself sound.

[19] Kuhn and those sympathetic to his ideas would still hold that pervasive conceptual confusion within a field indicates the absence of at least the necessary conditions for something being a science. One can ask, however, whether Kuhn could justify even that much – why, that is, should we insist on logic? Some commentators on an earlier version of this paper posed just that issue: why is the requirement of logic unquestioned herein. Certainly there have been those who have called this, however paradoxically, a 'Western superstition'. Indeed, I would argue that such a challenge is incoherent – there simply is no possible way to identify its meaning. Without logic it could mean anything whatever.

[20] See note 8.

[21] I suspect that I am offering a schema that owes more to features of Aristotelianism, with contemporary renditions and modifications, than to anything properly construed to be modern or progressive. Not being a historicist and Hegelian, I can rest at ease with the possibility that temporal development, even obvious accumulation of some kinds of knowledge, does not guarantee advance in all fields. I can imagine that even modern science owes more to earlier philosophy than to the philosophical views of its own times – at least when we consider the subtle fashion in which philosophical ideas have an impact on people's lives and ideas.

[22] Kuhn acknowledges the influence and place of metaphysicians but he is not concerned with whether some positions in metaphysics – i.e., whether some metaphysical claims – are true or false. He is making an attempt here to take a neutral or 'God's eye' position, as if he could escape the requirement to deal with whether his own metaphysical presuppositions are correct. Yet in the sort of discussion Kuhn carries on that requirement is central (1970:41).

[23] I.e., some value(s) may be axiomatic, rationally undeniable, while others may be contextually dependent.

[24] I have encountered this point in many circles, in many different types of discussions,

including in connection with ethics, political philosophy, and, of course, epistemology. In reference to the present topic it was brought to my attention by Professor Schagrin whose knowledge of the history of science is far more extensive than mine. Yet the point is actually more of a philosophical than historical one, dealing as it does with the problem of how borderline cases, anomalies, problems, etc., have to be faced by someone who offers an account of some aspect of reality. In ethics, for instance, must an ethical theory be capable of handling any and all so called 'desert island' cases? For a more detailed treatment of this matter see Machan (1974b).

25 For a full discussion of whether rationality requires that methodological directives produce 'unique substantive conclusions' see Wittgenstein (1969) and Stroud (1966). I should like to suggest that it is people who produce those conclusions, not directives. But that does not deprive us of objectivity – indeed, as I read Stroud, it renders it our responsibility to strive for it. See also Scheffler (1967).

26 Here again those who hold this view seem to have accepted a view of knowledge not entirely justified. See note 8.

27 Kuhn seems to speak of consensus without explaining whether it is the result of scientists having a common purpose, more or less, or because they all want to agree with each other. Keith Lehrer has recently offered this distinction as a very important one (in a yet unpublished paper read at the 1974 APA meeting, Pacific Division – 'Social Dissensus and Epistemic Agnoilogy'). Here again it would appear that the considerations that enter pertain in large measure to a scientist's character, morality.

28 Indeed, this is one of the difficulties in turning to scientists (their autobiographies, journals, letters, etc.) for accounts of what happened when they made discoveries. The scientist involved in his task, e.g., in identifying some feature of reality, cannot be taken as an expert about another task, identifying the process of identifying some feature of reality. He certainly does not have that as his professional task, regardless of how often he is tempted to turn to that issue. Of course his testimony can help – but it requires a philosophical analysis before we can make use of it for purposes of the *philosophy* of science.

29 The idea suggested that although we might know a great deal, we cannot know that we do, would require considerable discussion for a full treatment. Still, it seems clear that by rejecting the possibility of philosophical knowledge – e.g., "Human beings can have knowledge of reality" – one has made it very difficult to secure any other kind, also. I should note here that while developing this paper I often thought of what I would say to the comment that my efforts intend really in part to secure the prospect of being right about some of my own claims, e.g., that my own views about Kuhn, etc., are right. I finally discussed this issue with my wife, herself a student of Kuhn's works, and her response seemed to me entirely appropriate: "Quite right – that is *part* of the point of this paper. So what of it?"

BIBLIOGRAPHY

Agassi, J.: 1963, 'Towards an Historiography of Science', *History and Theory*, Studies in the Philosophy of History, Supplement 2.

Austin, J. L.: 1961, 'Other Minds', *Philosophical Papers*, Oxford U. Press.

Bastiat, F.: 1963, *Selected Essays in Political Economy*, Van Nostrand, Princeton.

Beardsmore, R. W.: 1969, *Moral Reasoning*, Schocken, London.

Boyle, J. M., Grisez, G., and Tollefsen, O.: 1972, 'Determinism, Freedom and Self-referential Arguments', *The Review of Metaphysics* **26**, 3–37.

Caton, H.: 1973, *The Origins of Subjectivity*, Yale U. Press, New Haven.
Chein, I.: 1972, *The Science of Behavior and the Image of Man*, Basic Books, New York.
Davidson, D.: 1971, 'Agency', in R. Binkley, R. Bronough, and A. Marras (eds.), *Agent, Action, and Reason*, U. of Toronto Press, Toronto.
Efron, R.: 1967, 'Biology Without Consciousness – and Its Consequences', *Perspectives in Biology and Medicine* **11**, 9–36.
Efron, R.: 1969, 'What is Perception?', in *Boston Studies in the Philosophy of Science*, Vol. IV (ed. by R. S. Cohen and M. W. Wartofsky), Dordrecht-Holland, pp. 137–173.
Efron, R.: 1970, 'The Measurement of Perceptual Duration', *Studium Generale* **23**, 550–561.
Feyerabend, P. K.: 1963, 'How to Be a Good Empiricist: A Plea for Tolerance in Matters Epistemological', in *The Delaware Seminar in Philosophy of Science*, Vol. 2 (ed. by B. Baumrin), Wiley Interscience, New York, pp. 3–39.
Fisk, M.: 1974, *Nature and Necessity*, Indiana U. Press, Bloomington.
Gerber, D.: 1974, 'On Argumentum Ad Hominem', *The Personalist* **55**, 27.
Jordan, J.: 1969, 'Determinism's Dilemma', *The Review of Metaphysics* **23**, 48–66.
Koestler, A.: 1964, *Act of Creation*, Macmillan, New York.
Koestler, A.: 1967, *The Ghost in the Machine*, Macmillan, New York.
Kordig, C. R.: 1971, *The Justification of Scientific Change*, D. Reidel, Dordrecht.
Kuhn, T. S.: 1962, *The Structure of Scientific Revolutions*, U. of Chicago Press, Chicago.
Kuhn, T. S.: 1970, *The Structure of Scientific Revolutions*, 2nd ed., U. of Chicago Press, Chicago.
Lakatos, I. and Musgrave, A. (eds.): 1970, *Criticism and the Growth of Knowledge*, Cambridge U. Press, Cambridge.
Louch, A. R.: 1966, *Explanation and Human Action*, U. of California Press, Berkeley.
Machan, T. R.: 1973, 'Wittgenstein and Metaethics,' *Journal of Thought* **54**, 252–257.
Machan, T. R.: 1974a, *The Pseudo-Science of B. F. Skinner*, Arlington House, New Rochelle, N.Y.
Machan, T. R.: 1974b, 'Kuhn's Impossibility Proof and the Moral Element in Scientific Explanations', *Theory and Decision* **5**, 355–374.
Machan, T. R. and Zupan, M. L.: 1975, 'Back to Being Reasonable', *Philosophy of Science*, in press.
Madden, E. H.: 1969, 'The Third View of Causality', *The Review of Metaphysics* **23**, 67–84.
Madden, E. H. and Harré, R.: 1973, 'In Defense of Natural Agents', *The Philosophical Quarterly*, **23**, 117–132.
Marx, K. and Engels, F.: 1968, *Selected Works*, International Publ., New York.
Merton, R.: 1962, 'Priorities in Scientific Discovery: A Chapter in the Sociology of Science', in B. Barber and W. Hirsch (eds.), *The Sociology of Science*, Free Press, New York.
Miller, E. F.: 1971, 'Hume's Contribution to Behavioral Science', *Journal of the History of the Behavioral Sciences* **7**, 154–168.
Miller, E. F.: 1972, 'Positivism, Historicism, and Political Inquiry', *The American Political Science Review* **66**, 796–817.
Polanyi, M.: 1969, *Knowing and Being*, U. of Chicago Press, Chicago.
Rand, A.: 1966, *Introduction to Objectivist Epistemology*, Objectivist, New York.
Rostand, J.: 1960, *Error and Deception in Science*, Harper and Row, New York.
Schagrin, M. L.: 1973, 'On Being Unreasonable', *The Philosophy of Science* **40**, 1–9.
Scheffler, I.: 1967, *Science and Subjectivity*, Bobbs-Merrill, Indianapolis.

Schoeck, H.: 1970, *Envy*, Harcourt, Brace and Jovanovich. New York.
Schoeck, H. and Wiggins, J. W.: 1961, *Relativism and the Study of Man*, Van Nostrand, Princeton.
Skinner, B. F.: 1953, *Science and Human Behavior*, Macmillan, New York.
Strauss, L.: 1971, *Natural Right and History*, U. of Chicago Press, Chicago.
Stroud, B.: 1966, 'Wittgenstein and Logical Necessity', in G. Pitcher (ed.), *Wittgenstein*, Anchor Books, New York, pp. 477–496.
Taylor, R.: 1966, *Action and Purpose*, Prentice-Hall, Englewood Cliffs, N.J..
Toulmin, S.: 1953, *The Philosophy of Science*, Cambridge U. Press, Cambridge.
Trigg, R.: 1973, *Reason and Commitment*, Cambridge U. Press, Cambridge.
Truzzi, M.: 1974, *Verstehen: Subjective Understanding in the Social Sciences*, Addison-Wesley Pub. Co., Menlo Park.
Wheatley, J.: 1969, 'Reasons for Acting', *Dialogue* **7**, 553–567.
Winch, P.: 1967, *The Idea of a Social Science*, Schocken Books, New York.
Wittgenstein, L.: 1969, *On Certainty*, J. and J. Harper, New York.
Wittgenstein, L.: 1972, *Philosophical Investigations*, Macmillan, New York.
Yolton, J.: 1973, 'Action, Metaphysics and Modality', *American Philosophical Quarterly*, **10**, 71–85.

NICHOLAS CAPALDI

THE MORAL LIMITS OF SCIENTIFIC RESEARCH: AN EVOLUTIONARY APPROACH

1. Introduction

The title of my paper is likely to cause some consternation among those who wish to preserve the purity of scientific research, and at the same time my title is likely to elicit a pleasurable sense of anticipation among politically activist sociologists. Both of these initial impressions are not only premature but contrary to my intentions. The general purpose of my paper is to argue for a rational science policy designed to preserve the purity of research, and my more specific purpose is to specify a criterion for judging what kind of research should be favored or encouraged and what kinds should be discouraged or prohibited. The criterion, as you shall see, is more likely to limit the existential activist research of social scientists than any other kind of research.

Simply to have raised the question of the possibility of a rational science policy is to presuppose a particular philosophical, or pre-scientific if you like, perspective. In fact, this and the related questions presuppose a particular philosophical outlook which when articulated makes the answers both clear and compelling.

To believe that it is meaningful to have a policy is to presuppose that we live in an intelligible universe, one that can be understood at least in part by man. A world that made no sense would make any policy meaningless. But of course it is not enough to say that our universe is intelligible, we must specify what kind of intelligibility we take it to have. Consulting the oracle at appropriate moments is a policy if we believe that the oracle has a special insight into reality. This brings us to the all-important qualifying adjective 'rational', for our policy is to be not just a policy but a rational policy. It is necessary, therefore, to specify how we are to understand rationality.

2. Rationality and science

Any philosophical analysis which is intended to answer methodological

Knorr, Strasser, and Zilian (eds.), Determinants and Controls of Scientific Development, 113–141.

and ontological questions will take as its paradigm the most extensive and coherent body of knowledge available to it at any given time. From the time of the ancient Greeks to the present the scope and coherence of Physical science has made it the basis for the evaluation of philosophical assertions on methodology and existence. But it has never been simply a matter of incorporating the latest fashionable opinion, for scientific opinions are known to conflict and to be modified in the light of new data. Hence, philosophers have tended to construct somewhat more stable models of science. The obvious danger of such models is that we tend to fall in love with them, so that any scientific practice which does not conform to the model is dismissed as the result of confusion and the promise is made that such a lack of conformity will be eliminated eventually by the discovery of new data or the working out of a new logical technique. In this respect philosophers of science can be as dogmatic and as authoritarian as some of their scientific and non-scientific colleagues. Despite the danger, it is generally recognized that if a comprehensive model can be articulated it would have enormous consequences for the organization of research.

I take it that for the members of this audience the paradigm case of rationality is provided for us by science in general and by natural science in particular. There are those who would not accept this paradigm, and for those protagonists the major part of my discussion will be irrelevant. However, at the conclusion of my discussion I shall return to them, for I believe that my position makes an adequate provision for dealing with those who do not accept this paradigm of rationality.

Having identified Science with a capital 'S' as a paradigm of rationality, one might imagine that our problem is easily solved. Unfortunately, this is where our serious intellectual difficulties begin. In order to make these difficulties clear, we shall introduce the distinction between science as the sum total of our present knowledge about the world and our understanding of what we are doing when we are being scientists. Put in a more traditional philosophical manner, we distinguish between reasoning about the world and reasoning about reasoning. We are thus creatures who not only reason but we are one of the objects about which we reason.

It is at this point that the history of science becomes a source of illumination. At various times, different sciences have dominated the intellectual world, and depending upon which science held sway, men have had different notions of what model to use for the understanding of science itself. In short, it has been our custom to take the predominant science and turn that science into the model for explaining not only the world but the nature of science as well. The situation has not been topsy-turvy, rather we can distinguish three traditions, each of which emerges during a particular period but which does not disappear with the emergence of the newer traditions. These three traditions have provided the main models for the understanding of the scientific enterprise, and corresponding to these three main models there have been three different conceptions of rationality.

I should also point out that rationality implies that the world as a whole is intelligible. Without this assumption science itself could not operate. Within a specific domain it is necessary for individual scientists to account for all the data, and any scientists who refused even to consider some data would be stigmatized as irrational. The same applies to the world as a whole. Failure to appreciate the general significance of this point has resulted in the fact that scientists themselves have failed to apply standards to questions of social science and value which they apply to physical science. There are, of course, some consistent individuals who are thereby noted in the history of ideas as the landmarks, and these individuals provide instantiation for the main models that I shall describe. On the other hand, among those who fail to make this assumption it is possible for an individual to hold one model of physical science, a model of social science based upon a second or different model of physical science, a model which I might add he would not accept for physical science itself; and, a theory of value based upon still a third and different model of physical science. Several combinations are possible, since an individual may be more advanced or retarded in one domain than in another. This is also a polite way of saying that some individuals are philosophically contradicting themselves.

3. A PRIORI MODEL

The first model, which I shall call the *a priori* model, took geometry as the standard of what constituted rationality. This geometric paradigm consists of self-evident axioms from which statements to be explained are deduced. To explain is to deduce from what needs no explanation. The first principles or axioms, it is assumed, are metaphysical in origin.

This model is well known and appeared among the Pythagoreans, was adopted by Plato, reappears again and again among such well known figures as St. Augustine, Descartes, and even at one point in the work of Bertrand Russell. To this day in the philosophy of science we find those obsessed with formal systems of explanation based upon the deductive or *a priori* model. Those colleagues who appeared distraught at the work or conclusion of Gödel and others exemplify this point of view.

Russell's *a priori* view may be characterized briefly as follows. Science is the key to understanding reality. Mathematics is the language of science. Further, mathematics is a formal system whose properties can be understood in terms of a finite logical structure. The *Principia Mathematica* (1910–1913) is the key to an understanding of that logical structure, and hence the *Principia Mathematica* is the key to reality. But even if we grant that science is the first key to reality we may still raise the possibility that the science of our time is not the final form of science, that even consequently new and different kinds of mathematics may be needed. If so, then there is nothing final about *Principia Mathematica.*

Not only is geometry generalized from a tool for the description of the world into a model for explaining science itself, but the *a priori* model of rationality has an analogue for generating answers to questions about values. It is this value-analogue which obviates the necessity for even having a separate conception of social science or even of history. All questions concerning man, his problems, values, etc. are to be explained and answered by deduction from self-evident axioms about the eternal values. These values are not the result of empirical investigation, but are given perhaps in some special act of intellectual vision or even implanted in our minds by God Himself. The *a priori* model allowed men concerned with intellectual rigor to harmonize traditional and religious value schemes with their conception of rationality.

Corresponding to each model of natural science there is a derivative conception of social science. When physical science is viewed in an *a priori* fashion, and when it is assumed that values are to be given *a priori*, the social scientist is indistinguishable from the social critic or prophet or visionary. He is content and in fact limited to the description of perfect worlds and ideal societies. He is, in short, a utopian thinker. Plato's *Republic* is a perfect instance.

We know the shortcomings of this *a priori* model. Self-evidence turns out to be in all spheres not an intellectual or logical category but a psychological category conditioned by a variety of social and historical forces. We are all familiar with the paradoxes generated in the minds of those who take self-evidence seriously. Moreover, and this is quite important, the *a priori* model is unable to make intelligible the conflict we observe between the self-evident principles and the world in which we live. Not only do the facts conflict with what we might expect if we take the axioms literally, for example, the planets do not move in perfect circular motion, but there is a corresponding conflict between the axiomatic values and what men actually practice. It is not a question of deviation or experimental error or the need for additional axioms. The contrast between the world of appearances and the world of reality becomes a serious intellectual problem. In such an *a priori* model it becomes quite unintelligible to us how men can act against their best interests. Even if we leave aside questions about man, there are serious intellectual problems. It seems paradoxical to a Greek that there can be such a thing as an *irrational number*. Further, to one who accepts the *a priori* view, the presence of alternative hypotheses each of which fits the data and is mathematically viable as in the simultaneous presence of Ptolemaic and Copernican hypotheses is cause for the gravest concern. Not only is this possibility unintelligible but it is viewed as a catastrophe for science as well.

Summary: First View of Science: A Priori (Platonic)

1. geometry is the model science, hence:
2. rationality involves self-evident axioms known independently of experience;
3. explanation is demonstration from first principles.
4. Social Science therefore consists of the description of perfect societies or utopias.
5. Values are given *a priori* in acts of intellectual vision.
6. There is no difference in status among first principles.
7. Failure of our experience either physically, socially, or morally to conform to the first principles is not viewed as a problem. As Steven Weinberg of Harvard University has recently expressed this view with respect to high-energy collisions: "nature, as we observe it, is but an imperfect representation of its own underlying laws."[1]

8. *Evaluation*:
   a. The notion of nature imperfectly representing itself is manifestly absurd and violates the basic stricture of total commitment to total intelligibility.
   b. When this *a priori* view is held in social science and value it is a tragedy and a catastrophe.
   c. Since this view is usually confined to physical science, it does little damage in practice. Those who recognize its shortcomings in social science and valuation usually rationalize their position by advocating a dichotomy of fact and value. This is also a violation of the principle of intelligibility. It is also what makes this view of science compatible with religious and totalitarian views of social and moral value.
   d. In practice, this attitude has been the most productive in the physical sciences and embraced by most of the 'greats'. It has none of the problems of the second model about fitting theories to the facts. It is also, in its disregard for the narrow second model, perfectly compatible with the third view I shall describe below.

4. Structuralist Model

The second model, which I shall call the *structuralist* model, took teleological biology as the standard of what constituted rationality. The emphasis on teleological should be noted, for later I shall mention a quite different conception of biology as a model. The structuralist model is opposed to any *a priori* view of how we obtain our knowledge, rather it emphasizes the empricial origin of all information. Nevertheless, it believes that such information when gathered reveals a rigid structure to reality, something totally objective in its independence of man. Insofar as it allows for change and development it believes that there are built-in ends or essences or final causes which explain why things happen as they do. These built-in purposes, even though they are discovered empirically, allow for deduction when once they are uncovered. To explain is still to deduce, but this time the axioms have an empirical origin.

Teleological biology is not only a set of categories for explaining the animal, vegetable, and mineral world, but these categories can be extended to explain the behavior of men in social groups and they can explain the categories of value. Here it becomes possible to develop a conception of social science but one which explains social behavior as designed to achieve certain ends through a taxonomy of institutions. Values are built-in and totally independent of human manipulation. Hence, any social diagnostic ultimately is construed in terms of deduction from empirically revealed built-in ends.

This structuralist model is first associated with Aristotle and by a peculiar metamorphosis is the foundation for Aristotelianism in the medieval period and subsequently. It still survives as a model among physicists, social scientists, and many philosophers of science who suggest that we explain by use of formal systems, only now the formal systems are based upon empirically discovered universal or nomic laws.

The difficulties with this model are worth noting in some detail. To begin with, insofar as it attempts to account for change and development it advances beyond the dichotomy of real and apparent found in the *a priori* model, but the best it can offer us is an account of change in the form of cycles. Since certain ends are to be realized by everything, if there is movement of any kind it must be movement toward those ends. Either this comes to an eventual halt, but this possibility is quite unintelligible if the model is taken consistently, or it simply repeats itself in an endless cycle.

If one takes this as a model for social science,[2] then one can develop a cyclic view of both history and of social forces, but here there would be little opportunity to make sense of the activities of men. What would it mean to learn from history if the cyclic pattern cannot be broken? The best one can do is to find his place in the cycle and at least enjoy the comfort of knowing what is next. Machiavelli, for example, does not know quite what to do with such a model of explanation which he implicitly accepts especially when he seeks to account for human effort, so he throws in the concept of fortune or luck, but this is hardly a rigorous intellectual category. Even Aristotle had recognized the insufficiency of this model to account for the world, for most acorns do not grow into oak trees, and even he is forced to speak of "births out of season." Conflict and competition, whether among theories or among men remain unintelligible on this model.

If one takes teleological biology not only as the master science for explaining the world but as the model of rationality and intelligibility, then it would seem that science is to be explained in terms of a human activity having its characteristic end to be achieved. Here the idealized vision of the scientist at work constructing a picture of reality for the sole purpose of contemplation emerges. This view does not accord either with what scientists themselves say that they are doing, not does it accord with what those who study science from the outside actually find, nor can it account for the theoretical and practical benefits which have been derived from the existence of intellectual conflict. It may be a comforting vision for some, but it does not square with reality. It is not simply the case of seeking to avoid conflict and the benefits of peace and tranquillity, but rather of offering an account of the existence of conflict.

The tendency of those who stress the importance of demonstration in the structuralist model is to analyze science taxonomically in terms of structures, usually logical or linguistic structures such as lawlike statements, theoretical statements, hypothetical statements, observation statements, etc. This kind of analysis ignores the intentions of the individual scientist and the influence of social surroundings. It is not only ahistorical, but it takes its ahistoricity as a virtue. By ignoring history it fails even to recognize that it is itself a view having a peculiar history. In any other field of intellectual endeavor its anti-historical bias would be taken and recognized as a flagrant form of anti-intellectualism.

A structuralist[3] tends to believe in a fixed structure to reality, and when he asserts that this structure is apprehended by observation, he is forced to adopt a naively empirical view about what one does when one observes. This naive empiricism generates the notion of observation statements. But clearly theoretical statements are not observational in this naive way. The structuralist must reconcile the two. He refuses to, and in fact cannot, surrender his notion of observation without surrendering his entire world view. Faced with two sets of entities, he must:

(a) Explain how a being who observes in a naively empirical way does not stop at laws but can generate theoretical statements. This he cannot do, and in fact, the structuralist cannot even offer a characterization of a law which is more than a generalization.

(b) Explain the relation between theoretical statements and observation statements. Since it is assumed that explanation is deduction, the relationship between the two statements must be construed as one of entailment.[4] This entailment relationship becomes as incomprehensible as the relation between Platonic forms and the world of appearance, and for an outsider trying to understand explanation by entailment it is like trying to understand 'participation' in the forms.

(c) Explain how observation can confirm or disconfirm a theory. This cannot be done either.[5] This kind of problem also has a Platonic analogue in the myth of reminiscence.

It should be made clear that these difficulties are not, as the structuralist would like to believe or pretend, purely technical problems. They are, rather, the direct consequences of a prior world view. But the structuralist is precisely the person who cannot recognize this because he would deny that methodology is in any way conditioned by ontology.

In the second model, the structuralist view of social science inevitably becomes a kind of classification scheme of social forms limited to institutions, wherein change and development are described cyclically. Here one speculates on how a given institution changes from one form into another. This view represents some kind of advance for at least it has acquired a historical dimension. But once again it is assumed that there are built-in values or ends for these institutions. This means that specific institutions might be criticized for failing to achieve the ends although it makes no sense for us to explain this failure. Alternatively, one is forced to argue that an institution always achieves its end because what it does is necessarily determined by its position within the cycle. This can become either a defense of the status quo or a prophetic apologetic for the next stage in the cycle. Hegel and Marx[6] exemplify these respective patterns. Whereas the first model allows for criticism of an institution for its failure to embody the ideal it cannot explain the failure. The second model does not really permit us to engage in meaningful criticism of an institution.

From another point of view, the second model of social science, by its taking for granted the existence of a built-in value scheme, is still similar to the first view of social science. Despite their differences, there is not and literally cannot be an appraisal of the values themselves. The values are simply given either *a priori* or empirically discovered in a relentless process. By its very nature this second view of the social sciences rules out the possibility of rationality, for rationality depends upon corrigibility. This explains the otherwise mysterious intolerance one finds among social scientists addicted to this view of their subject matter, and it explains as well the tendency of the sociology of knowledge to make a mystery of its own claim to rationality and objectivity. Further, this view of social science like its physical science counterpart insists upon a sharp dichotomy between theory and practice. Theory does not inform practice, but merely reflects it, and the social scientist is viewed more as an actor on the scene rather than either a viewer or director.

Structuralist social science leaves no room for individual effort, for individual reason is the product of external forces. Here the individual loses his identity in favor of the species as a whole whether that whole is called a class or a nation, etc. The moralist in Marx, for example, would not permit him to accept the ultimate logic of the teleological position, and that is why he turns it into an eschatology. Again, Marx himself sometimes suggests the later evolutionary view of an individual molding and interacting with his environment, but more often than not he lapses into the view that ideas do not shape social forces but are merely the epiphenomena of such forces. To the extent that we want to provide for the individual suggesting a new theory or finding contrary data, we must reject the Marxist theory of social science taken literally. Of course, there are 'marxist' social scientists who do adopt the evolutionary perspective.

Summary: Second View of Science: Structuralist (Aristotelian)

1. teleological biology is the model science, hence:
2. rationality involves first principles abstracted from experience;
3. explanation is still demonstration from first principles;
4. objectivity is a commitment to the correspondence theory of truth. This is the key factor of structuralism.
5. Social science is a description of the inherent ends of institutions which follow a cyclic pattern.
6. Values are the built-in ends of specific institutions.
7. The function of action or practice is to inform theory. Understanding is the final end. Hence there is a strict distinction between theory and practice.

8. *Evaluation*:

   a. Like the first view, it cannot account for deviations or "births out of season." It too violates the principle of total intelligibility.

b. Those who reject this as a view of social science and value, but nevertheless maintain it as a view of physical science, likewise invoke a distinction between fact and value; further, they embrace anti-historicity. The latter is manifested in strict dichotomies between the philosopher of science and what they consider mere history or sociology. They usually stigmatize holders of the third model of science as confusing logic and psychology.
c. Although they deny it, they do have an historical view of science as consisting of geniuses. The category of genius avoids the necessity for explaining the obvious failure of scientists to abstract theories from the facts.
d. The commitment to the correspondence theory of truth leads to the insoluble problems of theory invention, why it is that scientific explanations are never entailments, paradoxes of confirmation, and distinctions between theoretical and observation statements. Faced with these difficulties, they frequently revert to Platonism; a reversion preceded by numerous *ad hoc* admissions.

5. Copernican model

We come now to the third and most recent model of science and hence of rationality. I shall describe it generally and then distinguish two versions within the general framework. For historical reasons, I shall call this third model *Copernican.* It can perhaps be best construed as challenging the assumption made within the two previous models that science consists of categories for describing an obvious structure to reality. Explaining in science is not a simple matter of just describing what we see or envision, it is rather a much more complicated activity. Hence, any description or explanation of science itself cannot be a mere extension of categories used for describing the world. The emphasis should be upon the qualifying adverb 'mere'. To explain science itself we must take account of man, just as to explain the motions of the heavens we must take into account the movements of the observer on the earth.

This does not mean and should not be taken to mean that man is to be understood in a nonscientific or a scientific fashion. Nor does it in any way constitute a challenge to the goal of the unity of science. It can be argued that we may use some science to study man himself, and that the scientific study of man can then be extended to explain even science with both a big and a little 's'.

There are a number of ways of making clear what is new and unique about this third model. One distinction which we are now permitted to make is between the properties of the world in the nonhuman sense and the properties of peculiarly human entities. There has always been a psychological tendency to ascribe to the world things which might only be true even in a limited way of men or of our conceptual systems. Anthropomorphism is an example of this tendency. Believing that truth is a property of the world rather than a property of sentences is another example. This, again, should not be taken to mean that the properties of men or of conceptual systems cannot themselves be scientifically investigated and explained.

I do not believe that it is possible to describe or account for the activity of science itself and of scientific rationality by using only what we know to be true of a scientific description of the inanimate world. This is what vitiates the narrow positivistic theses and crude attempts to capture human rationality with a truncated logic. At a more basic level we can offer a scientific account for human rationality but it would obviously include more than is found on the first immediate level. This process is necessarily time-bound, so we may describe it as recognizing the unavoidability of infinite regress or even of a grand form of circularity. But it should be added that in this case circularity is a virtue and not a defect. I believe that this is what Harré has in mind by the necessity of transcendental argument.

This third model can be found in as early a figure as William of Ockham, but it is best represented in the modern (as opposed to the contemporary) period by David Hume and Immanuel Kant. This explains why the real beginning of the social sciences is a product of the eighteenth century. In its earliest form it was dominated by the success of Newtonian mechanics, hence it tended to explain even the activities of scientists from the point of view of a mechanical conception of human activity. Mill stigmatized this view as the geometric method. Popper has aptly described it in another context as psychologism. It is the failure of this mechanical view of explaining human activity in general and scientific activity in particular which has led a number of people to raise serious questions about this mechanical view of science. The kinds of issues which we are discussing at this conference I would attribute in large part to a recognition of the failure of the mechanical view of science.

From a purely scientific point of view there are serious doubts about the adequacy of the Newtonian view of physics to explain the inanimate world in other than a crude sense. There are now no doubts about its inability to explain the animate world. In an important sense there has not been a successor to Newtonian mechanics as a dominant science, and hence no dominant model for the explanation of science itself. Yet, having introduced the Copernican perspective, and having adopted the idea that human behavior can be scientifically explained, it was natural to look for some science of man in general and then generalize that science as the basis for studying science or scientific activity as one manifestation of human behavior.

The stage has thus been set for the second form of the Copernican model of rationality, and it is a form which takes biology as the key to understanding the activity of science itself. But it is not a teleological biology concerned with final causes and its consequent emphasis on classification. Rather, it is a form of *evolutionary biology*. If one looks upon human behavior in general and scientific behavior in particular from the point of view of a species evolving over a long period of time but not toward any final end we can meaningfully conceive, then rationality and objectivity take on an entirely new meaning. While I am not sure that these men would react kindly to my classifying them as evolutionists, it is nevertheless true that Feyerabend, Kuhn, Toulmin, and the late professor Hanson have by their historical studies made an evolutionary view plausible, and Popper by his emphasis on the methodological element of falsifiability has also contributed to making sense of scientific activity in this way. No doubt they would be more cautious, and in the light of my critique of classification, I should say that I have evolved my view from theirs.

By contrast with the structuralist or Aristotelian view which interprets objectivity as being free of bias and, more important, as having views which are directly translatable into a fixed reality, any Copernican view must argue that belief conditions observation. Perhaps this view sounds paradoxical because it is stated with the use of the term 'observation'. It might be more accurate to say simply and baldly that there really isn't any such thing as observation. By denying observation or by claiming that belief conditions observation, it is meant that any belief goes beyond all available data.[8]

From this point of view, theories present no special problem. Naturally, someone will ask how is it possible to explain changes in belief if there is no fixed reality against which to measure it, for it is obvious that beliefs do change. Avoiding the use of the term observation, perhaps we can say that there are such things as anomalous data which do not fit any previous belief or which are anomalous because they can be structured by two or more conflicting beliefs. The existence of such data gives rise to a search for a more comprehensive belief which resolves the conflict at a higher level.

From the Copernican point of view, objectivity is not explainable as the interaction between an individual and his environment but is instead an institutionalized form of intersubjectivity. Scientists are objective because of their membership in a community wherein each can cause the others to change or modify beliefs by either presenting new theories (beliefs) or by finding new data to challenge old beliefs.

By working with the old model of mathematical demonstration, the structuralist had come to believe that the development of new theories was somehow like the working out of the additional consequences of the old theories. Newtonians also think this way. This work is fine if we are talking about geometry, but it certainly does not apply to empirical science.[9] Actual theories we deal with both drop old parts, and this remains mysterious on the old views, and add entirely new elements which are not the mere consequences of the former. What we observe is a trial and error pattern which is best accounted for in terms of the Copernican model. But again this process must be viewed from the proper temporal dimension. If we look at too narrow a period of time, science certainly looks like a confused mess. If we choose too broad a spectrum, we are tempted to see overly-dramatic revolutions. If we choose a large slice but view it continuously and developmentally, then what we notice is that a theory retains an identity even in the face of apparently conflicting additions and subtractions. Change becomes intelligible as a series of cumulative discontinuities leading to a long term dramatic transformation. Surely, the notion of small changes leading gradually to big changes is an evolutionary perspective.

A theory, moreover, is not a univocal entity but has variants. Since theories are a series of beliefs going beyond all available data, this state of affairs is not only intelligible but to be expected. Not only is there competition between two different theories but clearly there is competition as well between two variants of the same theory. If we take our biological metaphor seriously, there is even an analogy with real species. The alternative variations in the same species constitute a healthy state for any given species in terms of adaptability. Small changes can be accommodated precisely because of the variants. Competition would even be healthy for the efficiency of theories in adapting to different data environments. Even error, which is the bane of existence for more rigid views, would be welcomed as contributing to efficiency by showing the ineffectiveness of one variant.

Not only does the evolutionary view of science seem to accord with the actual practice of science, but it helps to explain why scientists and philosophers of science keep old views of science even when the paradigm of science has changed. It can even explain why some scientists can operate with more than one paradigm at the same time. Any contradiction becomes a conflict when viewed from a different perspective, and conflict is part of the reality. Moreover, the evolutionary view can explain our concern with control over the environment, where control is the result of a successful adaptation. This concept of control completely undercuts the old distinction between pure and applied science. Finally, it seems to make sense of the relevance of science to human life.

By its very nature the evolutionary view is critical of the demonstrative conception of both science and scientific explanation. Hence, in seeking to generalize from scientific practice some conception of rationality, it will naturally be critical of any demonstrative view of rationality. A rational policy is not a demonstration from first principles *a priori*, nor is it a demonstration from the discovery of transparent and empirically discoverable eternal verities. Such a view, which I have labeled structuralist, has created bizarre theories of the formulation of scientific theories as well as unsolvable riddles about how one confirms the presence of a verity. The evolutionary view is concerned less with classifying laws and more with the actual life of theories. It is a view which is not bothered by the presence of theoretical terms as opposed to observation terms since it does not have a static view of structures, and rather than attributing the power of theories to vagueness (which sounds anti-rational anyway) it sees them as developing instruments of an organism coping with its environment. It allows for the influence of unconscious and external controls over the development of the process, and most important of all it makes intelligible the existence of conflict, the conflict within the species and the conflict among theories.

The work of Hanson, Toulmin, and Kuhn not only provides material for the refutation of simple-minded logical models, but it also indicates a number of recurring elements which are not so easily classifiable. Certainly, falsifiability is the major element, but falsifiability is not a mechanical logical process in a world which does not seem to come in easily identifiable pre-packaged units. A second element is the need for broad or comprehensive theories, so that frequently one finds two or more conflicting theories held at the same time. Thus, a naive or foolish consistency is never found in actual scientific practice. Another clearly detectable element is a conservative bent, a desire not to engage in wholesale revolutions. This conservatism is another version of comprehensiveness, for while it does not discourage new theories it views such theories sceptically until they alone are capable of explaining the data and the old views are no longer modifiable to accommodate the new data. The interplay of these forces must be viewed as either irrational or accidental on the demonstrative view. But to the Copernican mind it suggests instead that rationality has not been captured by simple-minded logical models.

What does rationality mean for the evolutionary perspective? The essence of rationality is that it relies upon persuasion and not force. As Whitehead remarked, the appeal to force is the mark of barbarism. To persuade is to convince someone that a conclusion follows from a set of premisses or that one set of premisses conflicts with another. The possibility that even our premisses may have to be modified rules out dogma or authority as a basis for rationality. The revision of premisses reveals the notion of corrigibility or falsifiability as underlying our conception of rationality.

If rationality is related to falsifiability, and if falsifiability is a temporal process, not a direct confrontation with 'reality', then rationality cannot be characterized as mere consistency. Further, if data does not come in pre-packaged units, then theories cannot be simply compared to the facts of observation. Rationality must then be viewed as the way in which beliefs are modified over time in the face of accumulating evidence. This conception of rationality allows for the obvious operation of individual judgment. Theories are not to be viewed as abstractions from the data but creative constructions of human intelligence. By creative I mean that human intelligence is not a tabula rasa whose operation can be explained as a simple mechanical and externally determined element. This will have important consequences for social science and rational control. It is not, of course, the creative process which is rational but what happens to the theory after it is articulated that determines whether rationality is present.

The third model of social science following the third model of science in general tends to be much more conscious of the activities of individuals rather than of institutions. In its mechanical form it has a tendency to assume that social facts and institutions are the mere mechanical results of the actions of human beings, and each individual acts in terms of certain drives or motives again viewed from a purely mechanical perspective. Among the proponents of this view, economics emerges as the dominant social science because it approximates most closely the model of the marketplace wherein each individual attempts to maximize his gain.

The shortcoming of the first version of the third model is that it either cannot account for cases where individuals seemingly act against their self-interest or reinterprets self-interest so that it becomes compatible with any state of affairs. As the latter, it fails the test of falsifiability. Moreover, its sense of social diagnostic has come increasingly under attack. It tends to view social problems as capable of rational resolution, and in this sense at least provides something not found in the previous models. Yet it tends to see human problems as problems of accumulation and distribution of wealth, so that all social problems are viewed as technical problems of a materialistic sort. In this view, the social scientist assumes the role of social engineer. At one end his theoretical function is to take polls or surveys to discover what particular mechanical motives exist and at the other end to instruct the natural scientist to secure a kind of vector resolution of social conflict. Rational control for this model means maximizing creature comforts. Insofar as it is self-conscious it tends to view social science as a research institution amassing more information, and its sense of usefulness begins and ends in the desire for more research funds. Those colleagues who are forever promising that all human problems are to be solved simply by increasing the size of their staff reflect this point of view. Unwittingly these men are partly responsible for our difficulties precisely because they promise more than we can deliver. They have no real sense of theory for the basic information is presumed to come from some mechanical or psychologistic view of human motivation.

The second variation of the third model, the model I have called Copernican, is evolutionary. It provides a rather different view of the social sciences and what they are able to contribute. The evolutionary view is naturally critical of utopian schemes, the first model. It is likewise critical of the structuralist model which seems to make no sense of individual behavior, least of all the behavior of social scientists. At the same time, the proponents of the evolutionary model of social science cannot but be impressed with the sense of historical development that one finds in the structuralist model of social science. That historical model is admired up to the point where it falls into the trap of believing in fixed or final ends.

With respect to the mechanistic view of social science, the evolutionary model is sympathetic to the emphasis on individual human beings. If I may use an analogy from biology again, a species may evolve but it does not evolve as a group. Rather it evolves through the mutations and adaptations of its individual members. The same seems to be the case for human beings when they function socially. However, the evolutionary model would reject the purely mechanical conception of human behavior. The evolutionary model seeks to explain what the mechanical model cannot explain, namely, the inability to resolve social problems merely by having more information and a debate. When one views different human beings in the context of varying historical and cultural backgrounds, then it is apparent why they do not immediately 'perceive' the information or data in precisely the same manner. This state of affairs should not, however, be viewed as ruling out rationality as a change in viewpoint over an extended period of time. The sociology of knowledge becomes a useful diagnostic technique so long as it does not degenerate into a self-contradictory ontological thesis.

The evolutionary model can also make sense of social science and its problems. Just as the individual whom it studies within given environments at given times can operate upon and change the environment, so individual social scientists can affect the course of development of social science. A given individual social scientist who articulates a new theory or discovers new data can participate in the process of rationality either by challenging previously held theories or providing a new way of looking at the data. Not only are institutions in this sense corrigible but one comes to see that social scientific theories are corrigible.

One of the consequences of this evolutionary approach is that social science is no longer to be viewed as the construction of utopias, not as prophecy, apologetic, social activism, or pure demography. It becomes a form of education when it is applied to the solution of practical problems. What does it educate us about? Education is conceived of in terms of the biological metaphor, an accumulated past experience interpreted and reinterpreted in the light of new experience and new theories about our own behavior. There are no rigid laws to be discovered here any more then there were in the natural sciences, rather we are concerned with the discovery of, at most, functional relationships or tendencies. Such knowledge by its very nature is not capable of leading to predictions. Even after we have provided an account of the past and present which takes into account every piece of data, we cannot show that this is the only account possible or that actual events are the only ones compatible with our theory. The limits of predictability have been admirably demonstrated by Peter Winch in his book *The Idea of a Social Science* (1958) and in W. V. O. Quine's theses of linguistic indeterminacy.[10]

The immediate reaction to such indeterminacy is to think that rational action is not possible. But just the opposite is true. If rationality is bound up with falsifiability, then it is precisely on this view of social science that rational action makes any sense. This view is well brought out by John Stuart Mill: "Knowledge insufficient for prediction may be most valuable for guidance. It is not necessary for the wise conduct of the affairs of society, no more than of any one's private concerns, that we should be able to foresee infallibly the results of what we do. We must seek our objects by means which may perhaps be defeated, and take precautions against dangers which possibly may never be realized. The aim of practical politics is to surround any given society with the greatest possible number of circumstances of which the tendencies are beneficial, and to remove or counteract, as far as practicable, those of which the tendencies are injurious. A knowledge of the tendencies only, though without the power of accurately predicting their conjunct result, gives us to a considerable extent this power."[11] The foremost contemporary exponent of this view is Sir Karl Popper. As Popper put it in the *Open Society and its Enemies*, "the main task of the explanatory social sciences" is "the discovery and explanation of the less obvious dependencies within the social sphere. It is the discovery of the difficulties which stand in the way of social action" (1950: 287). As a consequence, Popper has argued for a piecemeal conception of social engineering.

The knowledge that we might fail, that we might have to reconsider entirely our whole notion of what we are doing in social life follows from the recognition of the limits of predictability. Even if developments can at a later date and in retrospect be explained by some microtheoretic structure, the inability of such a structure to predict in advance provides the rationale for a developmental conception of social science. This is to incorporate the notion of corrigibility into our social science. It is the awareness of the possibility that we are fallible that necesitates as we shall see the policy of open research and its ultimate criterion of expanding research opportunities. The major element of social change which all previous models of social science fail to capture is the element of human intelligence itself. Reason is not merely a faculty of discernment, nor is it simply the byproduct of other forces however much they may shape it, it is also an active participant in the social process as well as in the intellectual process.

Summary: Third View of Science: Copernican (two versions)

1. Existence of alternative theories, both or all of which 'fit' the facts, is taken as the *reductio* of the two previous models.
2. We must distinguish between an understanding of the world and an understanding of understanding. Knowledge (science and rationality) is the product of the interaction of men and the world. Hence any analysis of science and rationality must take into account the activities of men. This does not exclude the possibility of a total and univocal view of reality, but it does rule out a description of science and rationality as mere correspondence to a structure independent of men.
3. This view controverts any distinction between facts and values and between theory and practice. The activities of the knower, including 'curiosity' and narrowly practical concerns, etc. are part of any analysis of knowledge.

4. Truth is to be understood either as some form of coherence or along the lines of the performative theory of truth as worked out by P. F. Strawson (1949).
5. Objectivity is to be strictly distinguished from absolute truth. It does no good to say that one believes in absolute truth but is perfectly willing to admit that all particular beliefs are corrigible. I fail to understand what the assertion of a belief in absoluteness means under these conditions. Objectivity is not correspondence of symbols to a fixed external structure but an institutionalized or socially oriented form of intersubjectivity. Thus objectivity is conditioned by the properties of the members of the community.
6. Rationality is the process in which conflicting beliefs (theories) are resolved into a new belief. The old beliefs are said to be falsified. However, different notions of falsifiability are involved, so we must distinguish between the two variants of this third model.

Variant A. Newtonian

1. Model science is Newtonian mechanics.
2. Falsifiability is the substitution of a more general or comprehensive theory, wherein generality means reduction to a common core. There is an emphasis on paradigm conservation and a tendency to view new theories as the working out of additional consequences of older theories.
3. Explanation is tied to prediction.
4. Social science is reduction to psychologism.[12]
5. Values are explained as the product of individuals acting to maximize self-interest usually conceived of in a materialistic sense. Economics thus emerges as the dominant social science. It claims to be value-free but only in the sense that it takes for granted that the problems all involve accumulation and distribution of wealth.
6. *Evaluation*: this view fails because of the lack of predictability. In the social sphere it fails to account for individuals who seemingly act against self-interest or self-interest is reinterpreted so that it becomes compatible with any state of affairs.

Variant B. Evolutionary

1. Model science is evolutionary biology.

2. Falsifiability is revision of basic premisses over time. Trial and error leads to the addition of new parts to a theory. Not only are there competing theories but variants of the same theory.
3. Social science is not utopianism, not a description of institutional essences, and not psychologism.[13] It sees itself as explaining the evolution of a species through the mutations and adaptations of its individual members not all of whom evolve as a group. This interplay between individual and species is what it seeks to capture.[14]
4. Values are the product of the interaction of men with the environment including other men. Value schemes are to be interpreted as theories and they can be modified or discarded accordingly. It makes sense for an individual to challenge the preexisting value structure just as individual scientists may challenge and falsify theories. Just as one theorist may be ahead of his colleagues so one individual may be morally more advanced than the other members of his community.

6. Values and the Science of Man

The problem we have set for ourselves is the formulation of a rational policy of science. We have determinated that science itself provides the model of rationality, and that the key element of rationality is corrigibility. Moreover, corrigibility has led us to formulate a view of science or the activities of scientists as best understood in an evolutionary pattern. We have now to establish the connection between these conclusions and our policy.

There is, however, a missing link in the chain, a link supplied by consideration of the question of value. Presumably, the formulation of a policy is to achieve a certain objective or set of objectives. Once we specify these objectives, then questions of means become appropriate. The objectives constitute our values, so we must inquire how our values are to be understood as part of this process.

In the first or *a priori* model, values are given *a priori* in the same way that our other first principles are acquired. Such values are not open to criticism or revision or falsifiability and hence are to be rejected as irrational.

In the second model, values are empirically discoverable as part of the permanent structure of the objective world and as such are independent of mankind. As part of that rigid external structure they remain impervious to human manipulation. One can merely note their cyclic pattern. Once more it is clear that such values are impervious to criticism and must be rejected as irrational. The pursuit of such values was what John Dewey continually stigmatized as the quest for certainty.

We come now to the third model in which it is clear that values function only as part of the interaction between man and his environment. Without man values would not exist. With different men or a different environment, the values will be different. This is not to advocate cultural relativism, for cultural relativism is a consequence of the structuralist model and not the evolutionary model. Within the evolutionary model, values like theories may be changed and challenged either by more comprehensive value claims or by the discovery of new data incompatible with old value schemes. Values may become as dysfunctional as theories and modified or discarded accordingly. One can on this view legitimately criticize a given set of cultural values as incompatible with the survival of the species. Moreover, it makes sense for given individuals to challenge the preexisting value structure just as individual scientists may challenge or falsify the views of the rest of the community. Just as one theorist may be ahead of his colleagus so one individual may be morally more advanced than the other members of his community.

Just as the evolutionary view rejects the dichotomy of theory and practice so in the value realm it must reject the dichotomy of value and fact. This means that values are themselves subject to scientific study. The possibility of a science of man carries with it the possibility of a scientific determination of value.

What I call the science of man is crucial for the formulation of a rational science policy. In a world which calls for the legitimization of scientific practice, that insists that the knowledge provided by the sciences both natural and social be used to solve human problems and not against man, we must provide some kind of answer. How often have we heard that more knowledge only makes our policies more efficient and not necessarily better. In short, we have the demand that scientific research not be abused or misused from society's point of view.

I presume that the information uncovered by scientists is abused or misused if it is employed in a manner contrary to human nature. Without a knowledge of human nature, it becomes impossible to decide what is and what is not compatible with human nature. Hence, without a science of man, no one is in a position to say what is and what is not a misuse of information.

The evolutionary view rules out the possibility of a fixed human nature. It does not rule out objectivity, for at any given point in time we may state in an intersubjective way the state of the species. Our problem now becomes to specify a policy which takes for granted the possibility of objective values and yet does not believe that those values are final. I believe that it is possible to construct a rational policy in the absence of final values or a final knowledge of values.

7. Policy

My policy may be constructed as a set of answers to a series of questions.

(1) Are there objective values for judging human behavior? Notice that I did not say absolute values, for that would imply no change over time. Objective is here to be understood in terms of an interpersonal structure of mutual criticism wherein individuals are at liberty to falsify the beliefs of others.

There are three possible answers: no, maybe, and yes. If there are no objective values, then there is no basis in terms of which we can judge, criticize, blame, or in any way evaluate the activities of individual scientists or the whole science community. No way would be provided to say that certain practices fail to achieve our goals. In short, there could be no policy. Therefore, the answer must be either yes or maybe, and as we shall see both of these answers lead to the same policy.

(2) Do we know now what these objective values are? There are two possible answers to this question: no and yes.

If the answer is no, then our situation is such that we believe such values to exist but do not claim to be in possession of the knowledge of what they are. This would necessitate a special kind of policy. And, as we shall see, this policy is the same as the one suggested by the answer 'yes'.

If the answer is yes, then it may be asked if we are infallibly correct about what we take those values to be. The assumption of infallibility is incompatible with the criterion of rationality we have adopted, namely, falsifiability. If we cannot assume for ourselves or grant to others the existence of infallible judgment, then we leave open the possibility that we might be wrong about our understanding of the values involved. A fallible yes is in practice no different from a 'no' answer, for both views must take the same precautions.

I might note that this policy is in principle adoptable by those structuralists who believe that there is a permanent value structure but that it has not yet been uncovered. The necessity for discovering the structure and taking precautions against error would dictate the same policy. Up to the time that the structuralist claims he has made the final determination, his policy would remain the same as that of the evolutionist. When the structuralist comes to realize that he will never grant the possibility of infallibility he will have falsified his own ontology. At that point we shall welcome his conversion.

(3) How do we act in the meantime, while we are trying to discover what those values are? It would seem that the first order of business is to discover the values or to engage in constant checks on their validity. Hence, the primary function of the entire community must be a commitment to the discovery and verification of those values. All institutions and practices are to be judged in terms of their contribution to that end.

A second consequence is to allow as much freedom as possible to individuals of the community to experiment. What might look like chaos should be viewed as alternative paths any one of which might lead to important discoveries or revisions. The only set of circumstances under which we are justified in interfering in the particular path of an individual is if we can prove in an objective way that he or she is doing harm to others or causing them to become dysfunctional. At first this may appear paradoxical since the existence of harm depends upon the kind of values we have in mind. But either we shall have already identified some values or we shall construe harm as incapacitating someone from either participating in the search for, or enjoying, the values we seek. Thus we can rule out the destruction of human beings, and we can especially rule out any behavior which incapacitates the rationality and reflectivity of other human beings.

Here an analogy may be drawn between freedom for research and censorship. Among rational men there is no justification for censorship, for the censor assumes some infallible standard. But does opposition to censorship rule out limitations on those who incite to riot? Is not a restriction of the rioter a form of censorship? The answer is a definite no because in a riot all possibility of rational communication is eliminated in favor of force. And force, as we know, need not always be brute. Riots are to be prevented or suppressed even by the use of force because that may be the only way to provide an atmosphere in which rationality may operate. What distinguishes the legitimate from the illegitimate use of force is precisely its contribution to an environment in which rationality may flourish.

There are two general points about rationality that I have made in this paper. The first is that rationality implies that the world as a whole is intelligible. The second is one that I should like to denominate as the *sanctity of dissent*. Rationality, as we have seen, means corrigibility. It is our recognition that we might be wrong which leads us to be tolerant. Tolerance is not an *a priori* value but a consequence of our view of rationality. It is this same sense of corrigibility which allows us to dismiss all forms of dogmatism, including most especially those which do not flow from science. It follows that we are entitled to take appropriate measures to protect the sanctity of dissent. If it is suggested, as sometimes happens when this kind of view is propounded, that a group of criminals may justify themselves as operating with a different paradigm, we may reply that such a paradigm is acceptable as long as it does not interfere with others who operate with a different one. Of course, part of what we mean by criminal behavior is that it engages in just such interference. But then that merely shows that my view does not condone criminal behavior.

It follows from what I have been saying that no form of research may be limited or prescribed unless it can be shown that such research interferes with the search for the truth on the part of either that individual or of other members of the community.

It follows as well that when for the most practical of reasons decisions must be reached about the funding of research under the prospect of limited resources, the primary value to be considered is which of the alternatives is likely to have as a consequence greater research possibilities. What we must look for are proposals which will lead to an even greater number of projects and proposals in the future.

I believe that the greatest threat to the freedom of research in the world today comes from *a priori* humanitarians regardless of the political system under which they live. Those who would seek to limit research because it is dangerous and who have made themselves the final arbiters of what constitutes the danger are to be found everywhere and especially in our universities. It may seem ironic that basic research has flourished best to date primarily where military rivalries exist. No doubt, political and military considerations of the most myopic kind are foremost in the minds of those who sponsor such research, but it is also true that fears for survival and the threat of giving an adversary an undue advantage lead to the sponsorship of projects that would otherwise go begging. It is also regrettable that research sponsored by the public for humane purposes is usually defined so narrowly that it rarely has residual benefits. Moreover, I suspect that if the world were at present to come under one government and if that government were controlled by *a priori* humanitarians, the effect on the sponsorship of basic research would be catastrophic.

An evolutionary view of theory development has shown that competition among theoretical proposals is both healthy and useful. In fact, one might say that it is necessary. But the fact remains that most people do not subscribe to the values of the free interplay of ideas either in the political-social sphere or even in the scientific sphere. Perhaps one can speculate how much an absolutist conception of science has reinforced moral absolutism. The freedom that does exist on the conscious level is the grudging concession of those who lack the power to impose their will. Like the origins of religious toleration, the toleration of ideas is the result of a multiplicity of weak sects and not a consciously held value. Our task is to make it a consciously held value.

Queens College,
Flushing, N.Y.

NOTES

1 *New York Times*, May 10, 1974, p. 56.

2 Kuhn's own account of science as an institution smacks of this kind of Aristotelian conception. The whole concept of normal science seems to imply a kind of institutional end. On the other hand, the notion of scientific revolution is an attempt to accommodate, in part, the idea of an individual reshaping the institution. This ambivalence may account for the dissatisfaction some feel about Kuhn's distinction.

3 I take Carnap and Hempel to be the best representatives of the structuralist view.

4 One of the earliest and best attacks on the deductive concept of explanation was in Dray (1957). For other attacks on this model, especially from a technical point of view, see the articles listed in the bibliography by Ackermann, Eberle, and Collins.

5 See Good's articles on the 'Paradox of Confirmation', mentioned in the bibliography.

[6] There is now a tendency to view Marx from an Aristotelian perspective. See the collection by Petrovic (1968). I would argue that Marx has an Aristotelian view of social science but a Platonic view of value.
[7] See my forthcoming book, Capaldi (1975).
[8] The critique of observation is to be found in Feyerabend (1965); Hanson (1958), pp. 5, 7, 8, 18, 20; Kuhn (1962), pp. 114–119; Toulmin (1961), 57, 95.
These views have been misunderstood and attacked by Kordig (1971).
[9] See Feyerabend (1962).
[10] Quine (1960).
[11] Mill (1843), VI, ix, 2.
[12] I am using 'psychologism' in Popper's sense as described in the *Open Society and Its Enemies* (1950). Unfortunately, Popper mistakenly ascribes psychologism to J. S. Mill.
[13] See Dewey (1929) for an attempt to develop evolutionary social science.
[14] A further consequence is the collapsing of internal vs. external social science or the dichotomy of the philosophy and the sociology of science. See Toulmin (1972), 505.

BIBLIOGRAPHY

Ackermann, R. J.: 1965, 'Deductive Scientific Explanation', *Philosophy of Science* **32**, 155–167.
Agassi, J.: 1963, 'Towards a Historiography of Science', *History and Theory*, Supplement 2.
Capaldi, N.: 1975, *David Hume: The Newtonian Philosopher*, Twayne, Boston, Mass.
Collins, A. W.: 1966, 'The Use of Statistics in Explanation', *The British Journal for the Philosophy of Science*, **17**, 127–140.
Dewey, J.: 1929, *The Quest for Certainty*, Gifford Lectures, Putnam's Sons, New York.
Dray, W.: 1957, *Laws and Explanation in History*, Oxford University Press, Oxford.
Eberle, R., Kaplan, D., and Montague, R.: 1961, 'Hempel and Oppenheim on Explanation', *Philosophy of Science*, **28**, 418–428.
Feyerabend, P. K.: 1962, 'Explanation, Reduction, and Empiricism', in *Minnesota Studies in the Philosophy of Science*, Vol. 3 (ed. by H. Feigl and G. Maxwell), University of Minnesota Press, Minneapolis, pp. 28–97.
Feyerabend, P. K.: 1965, 'Problems of Empiricism', in R. Colodny (ed.), *Beyond the Edge of Certainty*, Prentice-Hall, Englewood Cliffs, N.J., pp. 145–260.
Feyerabend, P. K.: 1968, 'On a Recent Critique of Complementarity', *Philosophy of Science* **35**, 309–331; **36** (1969), 82–105.
Good, I. J.: 1960, 'The Paradox of Confirmation', *The British Journal for the Philosophy of Science* **11**, 145–148; **12** (1961), 63–64; and **18** (1967), 322.
Hanson, N. R.: 1958, *Patterns of Discovery*, The University Press, Cambridge.
Hanson, N. R.: 1963, *The Concept of the Positron*, The University Press, Cambridge.
Hempel, C. G.: 1965, *Aspects of Scientific Explanation and Other Essays in the Philosophy of Science*, The Free Press, New York.
Kordig, C. R.: 1971, 'The Theory-Ladeness of Observation', *Review of Metaphysics* **24**, 443–485.
Kuhn, T. S.: 1957, *The Copernican Revolution*, Vintage Books, New York.
Kuhn, T. S.: 1962, *The Structure of Scientific Revolutions*, University of Chicago Press, Chicago.
Lakatos, I. and Musgrave, A., (eds.): 1965–1968, *Proceedings of the International Collo-*

quium in the Philosophy of Science, London, 1965, 3 volumes, North-Holland, Amsterdam.

Lorenz, K.: 1962, 'Kant's Doctrine of the A Priori in the Light of Contemporary Biology', *General Systems Yearbook* **7**, 23–35.

Mill, J. S.: 1843 (1961), *A System of Logic*, Longmans, London.

Petrovic, G.: 1968, *Marx in the Mid-Twentieth Century*, Doubleday, New York.

Polanyi, M.: 1958, *Personal Knowledge: Towards a Post-Critical Philosophy*, Routledge & Kegan Paul, London.

Popper, K. R.: 1959, *The Logic of Scientific Discovery*, Hutchinson, London.

Popper, K. R.: 1963, *Conjectures and Refutations*, Routledge & Kegan Paul, London.

Popper, K. R.: 1966, *Of Clouds and Clocks*, Washington Univ. Press, St. Louis.

Popper, K. R.: 1972, *Objective Knowledge: An Evolutionary Approach*, Oxford University Press, Oxford.

Quine, W. V. O.: 1960, *Word and Object*, The MIT Press, Cambridge, Mass.

Schilpp, P. A. (ed.): 1963, *The Philosophy of Rudolf Carnap*, Open Court, La Salle, Ill.

Schilpp. P. A. (ed.): 1974, *The Philosophy of Karl Popper*, 2 volumes, Open Court, La Salle, Ill.

Strawson, P. F.: 1949, 'Truth', *Analysis*, **9**, 83–97.

Taylor, C.: 1964, *The Explanation of Behavior*, Routledge & Kegan Paul, London.

Toulmin, S.: 1961, *Foresight and Understanding*, Harper & Row, New York.

Toulmin, S.: 1972, *Human Understanding*, Princeton University Press, Princeton, N.J.

Waddington, C. H. (ed.): 1968, *Towards a Theoretical Biology*, Edinburgh University Press, Edinburgh.

Winch, P.: 1958, *The Idea of a Social Science*, Routledge & Kegan Paul, London.

DAVID SILVERMAN

PRESERVING SCIENCE: VIRTUOSITY AS VIRTUE

Presence no longer reigns; it is the assault which prevails.

Martin Heidegger, *Holzwege*, p. 100

I

Both 'virtue' and 'virtuosity' derive from the Latin word 'virtus'. According to the dictionary, 'virtus' refers to 'goodness' moral perfection, high character, virtue'. 'Virtus' (virtue), we further learn, has its roots in 'vir' or 'man'. Hence the 'goodness' and 'virtue' of 'virtus' has a specifically human character. It is not things which may have 'virtus' but man alone. Nor is 'virtus' to be found only in man's relation to things. Man becomes 'virtus' when he expresses his specifically human character: to have 'virtus' is to exemplify that 'goodness, moral perfection, high character, virtue' is appropriate to men alone. It is the act of being human as a human being, not as a thing, nor in relation to a thing.

'Virtuosity' shares with 'virtue' the same root, but has a different usage. It is a term most frequently used in the sphere of music and it is from there that we will take our initial bearings. According to a musical dictionary, 'virtuosity' refers to "the manner of performance with complete technical mastery." An artist "who has achieved complete mastery over the medium in which he performs" is a 'virtuoso'.

Notice how the word 'mastery' recurs in both definitions. 'Virtuosity' is a quality which arises in a relationship with a thing. The 'virtuoso' is the one who, by the use of masterful technique, achieves "complete mastery" over some-thing. To have 'virtue' is to express the goodness appropriate to men; to have 'virtuosity' is to be master of a technique. To be 'virtuous' is to express the proper character of man; to be a 'virtuoso' is to be successful in controlling and dominating things. To have 'virtue' is to exemplify that which masters men (their humanity); to have 'virtuosity' is to have *complete* mastery.

The contrast between the usage of these terms is striking. One is tempted

Knorr, Strasser, and Zilian (eds.), Determinants and Controls of Scientific Development, 143–171.

to say that, despite their common origin, they now propose contradictory meanings. Yet this is to forget the all-encompassing character of 'virtus'. For, if 'virtus' (and from it, 'virtue') refers to the expression of man's specifically human character, then any activity of man must somehow express that humanity, if only in a partial, incomplete manner. So the fact that 'virtuosity' has come to mean what it has, defining goodness in terms of man's dominance over things, is not a *negation* of man's humanity but an *expression* of how that humanity makes itself heard in our time. If 'virtuosity' has almost replaced 'virtue' as the standard of the good life, we are then presented with an occasion for re-collecting our-selves as inheritors and co-producers of such a world.

The world where 'virtuosity' reigns is a world in which men recognize their humanity in their ability to dominate things. Such an ability is, of course, one of the possible modes of being human. Yet it is a limited and incomplete expression of 'virtue'. Mastery over a technique is, to be sure, an expression of a possibility of human existence. Yet, in revealing what man can do to things, it simultaneously conceals that which makes that mastery possible – the humanity, within whose grant men pursue their activities.

This essential incompleteness of 'virtuosity' is pointed towards in popular usage. So the musical dictionary while associating 'virtuosity' with "complete technical mastery," then goes on: "but the term is often used in a derogatory sense, suggesting that no other quality of interpretation is displayed by the performer." This "derogatory sense," to which the dictionary refers, implies that "complete technical mastery" is, in itself, somehow insufficient. What could be lacking? What "other quality of interpretation" does 'virtuosity' neglect?

Let us contemplate what might follow if, using the latest technical knowledge, a musical instrument could be constructed which would play automatically, by the turn of a lever. Of course, such instruments, like musical-boxes and player-pianos, do already exist. But let us suppose that this instrument was so finely made that it could give a technically perfect performance of a piece of music. In these circumstances, would it be altogether inappropriate to suggest that such an instrument had given a 'virtuose' performance or that its playing had exhibited great 'virtuosity'?

We now can see what is lacking in 'virtuosity'. For a man to be a 'virtuoso', he must achieve "complete technical mastery." Yet such mastery can also be attained by machines. Indeed, if one thinks of engineering principles, machines are constructed in order to operate efficiently on a given medium, by controlling and transforming it. It is, then, in the nature of machines to achieve "complete technical mastery": in the dictionary sense, to display 'virtuosity'. So the human 'virtuoso' expresses his humanity (his 'virtue') by acting as a well-constructed machine. And his ability to do so shows that one human possibility is for men to live their lives as machines.

Hence the incompleteness of 'virtuosity' arises insofar as it is an expression of only one aspect of being human. Moreover, it is an aspect which, by concealing the difference between men and machines, renders us forgetful of the man ('vir') which is at the root of 'virtue'.[1]

A machine *can* display 'virtuosity' – although it is an indication of the way in which the voice of our humanity still makes itself heard that we should prefer to say that a machine displays the 'virtuosity' *of its maker*. But a machine could never be 'virtuous'. So, in making 'virtuosity' rather than 'virtue' the central pre-occupation of our age, we identify the existence of men and machines. It now becomes appropriate to apply to machines a standard which, at its very root, exemplifies man's human (i.e. non thing-like) character. Equally, and more crucially, man is now to be addressed in terms of a standard which expresses the character not of man, but of machines. The essence of 'virtuosity' (and of machines) is exhausted by a striving for "complete technical mastery"; the essence of virtue (and of man) is concealed by it.

An irony: in finding themselves through their mastery over things, men become more like things. In dominating, they are dominated.[2]

A. *Virtuosity and Method*

There is an apparent confusion in this discussion of 'virtuosity' towards which we must turn. The musical dictionary locates such a quality in "technical mastery." Yet everyday usage tells us that the virtuoso's performance is, in some degree, different, new or superior to that of any-man. This apparent uniqueness of the virtuoso seems to clash awkwardly with technical mastery. After all, techniques are public and may be learned by any-man. So where is the 'difference' of the virtuoso's performance? How can 'newness' arise in practising a technique which is located in an accepted tradition?

This confusion between 'uniqueness' and 'tradition' as conflicting elements of 'virtuosity' can be re-solved if we recognize that the 'uniqueness' of the virtuoso is guaranteed entirely by the unquestioned standards of a tradition. Virtuoso playing achieves its distinctness, then, by suggesting new possibilities *of* and *for* a tradition. But it does not, indeed cannot, question the standards of that tradition. The man whose work brings into question those standards may, from the point of view of the tradition, be a lunatic, or merely an incompetent, but he cannot be a virtuoso.

For the virtuoso is the man, who, in exhibiting his technical mastery reveals the manner in which he is committed to (mastered by) his tradition. His concrete difference from other members of his community is the very guarantee of his analytic identity with them. So the recognition of 'difference' is only possible from within a tradition. And that tradition exemplifies the (forgotten) mode of production of a certain community.

In turn, what the community claims to unite it are its methods. These methods constitute, then, its concrete difference from other communities. It is true that it shares with other communities a tradition in which method and technique are the basis of knowledge and the means of technical control. However, the varying character of communally acceptable methods allow such communities to forget their analytic identity and to formulate their concrete differences. Each uses different techniques for warranting the work of its practitioners and each set of techniques constitutes the claimed site of its difference.

As a practice, science, like music, differentiated itself by its version of method – Scientific Method. As among musicians, one may recognize 'virtuoso' scientists: men who, by their complete mastery of Scientific Method achieve the acclaim of their community. The search for virtuosity in science, as in music, exemplifies an age in which 'virtue' is equated with the masterly practice of a technique for purposes of instrumental control. Hence, in understanding the role of method for the scientific community, we come closer to understanding the character of science.

What do methods do? Where method is equated with technique, it serves science as a means of gathering objective information about objective facts. By following communally warranted methods, we may feel secure that our accounts will be taken to have an objective character – to correspond to the way things are, rather than to our preconceptions.

By appealing to method, we acknowledge that we are constrained both in what we may study and in how we shall study it. Through our focus on 'proper' data, through our use of 'proper' methods of investigation, our community will come to warrant our account as an authentic report on Nature, for they will locate in it the method that any community member might have used. So they are interchangeable with us and we with them; if they, rather than we, had happened to carry out our study, there is little doubt that the use of a common method would have elicited the same information from Nature.[3]

As a consequence, for the scientific community, the acceptable account is the one which, through its display of communally warranted methods, is recognizably 'as objective as possible'. 'Goodness' becomes equated with impersonality for, untouched by human hands, the good account will be the very voice of Nature relaying its own message about itself. And the 'virtuous' scientist is to be a faceless man, interchangeable with other faceless men, applying techniques without bias, without fear or favour.

Viewed as a set of techniques applied in making sense, method is a means of producing accounts *and* accounters that will be 'as objective as possible'. Virtuosity is the very idea(l) of science. The scientist who displays complete technical mastery succeeds for the scientific community precisely because he seems to separate himself from his 'findings'. He is objective, like his account, because he has effaced himself behind communal techniques. He is simply the instrument of the scientific community; the medium through which a technique is employed. To the extent that he is any-one, he is technique.[4]

At each stage of Scientific Method, we are confronted with the idea(l) of object-like entities. The object of study must be established as an 'objective' fact, the account must be 'objective', while the 'objectivity' of the accountor will be apparent to the extent that he appears to be absent from what he writes. Unlike musical virtuosity, nothing appears to be missing from scientific virtuosity. For the greater the virtuosity of the scientist, the more both he and his account will approximate an object. And, the greater the objectivity of both, the more they are 'virtuous'.

Does method succeed? Invariably. Whatever the shortcomings of his data or the limitations of his findings, the scientist affirms that he is indeed a warranted member of his community through his display of being method-ical. Of course, method does not 'deliver' the goods' in two senses.

First, being method-ical cannot guarantee that a desired end will be attained. Chance happenings, unexpected developments, can frustrate even the most technically perfect attack upon Nature. Second, methods do not, in themselves, produce outcomes at all.

Consider the practice of 'working to rule' as used in industrial relations. By following the rule-book in an 'officious' spirit, the railwayman can regularly produce chaos on British Railways. This does not express some kind of failing of the rule-book: there is no way in which the book could be rewritten to prevent working to rule. For the rule-book does not produce correct outcomes. Rules, like railway rules, like sex-instruction manuals, like rules of scientific procedure, are always written by and for competent members of some community. Such people must already know what a 'proper' invocation of a rule looks like. They must already understand the 'spirit' of the rule-book and be able to identify 'proper' instances to which it can be applied. For the sense of any outcome rests upon a body of unformulated communal practices used to recognize contexts and to apply rules 'in the right spirit'.

The success of method resides, then, not in that it produces certain outcomes but in that it provides a communal rhetoric for displaying the good sense of *any* outcome. Method succeeds for members of a concrete community because in producing recognizably method-ical (i.e. rule-governed) accounts, they affirm their identity. And through the claims of method, rule and technique they preserve their difference.

Method, viewed as technique, reveals the good sense of outcomes while concealing the analytic identity of communities who affirm their difference through method. This identity arises in their treatment of the standard of good speech ('virtuosity') as an unquestioned means to an end ('technical mastery'). So the standard of artful speaking is always a resource but never a topic. Moreover, in a circular manner, the standard is justified through works and results which are further exemplifications of the standard.

So a difference grounded on method resides in an identity, exemplified in 'virtuosity' which is never addressed or perhaps address-able. Yet this must always be the case where method is viewed as a technique brought to bear upon speech. In making the grounds of speech some-thing (technique), the community of 'virtuosity' completes the assimilation of men to things.

For speech is always a dialogue in and with a human tradition. Yet 'virtuosi' fail to hear in any speech the openness created by the tradition which makes it possible, nor do they listen to how the speech re-collects that tradition. For their 'vision' fails to acknowledge the 'pro-vision' of an openness in which they can speak by the tradition in which their speech is grounded. Nor do they understand that both the tradition and its re-collection exemplify not some-thing but us. Our mode of production. Our expression of a possible way of living our lives together, of a possible society. Our confession of our-selves.

Science, then, cannot abandon technique, for it is technique. The members of its community 'find their feet' in technique; they confess themselves as technicians. They conceive of themselves after the model of objects which they encounter within their world and relate to things in their objectivity. Like their objects of study, their aim is to be objective, rule-guided and impersonal. The idea(l) is to reduce themselves to a "transparent medium" (Macomber, 1967) in which the quantitative structure of the world can manifest itself. The aim is always to be 'faithful to the facts', by seeking to eliminate bias, preconceptions and so on.

But how can men be faithful to some-thing ('facts')? Only if the 'virtuoso' forgets his humanity ('vir') and confesses himself as a thing too. For, in being 'faithful to the facts', he is faithful to no-thing but to a tradition, a mode of production, in which things become recognizable. And the tradition to which he commits himself (in which he finds himself) is that version of good speech through which it could be *faithful* to be 'faithful to the facts'. Such a tradition conceals its human character by effacing itself behind technique.

If science cannot abandon technique, neither can it (or any speech) abandon (some) method (of production). Ironically, despite recent attempts to attack science for its adherence to method (Feyerabend, 1970; Phillips, 1973), *it is science itself which warrants a conception of non-methodic speech.*

With its notion of technique as the ground of speech, science sets out to evaluate the standards of different speeches. Good speech is that which exemplifies communally acceptable methods; bad speech shows a poor grasp of technique, is disorganized – in short: is non-methodic. Yet a conception of non-methodic speech is only possible when method is addressed as technique. Because the text displays insufficient technique, it will be deemed non-methodic. Yet the intelligibility of both this evaluation and of the text itself is only made possible by some method of production (of sense). Making sense always exemplifies the dialogue of the sense-producer with a tradition. Indeed, the claim of a text to be taken seriously is grounded in its exemplification of some tradition (some method of production).

So an understanding cannot be other than methodic. Method is always already there in our speech. It is not brought with us as an aid to understanding, but it is always there in the understanding – it *is* the understanding.

Only because, as 'virtuosi', we have to believe in the independent intellect, free from tradition, applying techniques without fear or favour, can we conceive of non-methodic speech. Science, conceiving of method as a technique to attain its ends has thus been grasped by what it hoped to use to grasp. Method is not a mere instrument serving the sciences; rather "it has pressed the sciences into its own service" (Heidegger, 1971: 74).

B. *Preserving Science*

Both science and its critics share a conception of the possibility of non-methodic speech; they differ only over whether the presence of method is an indication of the goodness of speech. This criticism which is agreement should remind us of the parallels between the form of life of 'criticism' and of 'virtuosity'.

For criticism is an activity which concrete communities expect, even provide for. Indeed such communities base their claim to difference on their ability to evaluate and criticise their own work. So the critic is simply the community member in another guise. A member who produces his own work for others to judge and who, when the occasion arises, will judge the work of others. On both occasions, community standards constitute a resource rather than a topic. Whatever criticisms he makes, for instance, will be grounded in them and so will serve to preserve the community. Hence the critic's difference, like that of the virtuoso, arises from an identity – the analytic commitment to the version of good speech of his community.

Where we are impressed by that difference, believing perhaps that the critic's account stands out or is outstanding, we exemplify the manner in which criticism conceals its mode of production. For criticism succeeds precisely to the extent that it manages to score points off another work without addressing its own possibility. And this kind of success is only possible within the world of 'virtuosity' – a world in which criticism can display itself as reporting on some-thing rather than exemplifying a tradition which opens up a space in which things are observable and reportable. For, ultimately, the only object which criticism confronts is its own language (Barthes, 1966). Because criticism conceals the commitment that brings it to life, always speaking in such a way as to make its object-at-hand rather than that commitment its topic, to criticise is always, unthinkingly, to accept. Failing to point to his own possibility, the critic is reduced to a reader of the second degree: "a voyeur observing clandestinely the pleasure of the other" (Barthes, 1973: 31).

We do not seek to polemicise against science, then, for any kind of polemics conceals its own possibility. The aim, rather, is to preserve science by addressing its grounds and, thereby, our own. For science expresses the tradition in which we find our-selves. In re-collecting science, we locate it in a metaphysical tradition of which we are a part and a product. Rather than refuting it or turning our backs upon it, we must make our-selves open to this tradition. For we *are* the tradition and its illusions are our illusions. Any insight, any revealment, is always from within illusion. Hence we can only experience truth by a constant and resolute reflection upon tradition. In coming to understand that tradition, we come to understand our-selves (Macomber, 1967).

In "turning back to where we are in reality already staying" (Heidegger, 1971: 85), we seek to preserve science by respecting the historically-grounded character of its thought (and of our own) and by addressing its possibility (and our own). Such a re-trieval preserves (by pointing to the possibilities that provide for the intelligibility of its problem) and re-collects (by recognizing our own responsibility for the grounds of our formulation). No attempt is made to dominate and capture science – to discover its hidden 'deep structure' beneath its apparent 'surface structure'. Rather, in finding our way about our topic, we sue it as an occasion to renew a dialogue with the tradition that encompasses us both. As a consequence, our reading of science must necessarily be a *violent* reading. For to take science (and this text) seriously is to address not the specific propositions of which it is composed, but that which, although unstated as such, is made evident through these propositions (Deeley, 1971).

These, and other, attempts to preserve science, meagre in extent and in insight though they be, are doubly necessary. For criticism, as has been suggested, fails to re-collect itself. And so, necessarily, does science. Science grounds itself on a standard of good speech without asking what grounds that standard. Indeed, it cannot ask such a question without ceasing to be science. For science must take as its concern things-which-are. To enter into other spheres, by enquiring into the Being of beings, cannot be its task. If it did so, it "would constantly fall into the void" (Heidegger, 1968: 33). Science, as science, has no access to what makes possible its own being.

This is not to criticise science, nor, arrogantly, to claim a superior vantage point. 'Thinking', always knows far less than the sciences. Neither can 'Thinking' grasp the "essence" of science. Every revealing is always a concealing.

II

One point of entry into a dialogue with science is to consider the literature in which the scientific community seeks to define its own character. The most obvious area for this is the philosophy of science. Indeed, the existence of a philosophy of science presents apparently contrary evidence to our earlier assertion that science cannot understand itself. Here, after all, one might suggest, is the area where precisely such an attempt at self-understanding is being made.

Obviously, given limitations of space, the philosophy of science can only be touched in passing. Let us, therefore, focus, on a debate which, although confined to the pages of one book, has been regarded as quite influential. I refer to Lakatos and Musgarve (1970), especially the contributions by Lakatos, Kuhn and Feyerabend.

The debate focusses on Thomas Kuhn's theories of the 'structures' of scientific 'revolutions', in particular the implication that, if knowledge is rooted in a paradigm, there are no absolute standards to be applied in choosing between scientific theories. This appears to horrify Lakatos. For him, if there is no ideal of proven truth, then "the whole classical structure of intellectual value falls in ruins" (Lakatos, 1970: 92). To make truth a feature of the character of a community is, for Lakatos, to allow truth to devolve upon power. As a consequence, Kuhn's position vindicates what Lakatos refers to as the basic political credo of student revolutionaries (our "contemporary religious maniacs"), since they too would seek to have theories judged by "the number, faith and vocal energy of (their) supporters."

In response to Lakatos' argument that "In Kuhn's view, a scientific revolution is irrational, a matter for mob psychology" (Lakatos, 1970: 178), Kuhn stands firm by maintaining that truth must be regarded as an "intra-theoretic" category. Truth, then, for Kuhn, is always by and for members of the same community. Attempts to translate statements of another community are, however, always possible. But such translations merely transpose such statements into the language of the translator's community. For language is always a communal product and scientific communities are language communities. Consequently, since there is no 'neutral' languages "the choice of new theory is a decision to adopt a new native language and to deploy it in a correspondingly different world" (Kuhn, 1970: 277).

So far, so good. Albeit without questioning what grounds the possibility of a debate on 'relativism', Kuhn here seems willing to accept Lakatos' depiction of the relativistic implications of his position. Yet there is a curious ambiguity in Kuhn's paper. While persuasion is indeed involved in establishing the agreed premises of scientific activity, it would seem that there are always "good reasons" for choosing one theory rather than another. And these presumably non-relative, "extra-theoretic," "reasons," Kuhn has no difficulty in naming: "accuracy, scope, simplicity, fruitfulness and the like" (Kuhn, 1970: 261). Moreover, not *any* theory will do, even if scientists agree in its choice and enforce its acceptance. For, according to Kuhn, "nature cannot be forced into an arbitrary set of conceptual boxes" (Kuhn, 1970: 263).

Yet both these positions, suggesting as they do a resistance to the epithet of 'relativism', are immediately modified by Kuhn. While there are non-relative "reasons" for choosing between theories, it now seems that such reasons (rules) do not determine the outcome of all contests between scientific theories. "Though scientists share these values,' Kuhn comments, "...they do not all apply them in the same way." This retreat to the practices of a community as the basis of 'knowledge' is followed, two pages later, with a critique of "absolute truth" which weakens considerably his position that "nature" resists being forced in certain directions. Kuhn argues that to suggest that "a field theory... 'approach(es) more closely to the truth' than an older matter-end-force theory should mean, unless words are being oddly used, that the ultimate constituents of nature are more like fields than like matter and force. But in this ontological context it is far from clear how the phrase 'more like' is to be applied" (Kuhn, 1970: 265). Yet, if we cannot say that 'nature' is more like something than something else, what is the ontological claim of the position that 'it' resists being forced in certain directions?

So one is never quite clear to what extent Kuhn is committed to a conception of the community-based character of truth and knowledge. On the one hand, his historical studies seem partially to convince him that truth is whatever a scientific community values or tolerates. On the other, he writes as if there are (or should be) absolute standards by which to judge whether a theory is scientific or a science is mature. Indeed, he even feels able to write at one point: "I do not think the position just outlined departs very greatly from Sir Karl's" (Kuhn, 1970: 247).

Yet, however we judge Kuhn's stand in this debate, it is apparent the debate itself is located firmly *within* the scientific tradition. Like the virtuoso musician, the philosopher of science is able to suggest new possibilities of and for a tradition. Like the virtuoso, he is mastered by the standards of that tradition itself.

Notice how Kuhn will always defend his argument about the character of science in terms of the (unquestioned) standards of the scientific tradition. His own theory, he argues, has the virtues of economy and elegance; his confidence in it "derives from its ability to make coherent sense of many facts which, on an older view, had been either coherent or irrelevant" (Kuhn, 1970: 237). Similarly, the "refutability principle" is to be criticised for its lack of "sharpness" when applied to particular cases. Hence the scientific standards which are his topic come also to constitute his primary resource. And *the* central standard is that the grounds of the version of good speech inscribed in his text, should never be addressed.

So Kuhn will seek to persuade without addressing the persuasiveness of his version of good speech. For him, language, whether or not it should establish "empty boxes," does the work of picturing the world: hence we speak in order to persuade others of the character of that picture.

We now see why 'relativism' should be an important topic both for Kuhn and for his critics. If speech articulates a picture of the world, then relativism, even nihilism, confronts us head on because there are no absolute standards through which we can assess the 'truth' of such pictures. Relativism, then, is only a possibility within a 'picture' (or Tractatian) version of language under whose auspices speech seeks to correspond to the world. Speech which exemplifies these auspices seeks to conceal its own commitment – a commitment not to rendering a 'faithful' picture of the world but to the grounds of its existence. For speaking is faithful not to the facts but to the standard of artful speech that it articulates: it is faithful to its method, to its mode of production.

Whether ostensibly relativist or non-relativist, speech which seeks to 'picture' commits itself *absolutely* to a correspondence theory of truth. As a consequence, knowledge is conceived after an analogy with 'vision': it is primarily theoretical or speculative. The prior 'openness' in which it is possible for objects to manifest themselves is concealed. Before there is any-thing to which our thinking can correspond, the things-which-are must manifest themselves as 'on hand'. Hence truth consists not in the correspondence of speech to the world but in the activity of revealment which first renders things accessible. The revealment of 'vision' conceals its own possibility by concealing speech's 'pro-vision' of the openness in which we are brought into the presence of things (Macomber, 1967).

So correspondence theory is a logic of revealment alone. It forgets its own grounds and is forgetful that it has forgotten. When man seeks to "picture" the world, "what is it that man forgets", asks Vail (1972). He forgets: "The mystery which lurks within the openness; driven away and round and round the periphery of the mystery, man's attention is taken up with things in the open and he forgets the openness itself. He thus takes things for granted and does not see into the conditions for and the source of their presence" (Vail, 1972: 35).

As Vail points out, this forgetfulness does not reflect the limitations of an individual's thought or of a theory. Error ('*Irren*') is the way in which the tradition makes itself heard to us in our own era. So the first priority is to address where we already are. And we may re-collect that location as we consider the grounding of the 'scientific revolutions' debate in a correspondence version of truth. As a consequence, the debate exemplifies a mode of speech which completes the assimilation of men to things. For it requires a correspondence between two non-equivalent beings – things and the animating grounds of our speech (our mode of existence, our-selves).

It is true that some parties to the debate do give us a glimpse of the possibilities of a confusion between the address of the standards of science as a topic and their use as a resource. Feyerabend (1970), for instance, notes that it is not clear whether Kuhn's notion of "normal science" describes the course of scientific change or prescribes how a field should be turned into a "respectable" science. But the manner in which this 'criticism' itself is grounded within correspondence theory (which alone allows us to treat speech as descriptive or representational) is not considered. Similarly, Feyerabend's stand on the relativism issue owes its 'seriousness' to the very position that he wants to criticise.

Unlike Kuhn, Feyerabend seems to have no doubt that absolute standards of knowledge and truth are unavailable. Following one of Kuhn's own arguments, he suggests that each theory provides for what would constitute a test. Consequently, it is "incommensurable" with many other theories. "In these circumstances," he writes, "one can do one of the following two things. One can stop appealing to permanent standards which remain in force throughout history.... Or one can retain such standards as a verbal ornament, as a memorial to happier times when it was still thought possible to run a complex and catastrophic business like science by following a few simple and 'rational' rules" (Feyerabend, 1970: 215).

Since his preference is clearly for the former position, Feyerabend asks whether this means that the sciences must become "arbitrary" – part of the general "relativism" which Popper and Lakatos have attacked. Now arbitrariness arises if any speech is as good as any other. Correspondence theory re-solves such arbitrariness (its own creation) by positing absolute standards of good speech and, in doing so, posits speech as some-thing, as the product of a technique. As Feyerabend rightly points out, to question such standards is to re-collect the human character of our speech. Consequently, science can be less alienated because it can grasp itself as an historically-located mode of production.

From this it might follow that speech can never be arbitrary (non-methodic) because its claim to be intelligible locates its speaker by exemplifying his mode of existence (him-self). Yet instead of addressing the grounds of speech, Feyerabend immediately turns towards its ostensible topic – the theories that it speaks 'about'. Choosing between theories remains important for him and is to be a matter of "taste."

While this version of "taste" rather than 'absolute standards' is apparently 'different' from that of science, it is a 'difference' which is internal to the mode of scientific discourse. A discourse which must be concerned with what is said but not with what is unsaid but merely pointed towards; a discourse which is concerned to choose between theories but which is forgetful of the activity of theorising as a socially and historically locating production of sense (Blum, 1970).

While the goodness of speech is identified by the degree to which it corresponds to its topic, Feyerabend's position can appropriately be characterised as arbitrary. It may make sense to depict objects if there are absolute standards of speech. But why attempt this if good speech is merely a matter of "taste"? Moreover, the notion of 'taste' itself resonates with the version of the independent intellect, socially unlocated and making 'free' choices. And this depiction of the 'subject', as will be suggested, is precisely that which grounds not only utilitarianism but also the sciences. Hence, while from the point of view of correspondence theory, the consequences of Feyerabend's position may be arbitrary, its claim to be intelligible to us (who, like Feyerabend, live in this era of the 'independent intellect') *cannot* be arbitrary.

So Feyerabend's critique of the philosophy of science grounds itself on deeply scientific auspices. His final suggestion that there may be (absolute?) standards of taste is merely an additional indication that he is mastered by science's standard of artful speaking.[5]

The very human character of any activity, which Feyerabend wants to reassert, is, ironically, concealed by an assumption that the role of speech is to be persuasive and that there are standards of "taste" to guide us in assessing this persuasiveness. At the same time, the way in which this reflects a version of dialogue in the degenerate form of an evaluation helps to reveal how that humanity is lived in the present era.

The evaluator measures speech against the standards of his community. In doing so, he seeks to preserve two differences. First, the difference between "normal" and specially praiseworthy speech. Second, the difference between what is warranted as the speech of a community member and speech which is indifferent to the community or even constitutes an intolerable threat to community standards.

Yet both differences are grounded in an identity. The identity of all speech which speaks irresponsibly. Through its descriptions, explanations and evaluations, irresponsible speech claims to tell us about the character of some object other than its own mode of discourse. So its identity is that of 'virtuosity'. It seeks to persuade us by exemplifying, to the highest degree, community standards, while concealing the persuasiveness of its version of good speech. Indeed, its claim to be taken seriously, by stressing the presumed interchangeability of community members, grounds its speech as that of any-man.

Not all will have the ability to become virtuosi of their community. But each can seek to speak and to judge in a manner which exemplifies its warranted version of good speech. And, happily, the work of the virtuoso stands as a constant reminder of what the community can achieve. The virtuoso who is the very disembodied *instrument* of community standards ("economy," "elegance," "taste"). The master of technique who *is* technique. The virtuoso whose 'virtue' resides in his absence from his work – whose humanity expresses itself as non-humanity.

III

The self-understanding of the sciences, as exemplified by the Lakatos-Kuhn-Feyerabend debate, has a primarily 'protective' function. It serves to protect the sciences' standard of good speech by using that standard in its own work. It is philosophical insofar as that is necessary for the immunisation of the sciences against philosophy (Habermas, 1972). For science will have nothing to do with metaphysics. It seeks, through the philosophy of science, to provide its own beginning; to be its own authority. Yet, in delimiting itself from metaphysics, it grounds itself on a metaphysical position.

Knowledge, as understood by science, gives answers to questions concerned solely with particular determinate beings. Since 'nothing' is wholly indeterminate, science can have nothing to do with it – clearly 'nothing' can only be a topic for metaphysics. Yet science itself depends upon a negation. It is concerned with determinate beings and *nothing* else. It requires specific answers and none other. Moreover, since science must cut itself off from the tradition of metaphysics, it must establish itself as the grounds of this 'nothing'. Yet this is merely to take responsibility for that expression of the metaphysical tradition according to which something (in this case science's techniques) is the animating ground of speech and in which the openness through which things are re-cognised is concealed (Macomber, 1967). Consequently, science's attempt to cut itself off from the tradition merely serves to conceal its mode of existence as a retrieval of that tradition. By 'protecting' its standard of good speech, it provides a metaphysical end to metaphysics.

Science expresses our forgetfulness that we are always already rooted in the tradition. If there is a threat to man's rootedness today it is not because man has no roots but because he conceals his re-collection of them. For him, in Heidegger's terms, Being, the ground of ground, is only a "vapour and a fallacy."

Science's attempt to be its own author-ity is merely an expression of an era in which men demand to root themselves. In the guise of liberation from the constraints of the past, man has searched for a new certainty under the banner of 'freedom', 'individualism' and opposition to 'tradition'. Man was now to be the measure of all things and certainty was to be located in his own self-grounding. Consequently, human 'subjectivity' was to show itself as the ground of ground.

Contrary to this tendency to seek to establish a new rootedness, Heidegger asks us to allow to come into view the rootedness that is always already there in speech. Science is firmly rooted in what Heidegger will call the 'mathematical' project. According to such a project, man's sole concern is with a body of things of a pre-determined character. These things, apprehended as objects, seem to confront man, yet the hidden source from which the possibility of encountering a thing is forgotten. Man's sole concern becomes focussed on what he encounters *within* an area of openness whose mystery he does not explore. Moreover, the manner in which such things are to be investigated is prescribed: objects are to be studied in the mode of objectivity – method, viewed as technique, is given absolute priority over its possible objects.

Since it pre-supposes the character of things and the manner in which they are to be apprehended, the mathematical project presupposes itself. Consequently, it is incapable of interrogating its fundamental character: "Physics can, as physics, make no assertions about physics... Physics itself is not a possible object of a physical experiment" (Heidegger, quoted by Sallis, 1970).

Contrary to its attempt to ground itself, argues Heidegger, science cannot be the measure of itself. Hence we must seek in science the retrieval of the tradition which its mathematical project expresses. More specifically, Heidegger wants to locate the ground of science not in its willful separation of itself from metaphysics but as a culmination of modern metaphysics. It is the latter which alone provides the possibility of the mathematical project since it allows that project's distinction of subject and object. For modern metaphysics, man becomes a subject and things simultaneously become objects. Moreover, man becomes the *centre* to which beings are related. Regarding himself as the measure of all things, men can thus set out to organize and control the world in order to assure his position in the midst of the objects of his cognition. Yet this movement towards human domination and control conceals how man is unthinkingly dominated and controlled by the tradition his activities exemplify. "Relating solely to the things he can control – and conversely to the things which force him to take measures of control – man comes to regard himself as the 'measure of all things' ", writes Macomber. "Subjectivity and objectivity thus arise together in a movement which leads ultimately to science and technology.... In this movement the hiddenness out of which openness arises, and which binds subject and object together in a primeval unity, is concealed. The mystery is forgotten. To turn away from the mystery and to turn toward control are really two sides of the same movement" (Macomber, 1967: 111).

Committed to control, seeking to ground his activities in his own subjectivity, man approaches his very language with an instrumental orientation. Language, addressed as a thing among things, is, like them, to be subject to man's mastery – it too is to be an instrument. In this drive to control, then, language itself must become technologized and, since it has advanced furthest along this path built by the mathematical project, 'symbolic logic becomes the rightful heir of modern metaphysics" (Sallis, 1970: 157).

Yet, despite its continuous attempts to dominate and to control language, from operational definitions through to symbolic logic, science discovers and states its fundamental principles from *within* an already granted understanding of language. In addressing language as a technique of communication within its command, it provides for the re-cognition of good speech as a technical accomplishment and for the speaker as a technician. And this re-collection of 'virtue' as 'virtuosity' is securely located within the tradition from which it claims to separate itself. Yet, grounding itself on subjectivity, science remains forgetful of its auspices. Firmly located inside 'the house of language', science asks us to believe that it has constructed that house itself.

IV

What does all this 'represent'? Nothing. All the distinctions made in this paper (between 'virtue' and 'virtuosity', between 'method' as 'technique' and as 'mode of production') and all the parallels (between the mathematical project and the practice and philosophy of science) represent nothing because they *picture* nothing. Such distinctions and parallels are features of my theorising. There is nothing *outside* of my text to which they are to be compared because they are *products* of my text; they are constituted in and for my theorising.

Does this make me the perfect solipsist and give my readers no ground upon which to stand? Not at all. In producing my text (in reading it), we display our commitments to a mode of production of authoritative speech. Consequently, as people who 'find their feet' in such a mode, we display our-selves.

Of course, our commitment to a mode of production is always historically and socially located. There is no universal, unchanging tradition of authoritative speech because there is no universal, unchanging man. We live lives which are always already located in some era; we find our-selves as historically-rooted men (although, as 'virtuosi', we conceal those roots). In turn, through its claim to speak intelligibly, our speech re-trieves and re-collects that location and rooting – even if, especially if, we display our authority as 'technique'.

So, if a text represents nothing, it '*re-presents*' before our gaze the mode of authoritative speech, the mode of existence in and through which we live our lives. And that re-presentation is our election and our commitment for it re-presents the selves with which we can live. In producing our texts, we re-produce our-selves.

If writing exemplifies a mode of production (of sense, of selves), so, equally, does reading. Only as virtuosi do we criticise texts, or find them 'mistaken'. Secure in a communal version of technique and grasped by a picture theory of language, we seek to compare speech to its ostensible 'objects'. Whatever mistakes we find 'in' the text, whatever criticisms we feel bound to make 'of' it, our reading locates our-selves as entirely mastered by the certainties of a communal mode of production. Beginning as virtuosi, we conclude as functionaries.

This suggests that we read such debates as that between Lakatos, Kuhn and Feyerabend as re-presenting more than the ideas in which they are couched ('relativism', 'absolute standards', 'refutability' as a principle). A commitment to an idea is never grounded in an idea. We are 'certain' about some idea, or cannot bring our-selves to doubt it, because, ultimately that idea is rooted in how we elect to live our lives. 'Certainty' does not express an inner process or state. For instance, to be certain that *X* does not correspond to *Y* shows a certainty that language pictures 'things'. This latter kind of certainty is not the result of contemplation or test, it exemplifies a form of life.

Consequently, the debate about the character of science, while it may appear to take place in the elevated realm of ideas, cannot but re-collect the form of life of the debators. Their ideas are not free-floating creations to be judged on the intellectual marketplace; their ideas (and our reading of them) re-present and locate themselves (ourselves).

The kind of 'certainty' which speaking exemplifies has been treated by Wittgenstein. Certain convictions and assumptions, he writes, are "anchored in all my questions and answers, so anchored that I cannot touch them." They are not so much premisses or assumptions from which I *begin* to argue, "as the *element* in which arguments have their life" (1972, para. 103 and 105). In terms of how I live my life, I must accept some certainties. Indeed, those certainties express (are) my life, for they ground all I do. For instance, it makes no sense for me to continue doubting beyond a certain point that I have never been on the moon. "First and foremost," Wittgenstein argues, "the supposition that perhaps I have been there would strike me as *idle*. Nothing would follow from it, nothing be explained by it. It would not tie in with anything in my life" (1972: para. 117).

Certain suppositions are, then, in Wittgenstein's terms "idle" because they do not "tie in with anything in my life." We no more doubt that language pictures than that we have never been on the moon because we live in a representational, non-reflexive mode of existence. It is not that such certainties are necessarily thought through or intellectually appealing, but that they are held fast by the fabric of our existence. For instance, in assuming that my eyes do not routinely lie, I express what has to "stand fast" for me: "What stands fast does so, not because it is intrinsically obvious or convincing; it is rather held fast by what lies around it" (1972, para. 144).

When G. E. Moore asserts that he "knows" that "here is one hand, and here is another," then, he does not "prove" the existence of the "external world," he merely expresses his method of production (of sense and of self). "I should like to say: Moore does not know what he asserts he knows but it stands fast for him, as also for me; regarding it as absolutely solid is part of our method of doubt and enquiry" (1972, para. 151). In producing "convincing" ideas, in being "convinced," Wittgenstein implies, we are faithful not to the facts but to our mode of production ("our method of doubt and enquiry"). Certainty expresses what "stands fast for us" and the latter is not constrained by the "facts" which stand outside our form of life. "To say of man, in Moore's sense that he *knows* something; that what he says is therefore unconditionally the truth, seems wrong to me. – It is the truth only inasmuch as it is an unmoving foundation of his language-games" (1972, para. 403).

As Wittgenstein notes, 'truth' resides not in individual perceptions but in the form of life which, because it makes available the discoursive space in which notions of 'truth' and 'falsity' are possible and can be assessed in particular cases, is the "unmoving foundation of (our) language-games." This form of life makes possible individual perceptions; the latter are not the basis of truth but an expression of the grammar of our form of life: "It is not a kind of *seeing* on our part; it is our *acting* which lies at the bottom of the language-game" (1972, para. 204).

So, as Wittgenstein says, while 'I know' expresses the readiness to believe certain things, 'knowing' is more than a belief, it is a practice. For instance, the practice of scientific investigations excludes doubt about certain things: "... it belongs to the logic of our scientific investigations that certain things are *in deed (in der Tat)* not doubted" (1972, para. 342). And this failure to doubt certain things or to doubt after a certain point, is not some kind of limitation on the 'truthful' character of our assertions but an expression of the form of life in which 'truth' arises: "My *life*," Wittgenstein comments, "consists in my being content to accept many things" (1972, para. 344).

Hence 'certainty' arises because we cannot see where doubt would get a foothold. This kind of 'certainty' is not based on agreement on opinions but agreement on form of life. For, as we respect the grammar of our language-games, we commit ourselves to the form of life that they exemplify. Through playing these games, we confess our mode of existence and we confess our-selves.

For the virtuoso, work can never be self-confession. The virtuoso's products are the expression entirely of a technique and his own role is merely to ensure as close to a perfect rendering of that technique as is possible. The more the virtuoso is absent from his own work, avoiding 'bias' in his scientific writing or 'flaws' in his musical performance, the more it gains respect. Hence, in a parallel movement, 'virtuosity' separates the producer from his product as it separates his 'certainty' from the form of life it exemplifies.

Now the separation of the producer from his product is a familiar feature of the situation that Marx depicts as 'alienation'. Curiously, however, it is a concept that we can grasp most readily by distinguishing manual from intellectual labour and applying it only to the former. After all, the manual worker produces a clearly-defined product and his labour participates in a mode of production. Where he is separated from his product and where the mode of production, by using him as an adjunct of machines, makes him into some-thing, we can understand the alienated character of the manual worker's existence.

Intellectual labour, on the other hand, conceives itself as engaged in the production of 'ideas'. Rather than products of human labour, these ideas are understood as the reports of messages from Nature, conveyed by messengers who, by aiming to be 'as objective as possible', seek simply to be an efficient medium of message-conveyance. Further, these reports exemplify only a 'neutral' body of techniques, available to any-man, and quite unrelated to any commitment to a mode of production of sense or of selves.

So, where the manual worker is effaced by a technique, we have no difficulty in grasping his alienated condition. Yet, since the very standard of seriousness of intellectual labour is that of self-effacement, how can we conceive the possibility of its alienated character? Given this concealment of the potentially alienating character of *any* labour, we must ask: what warrants the distinction between intellectual and manual labour, with its ability to mystify human production?

The division between intellectual and manual labour expresses a form of life which is forgetful of itself – that is to say, forgetful of the socially located and locating character of human production. Such a form of life, argues Sohn-Rethel, following Marx, is that of commodity-exchange. Through the practice of commodity-exchange, bourgeois society appropriates the means of production and legitimates its action through the concepts of political economy. While these concepts have the appearance of pure thought which seeks to picture reality, they are a feature of the very mode of production which they would depict. Instead of representing reality, they exemplify (re-present before our gaze) the alienated mode of production which makes them possible.

Pure thought is conceived as the distinguishing feature of intellectual labours 'pure' because, seeking to exclude all trace of bias, it stands as the very voice of Nature; 'thought' because such labour is always within the wind, being neither socially located nor locating. And here we see the resonances between such intellectual labour and the form of life of commodity-exchange. As 'price' is the impersonal expression of market forces, speech is the impersonal expression of the character of Nature. The measurements of the market are analogous to the measurements of the scientist: as the market measures price, so the scientist records in measurement the voice of Nature. Both market price and market speech thus come to express the form of life of commodity-exchange. For both re-present a pattern of: "Abstract movement through abstract (and continuous) time and space of abstract substances which thereby suffer no material change and allow for none but quantitative differentiation" (Sohn-Rethel, 1973: 33). Consequently, the logic of the market exemplifies the logic of the mathematical project. Like the price of the market, the speech of the market presents itself as the outcome of an abstract process which, because it expresses the very voice of Nature, bears "no mark of any particular place, no sign of a particular time" (Sohn-Rethel, 1973: 34). All trace of its origin is effaced.

Grounding his cognition on a conception of the "independent intellect," the market speaker encounters the Nature re-presented by his senses in the mode of "pure thought" which is, at once, historically and socially unlocated and unlocating. Yet, because it cuts off the speaker from his human location, "pure thought" assimilates speech with the object-world. What thought is purer, after all, than the thought of the computer – the very ideal of market speech? (Silverman: forthcoming).

By proposing that the 'good' speech is the one that is most object-like ('objective'), market speech addresses itself as a commodity. A commodity which, like the commodities on the economic market, is to be priced (evaluated) through the activity of independent bargainers (intellects). Through and through, then, the fetishism of the "independent intellect" resonates with the fetishism of commodities. "What, then," asks Sohn-Rethel, "is the root of these fetishistic concepts of the 'subjects of cognition', the 'universal', the 'mind', of 'Nature' from which Man withdraws himself so as to make it the pure 'object world', and other concepts of this kind that provide the furniture of the world inhabited by pure intellects. The explanation lies in the fact... that the categories of the independent intellect are the socially synthetic functions by which a commodity-producing society forms a coherent nexus. Therefore any individual who thinks in these terms, thinks for society, and only for this reason is his thinking endowed with intellectual independence and rationality" (Sohn-Rethel, 1973: 34).

"Any individual who thinks in these terms thinks for society" because his intellectual mode of production re-produces its social mode of production. Consequently, while seeking to express the sense-perceptions of an independent intellect, such thought exemplifies an unthinking acceptance of the tradition as re-trieved through the form of life of commodity-exchange. 'Pure thought' hence becomes the synonym for the mode of existence of the functionary. A functionary who, like the market existence his speech expresses, possesses a determination to dominate and attack the world for purposes of instrumental control. In this process, language becomes a thing and, like every-thing else, is treated as a commodity to be dominated and, where necessary, to be used up, for purposes of profit. Hence nothing must be allowed to stand in the way of man's domination over language. Language must be got into shape, must be made into an efficient instrument with which to do description and explanation. Consequently, in the marketplace (which is the world), terms, like the prices of commodities, must be clearly defined and publicly available so that 'hidden' meanings are avoided and the real business of this project (reporting on Nature/making decisions about investment, production and consumption) can be accomplished 'without distortion' (Silverman: forthcoming).

The market speaker cannot re-collect his dwelling in the "house of language" for such re-collection, by pointing towards the tradition which speech exemplifies, would be disastrous to the picture of the "independent intellect." Instead, through his attempts to dominate language, his speech expresses the solipsism of commodity-exchange. For commodity-exchange is practical solipsism. It opposes people to each other as strangers whose sociality is limited only to an exchange where each seeks to gain. And the solipsism of modern metaphysics is an expression of the solipsism of commodity-exchange (Sohn-Rethel, 1973).

The thought of the "independent intellect" expresses its social blindness – a blindness to the mode of production which its thinking re-produces. Yet it is a blindness which, by effacing his mode of production from his mind, allows the marketspeaker to exert his control over a merely 'object' world through his frame of 'objectivity'. "In his 'theoretical reason'," Sohn-Rethel comments, "Man does indeed stand face-to-face with Nature as a mere object-world. This object-world is a universe of recurrent events for which the pattern of the exchange-equation provides the archetype" (Sohn-Rethel, 1973: 35).

Because "the pattern of the exchange-equation" grounds market speech, it cannot conceive itself except in terms of commodity-exchange. Its attempts to understand itself consequently *pre-suppose* that which must be re-collected (the persuasive character of speech, the need to evaluate different speeches and to search for standards – economy, elegance, taste – through which such evaluation may be performed). The "independent intellect" is thus rootless; not because it is not rooted, but because it is prevented, by its very nature, from grasping its roots. As Sohn-Rethel suggests: "It is an accurate description to say that the independent intellect is cut off from its root, at the root, and by its very nature. What the thinkers then seek by way of explanation of the forms they contemplate, has always already happened behind their backs. And they cannot step out of their light to see it" (Sohn-Rethel, 1973: 34).

Inevitably, thought grounded on a notion of the "independent intellect" merely re-produces the form of life of a commodity-exchange society: a society which understands itself in terms of the liberty of free men on the economic (thought) market. Such thought is condemned to express what "has already happened behind (its) back" because, while seeking to represent reality, it merely re-presents society in terms of the society's own self-understanding. And it is not possible for such thinkers to "step out of their light to see it" because their version of serious speech will not allow them to address the 'certainty' which grounds their knowledge, the sense (*Sinn*) in which arises their meaning (*Bedeutung*) (Silverman, 1974).

So the 'independent intellect' and its 'mathematical project', by attempting to distinguish 'intellectual' from 'manual' labour, is "the very corollary of a blind society." It cannot address the alienating nature of a system of commodity-exchange because it exemplifies that system. The light in which it lets beings shine provides for the vision of an objectified man: "The rationality of the independent intellect can never be more than the indispensable light to enable man to live in a world plunged into darkness" (Sohn-Rethel, 1973: 35).

V

The world of commodity-exchange is reflected in the political system of bourgeois society. Yet, as this paper has been implying, we need to grasp bourgeois society as more than a political system and as a mode of representation. To do so, we need to move from the language of politics to the politics of language. For the language of commodity-exchange has an alienating grammar – the grammar of market speech, the grammar of virtuosity. In terms of that grammar, it is proper to be faithful to the 'facts' (through mastery of technique), while remaining forgetful of the grounds of our speech. Indeed, 'virtuosi' *cannot* comprehend their commitment to the grounds of their speech, for those grounds, by providing for the speaker as a technician, assimilate men with the object-world.

The mathematical project, with its postulates of the "independent intellect," of Nature (not history), of *Ratio* (not mode of production), and of logic (not *Logos*), re-presents the form of life of commodity-exchange. The embeddedness of scientific discourse in the grammar of market speech, however, conceals the form of life that locates it. Like the bourgeois society it expresses, it can admit everything except its own character. Instead, it displays itself as a universally 'reasonable', 'logical' mode of speech – understandably enough since 'reason' and 'logic' exemplify its own self-understanding. Nowhere does (or can) it conceive itself as bourgeois: it is either universal or nothing. As Barthes reminds us: "The flight from the name 'bourgeois' is not... an illusory, accidental, secondary, natural or insignificant phenomenon: it is the bourgeois ideology itself, the process through which the bourgeoisie transforms the reality of the world into an image of the world, History into Nature" (1973: 141).

The crisis of the sciences, their forgetfulness of their human roots, of their location in the *Lebenswelt*, can now be addressed as a special case of the capitalist crisis. The abstract functionary of the mathematical project, practising his 'intellectual' labour, while absent from his text, is as alienated as any worker who loses him-self through a mode of work in which he is treated merely as a factor of production. As functionaries of a system of commodity-exchange, both present themselves as objects. In producing their products (texts), instead of re-collecting themselves they become estranged from themselves.

The decisive turn arises only when, by taking responsibility for the grounds of our speech, we re-collect our 'virtue'. Our speech differentiates and identifies, stratifies and confronts, in the context of a tradition of discourse (a mode of production) which it re-collects. Consequently, speech (or a text) participates in the movement and transformation of reality. The text, in Kristeva's (1969) terms, ties itself – reads itself (*se lie – se lit*) – doubly to that reality: to the signifying system in which it grounds itself and to the social process in which it participates as discourse.

Reading (scientific texts, this paper) only understands itself when the reader no longer seeks to be persuaded, no longer asks to be carried along by the 'story', but: "opens his eyes, at the risk of blinding himself, to the *source* of all stories, knowing that such an activity is and must be considered delinquent from the standpoint of the reigning stories and myths" (Sollers, 1968: 358).

University of London,
Goldsmiths' College

NOTES

[1] There may be a certain resonance between his usage of 'virtue' and the way in which we use the term 'virile'. Apart from their common root in man ('vir'), a 'virile' person is taken to be able to give expression to him-self without recourse to technique. Perhaps to the extent a characterisation of machines as 'virile' (for instance in car advertisements) makes sense, we display our non-virtuous existence. I am grateful to Barry Sandywell for this suggestion.

[2] As Kojève points out, the mastery of the Hegelian master over the slave turns out to dehumanise the master, to deny him the very recognition he sought through exercising his mastery. For, in treating the slave as an instrument of his Desire, the slave becomes an object. Hence the master is recognised by (re-cognises himself as) an object (Kojève, 1969).

[3] In its search to escape, through technique, the encumbrances of language, science seeks to avoid euphemism. Yet, since a commitment to method effaces both speech and the speaker, science's speech Says nothing. In this respect, it is curious that the Greek root of "euphemism" should be "fine-sounding" speech. Science wants to avoid "fine-sounding" speech (euphemism), yet because it wants to cut itself off from its own grounds, it merely produces speech that is "fine-sounding."

[4] As was the case with "euphemism," "technique" shows us the fateful way in which we re-collect Greek terms. According to the common translation, the Greeks use *techne* for "craft" and "art," *technites* for "craftsman" and "artist." Yet, as Heidegger suggests, *techne* was not "techniquee" in its present sense. *Techne* never signified a kind of practical performance but indicated a mode of knowing that breng forth beings out of their concealedness – a mode of *aletheia* (Heidegger, 1971).

[5] Polanyi's attempt to ground science on "personal knowledge" closely resembles Feyerabend's conclusions – Polanyi would seem merely to prefer "beauty" and "pure intellectual passions" to "taste." Indeed, Polanyi's references to "connoisseurship" (p. 64) and to "learning the art" (p. 125) indicate the manner in which this version of the philosophy of science makes the speaker a "virtuoso" and "virtuosity" the standard of good speech (Polanyi, 1964).

BIBLIOGRAPHY

Barthes, R.: 1966, *Critique et Vérité*, Editions du Seuil, Paris.

Barthes, R.: 1973, *Le Plaisir du Texte*, Editions du Seuil, Paris.

Blum, A.: 1970, 'Theorizing', in J. Douglas (ed.), *Understanding Everyday Life*, Aldine, Chicago.

Deeley, J. N.: 1971, *The Tradition via Heidegger*, M. Nijhoff, The Hague.

Feyerabend, P.: 1970, 'Consolations for the specialist', in I. Lakatos and A. Musgrave (eds.), *Criticism and the Growth of Knowledge*, Cambridge U.P., Cambridge.

Habermas, J.: 1972, *Knowledge and Human Interests*, Heinemann Educational Books, London.

Heidegger, M.: 1968, *What is Called Thinking*, Harper Torchbooks, New York.

Heidegger, M.: 1971, 'The Origin of the Work of Art', *Poetry, Language, Thought*, Harper and Row, New York.

Heidegger, M.: 1972, *On the Way to Language*, Harper and Row, New York.

Kojève, A.: 1969, *Introduction to the Reading of Hegel* (ed. by A. Bloom), Basic Books, New York.

Kristeva, J.: 1969, *Semiotiké: Recherches pour une Semanalyse*, Editions du Seuil, Paris.

Kuhn, T. S.: 1970, 'Reflections on My Critics', in I. Lakatos and A. Musgrave (eds.), *Criticism and the Growth of Knowledge*, Cambridge U.P., Cambridge.

Lakatos, I.: 1970, 'Falsification and the Methodology of Scientific Research Programmes', in I. Lakatos and A. Musgrave (eds.), *Criticism and the Growth of Knowledge*, Cambridge U.P., Cambridge.

Lakatos, I. and Musgrave A. (eds.): 1970, *Criticism and the Growth of Knowledge*, Cambridge U. P., Cambridge.

Macomber, W. B.: 1967, *The Anatomy of Disillusion: Martin Heidegger's Notion of Truth*, Northwestern U. P., Evanston.

Phillips, D.: 1973, *Abandoning Method*, Jossey-Bass, San Francisco and London.

Polanyi, M.: 1964, *Personal Knowledge*, Harper Torchbooks, New York.

Sallis, J.: 1970, 'Towards the Moment of Reversal: Science, Technology and the Language of Homecoming', in J. Sallis (ed.), *Heidegger and the Path of Thinking*, Duquesne U.P., Pittsburg.
Silverman, D.: 1974, 'Speaking Seriously: the Language of Grading', in *Theory and Society*, Vol. 1 (1), 1–15, (3) 341–359.
Silverman, D.: (forthcoming), 'Davis-Moore, Market Speech and Community', in P. Filmer, M. Phillipson, M. Roche, B. Sandywell, and D. Silverman (eds.), *Essays in Reflexive Sociology*, Routledge and Kegan Paul, London.
Sollers, P.: 1968, *Logiques*, Editions du Seuil, Paris.
Sohn-Rethel, A.: 1973, 'Intellectual and Manual Labour: an Attempt at a Materialistic Theory', *Radical Philosophy* **6** (Winter), 30–37.
Vail, L. H.: 1972, *Heidegger and Ontological Difference*, Pennsylvania State U. P., Trenton, Pa.
Wittgenstein, L.: 1972, *On Certainty*, Harper Torchbooks, New York.

SECTION B

COGNITION AND COMMUNICATION IN SCIENCE DEVELOPMENT

INTRODUCTION

Almost three decades have passed since Claude Lévi-Strauss first exemplified his concern with understanding facts about society by establishing facts which are true about 'the human mind' (1949). He had the bold originality of suggesting his approach as an alternative to the established functionalist perspective, which appears to be related to the fact that even now the critics among his professional colleagues greatly outnumber the disciples. Approximately fifteen years later, ethnographic anthropology was shaken by a debate as to the value of a cognitive as opposed to the traditional institutional perspective. 'Ethnoscience', as the new ethnography has been called, was to be concerned with the system of knowledge and cognition typical of a given culture rather than with cultural behavior, and systematic ethnoscience aimed at no less than an universal theory of knowledge (Sturtevant 1964; Werner and Fenton 1970). In the meantime, several additional specialities characterized by their focus on cognitive processes have evolved in different disciplines (e.g. cognitive psychology, artificial intelligence). However, it was not until recently that 'cognitive' factors have gained importance in discussions belonging to the field of sociology of science (e.g. Nowotny, 1973).

The long delay with which a field, intrinsically related to knowledge production, has now begun to emphasize cognitive variables is in itself interesting. We may perhaps link the neglect of attention paid to cognitive concepts to the fact that variables generally subsumed under the notion 'cognitive' had long been dealt with extensively in philosophy of science, so that there obviously was no need for dealing with them in sociological studies. However, there are two points which should be noted here: First, a cognitive approach as opposed to the methodological has the advantage of guarding the focus on the conditions of the development of scientific knowledge without withdrawing into the 'third world' (Popper) of ideas. We have seen in the Introduction to Section A that a 'normative' or 'third world' conception of philosophy of science generates difficulties of understanding. Here we want to state that whatever the logical status of phil-

Knorr, Strasser, and Zilian (eds.), Determinants and Controls of Scientific Development, 175–183.

osophy of science approaches may be, a cognitive perspective as exemplified by several disciplines may lead to results different from those exposed by philosophers and certainly to empirical results. That is to say, an argument which concedes 'cognitive' matters to the realm of the philosopher and logician is not conclusive.

This brings us to the second point we wanted to mention: What exactly do we mean by the notion of 'cognitive' in science studies and is it the same as was meant in the original disciplines showing concern for cognition and cognitive processes? If we define cognition in accordance with other disciplines we have to define it in terms of socially constructed models of the outer world and cognitive processes by referring to the production of cultural knowledge through the operation of individual information processing mechanisms (cf. Neisser, 1967). Then we can see that the notions of cognition and cognitive processes have undergone a slight shift of meaning when being applied to science studies, in that the aspect of individual information processing and its relationship to the growth of knowledge has largely been ignored. The term 'cognitive' has been used synonymously with 'intellectual' (e.g. theoretical and methodological) components of research processes and research results. The limited conception of 'cognitive' enhances an outlook which, for substantial results, is again referred back to the logician and philosopher of knowledge. However, papers introduced here, attempt to transcend those limits.

Let us now pose the question of where cognitive approaches do fit into sociological undertakings. Here we immediately see that the positive stance which is maintained associates cognitive theories with social approaches defining science in terms of communities, communication networks and institutions. Beyond that, the two approaches differ markedly, not only in relation to the central topics under examination but also regarding basic methodological presuppositions. For a cognitive approach, understanding science, by definition, consists of the establishment of conceptual relations between *different types* of realities: between the (logical) context of justification, the (social) context of conventional acceptance, and the (psychological) context of knowledge processing. In contrast, social explanations are generally much more limited in perspective.

Let us illustrate this by pointing to the difference between the approaches called structural in sociology and cultural anthropology. In sociology, 'structure' has been defined for decades in terms of an observable social order at the phenomenal level, an order which is based mainly upon normative regularities. Hence structural studies have sought to establish empirical patterns among phenomena by means of which people maintain stable social relationships among themselves. The structural perspective in cultural anthropology (and linguistics), by contrast, rests on the assumption that an empirical process is inconceivable without some ordering system lying *behind* it, and that it is the hidden system of latent relations or structures *generating* observable phenomena which has to be attained. For Lévi-Strauss (1955), for instance, understanding consists of the reduction of one type of reality to another, and the fundamental problem is one of exposing the relation between reason and sense-perception. In short, the difference between structural studies in sociology and anthropology is one of the level at which explanation of socio-cultural reality takes place.

It follows that the general sociological approach has been structural only in the sense of elaborating surface structures, and sociology of science studies do fit rather neatly into that pattern: The most pronounced examples going beyond a purely institutional or organizational perspective speak of gift-giving of information in exchange for recognition as the organizing principle of science (Hagstrom 1965), or still lean heavily on the work in the Mertonian tradition (Storer, 1966). Kuhn has never relied primarily on social variables, although he has often been praised for providing sociological tools in dealing with philosophy of science issues. His criterion of demarcation between mature sciences and preparadigmatic ones refers to the consensus over a *paradigm* or a disciplinary matrix. The entries of the matrix are made up of cognitive components of the research procedure, which define the *intellectual* state of affairs in the field in question. As an example consider the rules embedded in a paradigm, which, according to Kuhn, do not have the character of social norms maintained through the operation of social control. Rather, they must be described as unconsciously shared commitments, providing a background on the basis of which normal science can proceed. Most importantly, Kuhn's central idea of the incommensurability of paradigms is backed by a cognitive learning theory, which states that science is learned by doing shared 'exemplars' rather than by acquiring rules for doing it. The requirement that learning must not be reconstructed as something we manage by abstracting and applying certain criteria rules out the possibility of rational choice between theories. It is because his fundamental propositions are thus not conceived in sociological terms that Kuhn's theory cannot be put on a par with the work of many of his sociological contemporaries.

We have been trying to sketch some of the substantial and methodological features of cognitive versus social approaches to science. Yet we feel that it is still not easy to decide in favor of one or the other brand. The problem can be illustrated by turning to linguistics. Here, as in science, language is a social phenomenon in the sense that it depends for its existence, maintenance and development on the use which is made of it by natural speaker-hearers. Nevertheless, it has proven to be a successful move on the part of linguistics to treat language (*la langue*) as a system of its own, interdependent with but nonetheless delimited from the social sphere. The difficulty to conceive adequately of the interdependency relationship has, however, led to numerous controversies as regards, say, the significance of language 'performance' as opposed to language 'competence'. The difficulty lies in the fact that language as a system of rules includes the principles validating the acceptability of utterances, and yet it is these utterances which construe the rules in that they allow for rule changing creativity.

In science, as compared to language, there exists a similar difference between the set of communally warranted methods and results supposedly internalized by each member of a scientific community, and the application and modification of this knowledge when actually producing scientific outcomes. More important, however, is the fact that in science as opposed to language the rules governing scientific behavior constitute a set of (again communally warranted) *norms* endorsed by scientists in the course of their professional socialization, where the intellectual conformity produced by scientific education is being maintained by social control largely operating through a system of rewards and professional recognition. On the other hand, rule-governed linguistic behavior is part of a faculty seemingly stimulated by, but not originating from, socialization processes, a faculty which depends on social control only for surface elaboration. Hence, social variables should play a more essential role in scientific development than they play in language.

When trying to assess the differential importance of social versus cognitive factors it must be remembered that it is the fundamental aspect of the *growth of knowledge* which has to be rendered intelligible. It should be noted that the notion of growth as introduced here does not presuppose a linear conception of progress, nor does it imply a continuously better approximation of truth. Even if one advocates a circular notion of progress towards increasing complexity of thought one cannot ignore the fact that science today presents us with more (in terms of the range of phenomena covered) and better (in terms of predictability and illumination) stories about reality than some time ago. Generally speaking, social approaches succeeded in showing that science shares the distinctive features currently thought to characterize social systems, and that the specific quality of these features seems to be sufficient in order to delineate scientific *communities* from other societal subgroups. Social control and exchange mechanisms amply account for the maintenance of a supportive surface structure which sustains the ground on which science as a knowledge producing and knowledge processing activity can proceed (cf. also Introduction to Section C). However, to explain the cumulative aspect of science, more is required than an explication of community structures. We are usually left with a reference to the 'rationality' of scientists, a concept which in itself needs cognitive investigation.

The cognitive approach seems to have several advantages, the first and most important of which we have already mentioned. It consists of maintaining an intrinsic relationship with the individual person as the undisputed unit of knowledge processing, without ignoring the social and historical construction of the cognitive map the individual draws on. Consequently, the cognitive approach precludes the reification of social aspects of the system, and yet does not stumble into the pitfalls of psychological reductionism. Second, the cognitive approach has the advantage of enhancing a differentiation between socially controlled imperatives of behavior internalized to varying degrees by the individual, culturally defined rules of the game, which set the stage for communally warranted action, and universally valid quasi-laws, which can still be aroused to consciousness and possibly modified, which, however, constitute indispensable requirements for the process of human symbolic interaction itself. Third, the cognitive approach promotes the search for infrastructures, since the transition from the social to the cognitive sphere constitutes in itself a change to a more fundamental layer of explanation. The latter is true because a cognitive process as defined above potentially takes care of the *cultural* and linguistic organization of scientific as well as everyday knowledge, and of logical as well as empirical preconditions of knowing.

The search for deep structures at a cognitive level of reality should be discussed in the light of the search for economic infrastructures. We believe that economic infrastructures do play a key role in shaping the cognitive map of participants in a culture. Furthermore, it is granted that economic infrastructures may have a strong influence on the direction which scientific processes may take. However, any attempt to go further than that and to establish direct relationships between the structure of the economy and conceptual developments in science up to now has been doomed to failure (cf., e.g. Galtung, 1972). It seems as if we have to wait for a much broader understanding of cognitive mechanisms before we can expect to establish the transformation rules connecting cognition and its alleged economic basis. Be that as it may, the mode of production of knowledge is attained most directly by the cognitive code which generates scientific information, and it is this code which has to undergo a most scrutinized examination.

Finally we have to ask ourselves where the papers presented in this section fit in the scheme of social and cognitive approaches to science. None of the studies exemplifies a narrowly defined sociological perspective. However, two of them focus on scientific communities conceived in terms of communication processes, whereas the others concentrate on conceptual development and differentiation in the social and natural sciences.

Let us begin by briefly sketching the argument as presented by Nicholas C. Mullins 'A Sociological Theory of Scientific Revolutions', who most extensively relies on social conditions of scientific change. By slightly modifying Carlsson and Karlsson's model of 'generations' based on differential acceptance of innovation by different age cohorts, Mullins attempts to explain the Kuhnian revolutions in science solely by normal science processes. Hence, he specifies differential access to information and resources resulting from the normal process of communication and differential susceptability to change by age as the basic mechanisms guaranteeing the 'natural summation of small changes' that characterize paradigm change.

Mullins' approach is genuinely sociological in the sense of taking care of the ranking and reward system in science, the role of socialization, and group coherence and communication networks. Its sociological strand will become especially obvious if we contrast Mullins' model with that of Gernot Böhme 'The Social Function of Cognitive Structures: A Concept of the Scientific Community Within a Theory of Action', who explicitly strives for delimiting social as opposed to cognitive determinants of scientific development. Böhme, in contrast to Mullins, directs our attention to the *content* instead of the structural and functional aspects of scientific communication, arguing that paradigmatic change in science can only be understood in terms of cognitive factors. Consequently, he holds the essence of scientific communication to lie in 'argumentation', referring to the requirement of adducing reasons for what is presented as true. It should be noted that there is no necessary contradiction between the characterization of science as exchange of information and the pointing to the fact that there needs to be an explicit legitimation for what is proposed. Rather, the latter approach specifies an additional (and allegedly more important) cognitive function which communication provides for. It should also be noted that depicting scientific communities as organized by the state of argumentation in their respective specialty is linked to the Popperian context of justification and to the original Kuhnian attempt to define these communities by a shared paradigm. This becomes especially clear if we conceive of the 'state of argumentation' in terms of the components of a specialty matrix. Hence, the question yet to be solved by Böhme is somewhat similar to one of the main problems encountered by Kuhn. It consists of either *defining* a scientific community by its state of argumentation in which case one would have to elaborate what exactly belongs to this state in order to possibly arrive at a classification one would hope to be more systematic than the classification into specialty groups currently used by scientists themselves. Or of treating the state of argumentation as a substantial hypothesis about what holds a scientific community together, in which case one would have to suggest an independent criterion of what constitutes those communities, such that the hypothesis remains open to test by empirical evidence.

The study presented by Karin D. Knorr, 'The Nature of Scientific Consensus and the Case of the Social Sciences', relates easily to what has been presented so far, and yet provides a different – albeit still cognitive – perspective. By slightly modifying an idea by Stephen Toulmin (1958) the structure of arguments is connected with specific methodological approaches and the latter with the amount of consensus found to exist in social science specialty groups. It is shown that, contrary to Kuhn and many other commentators upon the state of social science, there exists a fairly high degree of consensus depending upon the methodology used in research work. Hence, the author maintains that it is not necessarily the paradigmatic maturity of a discipline which accounts for consensus. Rather, the latter is primarily enhanced by distinctive features of the methodology and the structure of arguments upon which it is based. What is important in terms of cognition and communication is that the social-communicational element of consensus is, by empirical evidence, shown to be related to specific cognitive factors, thus bridging the gap between the two levels of explanation exemplified so far.

The last paper included in this section is by H. Rom Harré, 'Images of the World and Societal Icons', and represents the most pronounced effort in the direction of a cognitive approach, exposing the features of a structural theory of conceptual understanding. The special flavor of the work presented here rests with the attempt to reconcile conceptual explanation in science with an ethnomethodological theory of society, which results in the tendency to deny any causal effects flowing from societal factors into the thinking of scientists. If social 'facts' are conceived of as *no thing* but icons generated by and at the back of our accounting rhethoric, it is logically conclusive to do away with 'external' independent influences on scientific accounts. However, Harré, too, encounters the problem of validation, that is, of finding independent criteria for what constitutes a social fact.

It has been our contention that science cannot be rendered intelligible by exclusively drawing from social or communicational variables in the narrow sense of the word. On the other hand, it is obvious that there need to be behavioral correlates on the side of scientists for each aspect of system development. The fact that there have to be such correlates may easily lead one to advocate a social explanation approach to science. The papers introduced here represent a series of arguments progressively inclined to leave that level of explanation. The reader should watch the problem shifts which go together with it.

BIBLIOGRAPHY

Galtung, J.: 1972, *Social Structure and Science Structure*, Paper presented at the Institute for Advanced Studies and Scientific Research, Vienna.

Hagstrom, W. O.: 1965, *The Scientific Community*, Basic Books, New York.

Lévi-Strauss, C.: 1949, *Les Structures élémentaires de la parenté*, PUF, Paris.

Lévi-Strauss, C.: 1955, *Tristes Tropiques*, Presses Universitaires de France, Paris.

Neisser, U.: 1967, *Cognitive Psychology*, Appleton-Century-Crofts, New York.

Nowotny, H.: 1973, 'On the Feasibility of a Cognitive Approach to the Study of Science', *Zeitschrift für Soziologie* **2** (3), 282–296.

Storer, N. W.: 1966, *The Social System of Science*, Holt, Rinehart and Winston, New York.

Sturtevant, W. C.: 1964, 'Studies in Ethnoscience', *American Anthropologist* **66**(2), 99–131.

Werner, O. and Fenton, J.: 1970, 'Method and Theory in Ethnoscience or Ethnoepistemology', in R. Naroll and R. Cohen (eds.), *A Handbook of Method in Cultural Anthropology*, The Natural History Press, Garden City, New York.

NICHOLAS C. MULLINS

A SOCIOLOGICAL THEORY OF SCIENTIFIC REVOLUTION*

The intellectual structure of science is affected by both large and small changes. In his 'postscript' Kuhn (1970) moved his conception of 'revolutionary' science from its prior focus on the massive and heroic shifts of the Galileian, Newtonian and Einsteinian examples to one emphasizing the aggregation of changes in the 'paradigms' held by groups of scientists. This paper investigates the intellectual and social preconditions for the aggregation of changes in modern science. It also discusses the processes that allow aggregated, small changes to be perceived by most scientists as a single block of changes – a revolutionary change in paradigm.

Kuhn's original model for the diffusion of a new paradigm throughout science was a cohort fixation model (Ryder, 1965). He argued that production of a new exemplar influenced a whole cohort of new scientists in training to see the world in a new way; that is, a new paradigm developed. Carlsson and Karlsson (1970) provided a mathematical formulation for describing such apparent cohort effects as Kuhn's model postulated. Also important to thinking about scientific change has been the developing theory of small coherent groups which act as an environment in which initial aggregation occurs. Griffith and Mullins (1972) concluded that small, highly coherent, consciously organized social groups have been consistently associated with major changes and advances in the directions of science across a variety of disciplines, periods of time, and types of research.[1] These groups have the following characteristics: (1) a distinctive theoretical orientation and substantial scientific achievements, usually the result of attention focused on a restricted set of phenomena; (2) conscious effort by a member (or members) to develop the group socially and intellectually; (3) recognition of these members as social and intellectual leaders; (4) an active and successful program for recruiting new members.

At this point, sufficient research has been done to outline the aspects of a theory that both describes and explains the normal processes of science –

Knorr, Strasser, and Zilian (eds.), Determinants and Controls of Scientific Development, 185–203.

communication, publication, status-giving and receiving, etc. We can then examine this theory to determine whether it also explains the fundamental conditions associated with specialty group developments and revolutionary changes. A satisfactory sociological theory of the development and demise of specialty groups' extraordinary surface structure must either connect it to the normal social processes of scientific research or else account for deviations in terms of other sociological processes.

1. The model

Carlsson and Karlsson (1970) proposed a mathematical model for the development of 'generations' (i.e., the generation of generations). This model is based on the differential acceptance (in a smooth function) of innovations by cohorts of different ages. By definition, each cohort has experienced a different set of events. The striking aspect of the authors' mathematics is their conclusion that if a series of random stimuli impinge on a whole population over time, they will break the population into generations with sharp breaks between them.

Carlsson and Karlsson modeled the total behavior of a group assuming (1) only random variation in the underlying 'stimulus pressures' and (2) differential response by age. Their basic equation is:

$$(1) \qquad Y^h(h) = S_h$$

where the behavior Y (proportion new style) in cohort C^h at time T_h results from a stimulus, S_h, to the new behavior. For scientists, S_h is the current literature of the discipline. Accordingly, a cohort trained between 1960 and 1965 is completely responsive to the literature of that period; however, as that cohort ages, its responsiveness to innovation is smaller. Carlsson and Karlsson also showed that if we assume a change in style to be irreversible, and a smooth declining curve in change proneness, for a specified time period (theirs was 25 years) the result will be a swing from old to new style in the population.

Carlsson and Karlsson (1970: 715) also proposed an equation for the population total of new style behavior at time h. However, I cannot assume, as they did, that each cohort remains the same size for all time periods. I therefore changed this equation, adding a weight, w_h^T, which varies from 0 to 1. If h is the cohort index and T is time, the equation defines the proportion still active in cohort $C^{(h)}$ at time T. My equation for $Y_h^{(T)}$ is:

$$(2) \qquad Y_h^{(T)} = W_h^{(t)} \, Y_h^{(h)} + W_h^{(t-1)} \, Y_h^{(h)} \ldots W^{(0)} \, Y_h^{(h)}$$

This equation, like the original formula, is also a set of moving averages and is therefore subject to the same Slutzky-Yule effect. The major difference between the two formulas is that the differential weights, which make new cohorts much more pronounced in size than old ones, lessen the smoothing function. The behavior of older scientists, in the aggregate, will still lie close to the long-term average, but they will constitute only a small percentage of those involved in revolutionary scientific changes. (As Carlsson and Karlsson pointed out, their model describes only oscillations in behavior, assuming a random variation.)

The Carlsson and Karlsson model depends on a natural summation of small changes and differential susceptibility to change by age rather than on an artificial time break or on events that compel a change.

To apply this model to science, it must be shown that (1) there is a source of random fluctuations (2) that impinge on much of science. It must also be shown (3) that different cohorts in science are differently susceptible to change, and (4) how these three factors act jointly to create new specialties from normal science activities.

The model permits us to understand why the scientists involved frequently perceive revolutionary change as simply the result of small changes: it is only in the aggregate that changes become sufficiently influential that older generations see them as revolutionary. In short, when applied to science the model is empirically indistinguishable from the model of 'generations' that strongly choose different styles, even though the social and social psychological processes involved in the two models are quite different. It incorporates the phenomena that interested Kuhn, but within an empirically based picture of science. Furthermore, this model cumulates the results of past theorizing and research.

2. The theory of normal science

The following are definitions and assumptions together with documentation of their applicability to science. The topics of importance are innovation and communication, and the ranking system and research.

2.1. *Innovation and Communication*

DEFINITION 1. *Innovations* are changes in procedures, materials, and/or theoretical orientations.

I assume that any scientist may innovate. The available data indicate that innovations occur randomly among scientists in all status rankings and groups throughout science. Merton (1961) noted that many discoveries (innovations) are not unique to one man but are developed simultaneously by several. Price (1963:66–68) generalized that insight by noting that the number of discoveries follows a Poisson distribution (i.e., the process can be treated as if the discovery were occurring randomly to rare individuals in the population of scientists).[2] Small innovations do not require explanation; their adoption and collection into greater changes does (Hagstrom, 1965; Mulkay, 1972).[3]

DEFINITION 2. *Direct Connection* from scientist A to scientist B exists when A discusses his research in detail with B: $A \rightarrow B$.

This relation may or may not be reciprocated (Mullins, 1966). Direct connection is important because it conveys much of the information flow within science. The number of scientists *directly* connected to any given scientist is probably less important than the *pattern of indirect connections* (i.e., A is indirectly connected with C if A is directly connected with B, and B with C, e.g., $A \rightarrow B \rightarrow C$). Research has not yet determined the average number of direct contacts (either outgoing or incoming) maintained by individual scientists, but there is some evidence (Mullins, 1966) that this number is small, averaging between two and three per scientist.

Numerous studies have shown that informal communication, of which direct connection is only one kind, is not the largest part of the total communication picture. More information, and of higher quality, is communicated through journals (see, e.g., Garvey and Griffith, 1963), but the time lag in publication makes the 'early-warning system' of informal communication quite crucial.

DEFINITION 3. *Adoption* requires (1) incorporation of an innovation into the research of a scientist not involved in the initial innovation or (2) its synthesis into his description of research in the discipline involved.

Direct connection affects adoption. A fundamental assumption here is that innovations are most frequently adopted by those who have learned of them by direct connection. Adoption is probabilistic with respect to communication. Not every receiver of a message adopts the relevant innovation in his own work, but the probability of adoption is increased by (1) the adoption of that innovation by the sender (if he is not the innovator), and (2) the frequency of communication between sender and receiver. These assumptions are supported by Rogers' (1962) study of the diffusion of agricultural innovation, Coleman *et al.*'s (1968) analysis of medical innovation; and Katz and Lazarsfeld's (1964) assertion that 'opinion leaders' operating in informal settings set trends in fashion, politics, etc. If the average probability of communication for a given message drops sufficiently that the expected number of scientists reached is smaller than the number sending (e.g., for the above estimate of three direct contacts, below 0.33), the message will shortly die out; similarly, if the probability increases, the number reached will also increase. Easily adoptable methodological innovations will pass rapidly through the network to all affected areas of science if most researchers who learn of them try them and find them useful.

DEFINITION 4. *Centrality* defines scientist *A* as more central than scientist *B* if more scientists are connected directly and through close intermediaries with *A* than with *B* (Harary *et al.*, 1965). Figure 1 shows the difference between centrality and number of direct connections received.

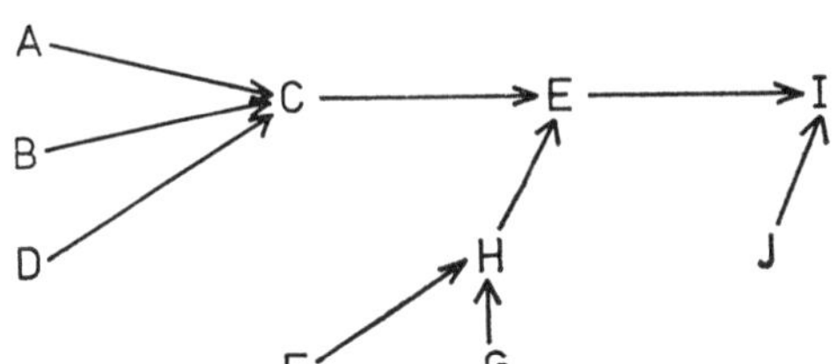

Fig. 1.

I is most central even though *C* receives more direct connections.

Indirect connections permit innovations to pass through several scientists, thus increasing the possibility that they will be tried and checked for accuracy or usefulness. If an innovation is adopted, it will be passed on; if not, it will be dropped. Central people, then, receive information that has been better filtered. Empirical support for this assumption can be found in Griffith *et al.* (1973b), who report that 40 percent of the variance in publication rates for scientists in two biomedical specialties is accounted for by sociometric centrality within the specialties.

From Definitions 2 and 4 and their related assumptions 2 and 3, we can state that: The probability is higher that central (rather than non-central) persons receive messages that have been (1) adopted by more people and (2) proven acceptable in terms of having facilitated present work. However, because a scientist does not necessarily force all of his work into logical consistency, inconsistent messages can easily be passed on. This fact does not imply mindless acceptance of any results, but it does indicate that hard thinking and analysis are required to integrate the incoming messages.

The communication pattern is dynamic; therefore, it slowly shifts direct connection patterns. Mullins (1966) found that one-half of the connections in a system were less than eight years old. Part of this phenomenon is caused by the entry of new investigators (potential new communicators): if the number of scientists is doubling every 15 years, then half of all scientists must have been in the field less than 15 years. Furthermore, attrition produces a situation in which roughly $\frac{1}{3}$ of all scientists are between 30 and 40 (i.e., are in their first 10 to 15 career years; NSF, 1970). Finally, the dynamics vary somewhat from area to area because specialties differ greatly in their average age of practitioner.

DEFINITION 5. *Actively involved* scientists routinely do research and report their results in the archival literature.

Price (1963) estimated that the number of actively involved ('solid') scientists (consistently reporting at least one paper per year) equals the square root of the number of total scientists; if, for example, there are 450000 total scientists in the United States, then 671 are *actively* involved. (This number seems small, but certainly not by a factor of ten).[4]

Furthermore, all actively involved scientists in any field are joined, however indirectly, with all other actively involved scientists in all other fields. Evidence for this assertion is found in Mullins (1966, 1968; see also Crane, 1969), which showed that scientists' communication is not limited by field or locality and that all active scientists seem to be involved in some communication of this sort. In addition, Rapoport (1963) showed that if a network is formed by as many as two connections per node, purely random connections will fully join it, regardless of size.[5] Since all active scientists regularly discuss their research with at least two other scientists and the communication pattern is not strongly bounded by discipline (see Mullins, 1968; Gaston and Rose, 1970), then all active scientists must be connected (some connections may be quite long, but all are still connected). Any given piece of information may not necessarily pass completely from one point to every other over the entire network; if formal publication occurs, it frequently short-circuits informal communication.

To summarize: An all-encompassing communication network spreads important research information throughout science. Scientists central to the network receive information sooner because they are closer to more potential innovators; they also receive more useful information, since others have tested it before passing it on. The average probability of a given scientist's passing on a given piece of information is an important parameter because it determines the rate of spread (or lack of it).

2.2. *Ranking System and Research*

Fundamental to the conception of a ranking system is the presumption that most scientists prefer higher ranking to lower. This situation is presumed necessary to motivate the operation of the ranking system (Cancian, 1967, made a similar assumption).

DEFINITION 6. *Productivity* is the publication of papers in the archival literature.[6]

Zuckerman (1970) showed that (1) high productivity is associated with high rank, and (2) age is associated with rank.

DEFINITION 7. *Resources* are whatever is needed to do research (the actual items differ from field to field): equipment, assistants, computers, books, and articles (operationally, the research budget).

Productivity confers the ability to attract resources. This assumption is hard to demonstrate in detail. It is believed by scientists, but the research needed to prove it has not been done. However, for the theory being developed in this paper, perfect correlation is not necessary. In support of this assumption, it should be noted that resources are distributed so that the top ten universities receive over a third of all research funds

(Hirsch, 1968: 106). Blume and Sinclair (1973: 129) report that studies of the reward system in science have "largely involved juxtaposing... indices ...measuring the contributions of individuals on the one hand [with] those measuring the recognition received on the other." Yet it should be noted that Hagstrom (1971) found only a small relation between status as measured by funds, rank, age, or a combination of these three variables, and productivity. In fact, he noted that no sociological variables, either singularly or in combination, account for much of the variance in productivity.

From this assumption we can state that those with higher productivity, and hence more resources, can use them to produce even more published papers. This statement oversimplifies the actual operation of the system. Nevertheless, there is no denying the advantages that result from available resources; if research encounters difficulties, these can often be circumvented if sufficient resources are available.[7] In extreme cases (e.g., the World War II atomic project or the space program of the 1960's), the application of multiple approaches can be used to assure that at least one line of research pays off within a relatively short time period.

Some innovations require more resources than others. This assumption is easier to demonstrate. It is well known that some innovations require equipment that has a large fixed cost. Some require a much longer period of work than others. A scientist with more resources has an advantage over the scientist with fewer resources in that he may therefore be able to incorporate into his work either more innovations or innovations that require large resource outputs (e.g., a new piece of equipment). Therefore, those who are central and who have resources are more likely to be productive. If they *are* productive, they subsequently gain more resources and hence can become more successful. This 'compounding of interest' results in the observed distribution of papers in science: those who publish frequently publish many more papers than those who publish less; in fact, the distribution of paper publication relative to the number of scientists is close to exponential (Price, 1963; Menard, 1971).

The reason for success, then, is fairly straightforward. The communication network cmulates innovations, particularly important ones. The uneven resource distribution enables those who have successfully capitalized on previous information to capitalize again, even though they have no assurance that they will succeed again (Allison and Stewart, 1974).

Indeed, the processes underlying failure after initial successes are not hard to discern. A scientist with substantial resources can choose either to support more small projects or to increase the amount he spends per project and do fewer of them. The latter route increases the scientist's risk (i.e., the amount of resources that might prove unproductive if a project fails). Unlike outcomes as posited by economic theory, though, the scientist's rewards are not necessarily correlated with risk. Even if successful, his work may be perceived simply as large and routine rather than innovative and important. Hence, increasing the number of small projects spreads the risk and increases productivity, particularly when collaboration is involved; unfortunately, though, it may also produce a situation in which marginal returns (relative to resources) become small as numerous, similar papers are produced. To understand these processes in more detail, we must distinguish between current and total rewards.

DEFINITION 8. (a) *Current rewards* flow from the utilization of a scientist's work by other scientists currently doing research. A good indicator for this concept is citations in the current literature.

(b) *Total rewards* are all items believed by scientists to be rewarding. A good indicator is the total of prizes, lectureships, citations (including those in review articles, monographs, and (textbooks), salary, etc.

Highly rewarded scientists may lose their current rewards due to (1) loss of centrality, which results in less information and, eventually, failure of productivity; (2) failure of productivity despite continuing good information; or (3) diminishing returns. The 'ratchet effect' (Merton, 1968; Cole and Cole, 1972) maintains the total-reward status of the few, high-status, highly productive scientists. However, it does not assure their continued productivity; and if they stop, their current rewards will eventually stop, and their reputations will be eroded as those who knew them gradually become a minority in their discipline (see Westie, 1972, for an interesting example); only a few exceptional scientists will retain their reputations as their work is condensed in the literature.

At this point, the packing-down process becomes important. If a scientist has received current rewards, his work has been utilized at the research front; it thus has some chance of surviving the next process – review articles – and yielding total rewards. In the hard sciences, as Griffith (1972) pointed out, review articles cover only those results that are central to a line of research and have also been judged good. At each step in the packing-down process, every paper – and the reputation of its author – passes through a finer and finer filter. At the final enshrinement (in textbooks, ten to fifteen years later), only a few reputations, based on a few papers, will have survived. A single crucial paper may produce a fifteen-year string of total rewards. Interestingly, in the hard sciences (where the average citation is less than two years old), current-reward status is gone in two years or so. In sociology, in contrast, a paper can be as much as ten years old and still be yielding current rewards (see Garvey *et al.* 1970a). Despite much speculation, research has not yet determined whether the sheer number of papers or some specific aspect of structure in the hard sciences produces their quicker pack-down.

Students are a major factor in both success and failure.

DEFINITION 9. *A student* is a person working for some limited period of time on a senior scientist's research and under his direction. Initially, every student has the potential to become a senior scientist.

Successful scientists have students, and students from the recent past (i.e., within 15 years) constitute over half of all scientists. Only a few scientists have many students, and they usually appear in clusters rather than evenly spread over time. We can reasonably assume that scientists communicate with their students; furthermore, that successful scientists provide good information to their students, some of whom take advantage of it.

When scientists are successful on projects, their students receive good starts; when projects are unsuccessful, the students suffer. Top scientists tend to be the students of top scientists (Zuckerman, 1967). All students of top scientists should be able to assess risks better than students of lower-status scientists, because they have benefited from seeing an expert work. Nevertheless, research shows that while top scientists often have groups of top students, these are separated in time by groups of not-so-successful students (see, e.g., Mullins, 1973b: 49–70). We can hypothesize that when scientists are working on successful projects, they give a better education in risk-taking than when they are involved with unsuccessful projects (and virtually all scientists, at some time, have unsuccessful projects). When successful projects and groups of students coincide, the result is a group of 'followers'. After a successful project, highly productive scientists usually continue to innovate, but only some of their students retain contact with them, take risks themselves, change in the same direction, and publish, thus moving toward very high status themselves.

In short, (1) those with high productivity are able to attract resources. However, some projects require more resources (are more risky), and some continuing research is subject to diminishing returns, even though productivity remains high. (2) Scientists change their theoretical positions according to status. Some active, highly rewarded scientists change more rapidly than anyone else as, over time, they adopt innovations that may have started at many different places in the field. Others do puzzle-solving but retain their high status through operation of the rachet effect. Middle-rewarded scientists generally remain locked into normal science patterns. Low-rewarded scientists have little to risk and usually lose when they do; they therefore drop out. Students begin their careers by reflecting their professors' status, but most soon lose it. As those with high rank and re-source levels innovate more and, in innovating, move away from the conceptions of those with lower innovation rates, instability is created.

The effect of normal science processes, then, is to generate generations, as in the model proposed by Carlsson and Karlsson (1970). If we divide the population of scientists into cohorts based on time of training, we will find a few older scientists (high in rank and re-sources) changing from one perspective to another. However, we will also find that most of the older and middle-aged (who are middle- and lower-ranked, some because they have not yet made it, some because they never will) have remained faithful to the perspective in which they were trained. Finally, the newest group will be quite large (because science expands rapidly), quite active (because attrition has not yet eliminated them), and faithful to the newest perspective(s) in greater percentages.

Even though the social and social psychological processes involved in the two models are quite different, this situation is empirically indistinguishable from that of Carlsson and Karlsson's (1970) 'generations', whose members strongly choose divergent styles. Carlsson and Karlsson's model depends on a natural summation of small changes and differential susceptibility to change by age rather than on an artificial time break or on isolated events that compel a change. As was shown above, these conditions are met by the conditions of science. Moreover, the model permits us to understand why the scientists involved frequently perceive revolutionary change as simply the result of small changes: only in the aggregate does change produce influences that older generations perceive as revolutionary.

The sociologist of knowledge might ask where the leaders of scientific groups obtain their insights. The theory presented here suggests that the leaders' insights derive from the same source as everyone else's: the communication network. Also, like all other positions, their locations in that network have access to a wide range of information of varying composition, quantity, and quality. Only their long experience with quantity and quality in information enables them to make decisions that may succeed in the face of high risks.

3. The development of new paradigms

At this point, the general phenomenon of revolutionary paradigm changes becomes understandable in terms of Carlsson and Karlsson's model. What Kuhn termed 'revolutionary' change in science is equivalent to the fads and changes in generations that sweep society as a whole. A summary of small changes simply, but inexorably, adds up to a new paradigm.

This pattern explains several aspects of coherent groups. (1) Development begins when two or more established scientists recruit their students and, changing previous intellectual positions a bit at a time, incorporate new ideas. (2) The mutual absorption of the principal scientists' direct ties by each other and their students seals their work off from the rest of the discipline until the new development(s) has matured.[8] This isolation virtually guarantees that news of the new research will spread only with the graduation (or at least departure) of the first student group, the development of a second group, and/or the redevelopment of outside contacts by the senior scientists. (3) The distinctive theoretical orientation develops because the senior scientists' innovations feed into each other's research, gradually causing their work to become more and more unlike that done by the rest of the parent discipline. The rate and or type of change for any one scientist may not seem at all strange to him, but his lack of contact nevertheless makes his work unknown to most in his discipline. (4) The responsiveness of innovative scientists and their students to a specific set of research problems increasingly focuses their attention on innovations from one or a very few sources. (5) The actual innovators become recognized as intellectual leaders; those organizing the social aspects of the group (jobs, fellowships, etc.) are recognized as social leaders. The active recruiting of students becomes easier because the researchers' activity levels, concentrated in a small area, convince, students that the 'action' is in that area.

The discussion thus far sounds overly social; nevertheless, the intellectual check on whether various innovations produce new understandings and make sense of specific data is quite real. Both leaders and followers must produce *results* from their innovations (i.e., an improved understanding of their data's order). Groups can fail to produce the necessary results, and many have. In such a case, however, lack of intelligence is hardly ever the reason. The high-status people in an area are bright. They have long taken advantage of the innovations coming to them and have made numerous correct inferences. They themselves have innovated constantly and had a fair measure of success. Indeed this experience may help them to determine whether a disappointing line of research has been stopped cold or just temporarily sidetracked. It sometimes happens that the data of interest are unresponsive to the methods/concepts of the group. In that case the high-ranking scientists involved have two choices open to them: (1) they can move to another line of research and communicate with high-status scientists in other areas, or (2) they can cease being very productive and become middle-status persons. Finally, budding groups sometimes die aborning because one or more necessary social and/or intellectual characteristics fail to develop (e.g., a program statement; for discussions of failure, see Mullins, 1973b: Chapters 5 and 7).

As Griffith and Mullins (1972) point out, a coherent group is inherently unstable. It inevitably dies of either failure (the lack of significant work) or success (the productivity yields high-status scientists, who move off into other areas as they become numerous and less directly connected with the rest of the group). In either case, the break-up of the group returns most of its members to normal science activities.

4. Summary

This paper argues that the normal process of communication leads to differential access to information, which creates differential probabilities that discoveries will be combined into major developments. Similarly, the normal ranking process causes differential probability of and desire to risk development: only high-status scientists (and their students) with substantial resources are able to take the risks demanded by paradigm change; low-status scientists (lacking resources) cannot innovate, and the middle-status will not. The normal process of student distribution sometimes concentrates students where new developments are occurring. Students constitute the next generation of researchers, and as the low ranking of the previous generation become silent, the new generation defines the new style of work. We sometimes see these occurrences as specialty developments, but they are not, strictly speaking, 'developments'; they are rather the cocoons left behind by bright butterflies that have flown on. In short, normal science processes, viewed in the context of Carlsson and Karlsson's model, are sufficient to explain both normal and revolutionary science.

Indiana University,
Bloomington, Indiana

NOTES

* This paper is a revised version of a paper prepared for the Conference on Determinants and Controls of Scientific Development. Graz, Austria, June 16 to June 22, 1974. Another version of this paper was given at the American Scoiological Association meetings, Montreal, Canada, August 29, 1974. Parts of this paper were previously delivered as an IBM Lecture at Williams College, Williamstown, Massachusetts, U.S.A., on May 9, 1974. My thanks to Robert Althauser, Belver Griffith, Lyle Groeneveld, Lawrence Hazelrigg, Derek deS. Price, and Austin T. Turk for helpful comments on a prior draft of this paper. Editorial services were provided by Carolyn Mullins. Support was provided through NSF Grant GS 32769.

1 Other important studies of coherent groups have been done by Ben-David and Collins (1966), Crane (1969), Crawford (1972), Edge and Mulkay (1973), Krantz (1965, 1969, 1971), Law (1972), Mulkay (1972), Mullins (1973a, 1973b), Price and Beaver (1966), and Stoddart (1967).

2 Apparently Ben-David's (1960) argument – that 'outsiders' are the most innovative – was true of his historical case, but not true generally.

3 The model presented here differs fundamentally from that presented by Crane (1972). Crane's model assumed that the innovators are central and that researchers must therefore identify a diffusion process to explain the spread of an innovation from one scientist to others. I propose rather that *innovations* – not innovators – are widely scattered, and that the *combination of diffused innovations* at central points in the communication process causes differences in publication rates. To say that " it is hypothesized that research areas develop as a result of important innovations that appear prior to the period of exponential growth" (Crane, 1972: 67) is (1) to misinterpret the data on innovations and, consequently, (2) to misconceptualize the way science works. First, Crane (1972: Table 16) showed that for five-year time periods, the percentages (of 201 total innovations) appearing in each time period are 21, 10, 20, 31, and 17 Crane described the last five-year period as one in which "only 17 percent of the innovations

were produced in spite of the fact that the field was very productive at the time"; she further stated that 31 percent appeared in the first two five-year periods combined. She thus hid the small 10 percent figure by combining it with the larger 21 percent figure. This manipulation, whose apparent intent is to make the reader ignore the 10 percent and concentrate on the 17, does not constitute proper use of the data. At best, no sensible conclusion can be drawn. Furthermore, Crane provided no information on the source of these innovations (e.g., what scientists published them; whom they cited as having helped to develop them).

[4] Small and Griffith (1974) find only 1819 papers cocited with each other as many as six times in a 3 sample of the first quarter, 1972, *Science Citation Index* (809 papers constitute the biomedical science sector of their highly cocited set). They propose and proceed to demonstrate that this small subset of the 1.3 million papers published in the same period produced a reasonable and complete enumeration of the large active area of science. This indirectly supports Price's conception that only a few highly active persons are at the core of science.

[5] 'Small World Experiments' (Milgram, 1967; Travers and Milgram, 1969) have suggested that the population of the entire world is tenuously connected; if the connections are known, the distance from any given person to any other is surprisingly short. If we accept the estimates above (3 persons per step, 671 total persons) and assume no overlap of choice (clearly an extreme assumption), we can calculate that all active scientists are between five and six steps apart.

[6] Throughout this discussion, paper counts are used as the measure of productivity. There are two problems. First, as Cole and Cole (1971) have shown, the quantity and quality of papers are highly related; thus, despite some exceptions, highly productive scientists usually also produce high-quality papers. Second, while in the 'softer' sciences (e.g., sociology), books are as important as papers, such is not the case in the hard sciences (hence the *Science Citation Index* – *SCI* – counts citations *to* and *within* papers, but only *to* books). The discussion here focuses on a measure only for papers because much of the relevant research has been on physical and biological rather than social science groups. In addition, the *SCI* is currently the basic data source, and it focuses on papers. As research on social science groups increases, measures may have to be expanded to include books.

[7] Zuckerman and Merton (1968) found that the review system for publication has some bias for eminent people, which increases their advantage in publishing. However, Price (1967) holds that with sufficient persistence, almost all work will be published sooner or later. Even within sociology, a field characterized by rejection rates averaging 85 percent (Zuckerman and Merton, 1968), almost 85 percent of the papers given at one national meeting – many of them extremely trivial – were published within the next five years (Garvey, 1971).

[8] Students rarely develop contacts outside their departments. Also, groups vary on the degree of their openness. Bohr's quantum mechanics group (which could be described as an elite specialty) was quite open. In contrast, operant conditioning (which could be described as a revolutionary specialty) was tightly closed against outsiders (see Griffith and Mullins, 1972; and Mullins, 1973b, for further discussion of these groups).

BIBLIOGRAPHY

Allison, D. and Stewart, J. A.: 1974, 'Productivity Differences Among Scientists: Evidence for Accumulative Advantage', *Amer. Sociological Rev.* **39** (August), 596–606.

Bayer, A. E. and Folger, J.: 1966, 'Some Correlates of a Citation Measure of Productivity in Science', *Sociol. Education* **39** (Fall), 381–390.
Becker, M.: 1970, 'Sociometric Location and Innovativeness: Reformulation and Extension of the Diffusion Model', *Amer. Sociol. Rev.* **35** (April), 283–296.
Ben-David, J.: 1960, 'Roles and Innovations in Medicine', *Amer. J. Sociol.* **65** (December), 557–568.
Ben-David, J.: 1964, 'Scientific Growth: a Sociological View', *Minerva* **1** (Summer), 455–476.
Ben-David, J.: 1965, 'The Scientific Role: Conditions of Its Establishment in Europe', *Minverva* **4** (Autumn), 15–54.
Ben-David, J. and Collins, R.: 1966, 'Social Factors in the Origins of a New Science: the Case of Psychology', *Amer. Sociol. Rev.* **31** (August), 451–464.
Blume, S. S. and Sinclair, R.: 1973, 'Chemists in British Universities: a Study of the Reward System in Science', *Amer. Sociol. Rev.* **38** (February), 125–138.
Cancian, F.: 1967, 'Stratification and Risk Taking: a Theory Tested on Agricultural Innovation', *Amer. Sociol. Rev.* **32** (December), 919–927.
Carlsson, G. and K. Karlsson: 1970, 'Age, Cohorts and the Generation of Generations', *Amer. Sociol. Rev.* **35** (August), 701–717.
Cole, J.: 1970, 'Patterns of Intellectual Influence in Scientific Research', *Sociol. Education* **43** (Fall), 377–403.
Cole, S.: 1970, 'Professional Standing and the Reception of Scientific Discoveries', *Amer. J. Sociol.* **75** (September), 286–306.
Cole, S. and Cole, J.: 1967, 'Scientific Achievement and Recognition: a Study in the Operation of the Reward System in Science', *Amer. Sociol. Rev.* **32** (June), 377–390.
Cole, S. and Cole, J.: 1968, 'Visibility and the Structural Bases of Observability in Science', *Amer. Sociol. Rev.* **33** (June), 397–413.
Cole, J. and Cole, S.: 1971, 'Measuring the Quality of Sociological Research: Problems in the Use of the *Science Citation Index*', *Amer. Sociol.* **6** (February), 23–69.
Cole, J. and Cole, S.: 1972, 'The Ortega Hypothesis', *Science* **178** (27 October), 368–375.
Coleman, J. S., Katz, E., and Menzel, H.: 1968, *Medical Innovations: A Diffusion Study*, Bobbs-Merrill, Indianapolis.
Crane, D.: 1965, 'Scientists at Major and Minor Universities: a Study in Productivity and Recognition', *Amer. Sociol. Rev.* **30** (October), 699–714.
Crane, D.: 1969, 'Social Structure in a Group of Scientists: a Test of the 'Invisible College" Hypothesis', *Amer. Sociol. Rev.* **34** (October), 335–352.
Crane, D.: 1972, *Invisible Colleges: Diffusion of Knowledge in Scientific Communities*, University of Chicago Press, Chicago.
Crawford, S.: 1972, 'Informal Communication among Scientists in Sleep Research', *J. Amer. Soc. Information Scientists* (September-October), 301–310.
Edge, D. O. and Mulkay, M. J.: 1972, *A Preliminary Report on the Emergence of Radio Astronomy in Britain*, Vol. 1, Department of Engineeriqg, Cambridge, England, (mimeo).
Fisher, C.: 1966, 'The Death of a Mathematical Theory: a Study in the Sociology of Knowledge', *Arch. History Exact Sci.* **3** (2), 137–159.
Fisher, C.: 1967, 'The Last Invariant Theorists: a Sociological Study of the Collective Biographies of Mathematical Specialists', *European J. Sociol.* **8** (2), 212–244.
Garvey, W. D. and Griffith, B. C.: 1963, Vol. 1; 1965, Vol. 2; 1966, Vol. 3; *Reports of the American Psychological Association's Project on Scientific Information Exchange in Psychology*, American Psychological Association, Washington, D. C.

Garvey, W. D. and Griffith, B. C.: 1967, 'Scientific Communication as a Social System', *Science* **157** (1 September), 1011–1016.
Garvey, W. D., Lin, N., and Nelson, C.: 1970a, 'Communication in the Physical and the Social Sciences: the Process of Disseminating and Assimilating Information Differs in These Two Groups of Sciences', *Science* **170** (11 December), 1166–1173.
Garvey, W., Lin, N., and Nelson, C.: 1970b, 'Some Comparisons of Communication Activities in the Physical and Social Sciences', in C. E. Nelson and D. K. Pollock (eds.), *Communication among Scientists and Engineers*, D. C. Heath, Lexington Mass.
Gaston, J.: 1970, 'The Reward System in British Science', *Amer. Sociol. Rev.* **35** (August), 718–732.
Gaston, J. and Rose, D. S.: 1971, 'A Reconsideration of Communication among Biological Scientists', paper presented to the American Sociological Association meeting, August (Denver).
Goffman, W.: 1966, 'Mathematical Approach to the Spread of Scientific Ideas – the History of Mast Cell Research', *Nature* **112** (29 October), 449–452.
Goffman, W. and Newell, V. A.: 1967, 'Communication and Epidemic Processes', *Proc. Royal Soc.* **A 298**, 316–334.
Griffith, B. C.: 1972, 'On the Nature of Social Science and Its Literature', Drexel University, Philadelphia (mimeo).
Griffith, B. C. and Miller, A. J.: 1970, 'Networks of Informal Communication among Scientifically Productive Scientists', pp. 125–140 in C. Nelson and D. Pollock (eds.), *Communication among Scientists and Engineers*, D. C. Heath, Lexington, Mass.
Griffith, B. C. and Mullins, N. C.: 1972, 'Coherent Social Groups in Scientific Change: "Invisible Colleges" May be Consistent Throughout Science', *Science* **177** (4053, September 15), 959–964.
Griffith, B. C., Small, H., and Stonehill, J. A.: 1973a, 'The Structure of Scientific Literature II: the Macro and Micro Structure of Science', Drexel University, Philadelphia (mimeo).
Griffith, B. C., Meier, V., and Miller, A. J.: 1973b, 'Describing Communication Networks through the Use of Matrix-Based Measures', Drexel University, Philadelphia (mimeo).
Hagstrom, W.: 1965, *The Scientific Community*, Basic Books, New York.
Hagstrom, W.: 1971, 'Inputs, Outputs and the Prestige of University Science Departments', *Sociol. Education* **44**, (Fall), 375–397.
Harary, F., Norman, R. Z., and Cartwright, D.: 1965, *Structural Models*, John Wiley, New York.
Hargens, L. L. and Hagstrom, W. O.: 1967, 'Sponsored and Contest Mobility of American Academic Scientists', *Sociol. Education* **40** (Winter), 24–28.
Hirsch, W.: 1968, *Scientists in American Society*, Random House, New York.
Homans, G. C.: 1950, *The Human Group*, Harcourt Brace, New York.
Katz, E. and Lazarsfeld, P.: 1964, *Personal Influence*, Free Press, New York.
Krantz, D. L.: 1965, 'Research Activity in "Normal" and "Anomalous" Areas', *J. History Behavioral Sci.* **1**, 39–42.
Krantz, D. L.: 1969, 'The Baldwin-Titchener Controversy', pp. 1–19 in D. L. Krantz (ed.), *Schools of Psychology*, Appleton Century Crofts, New York.
Krantz, D. L.: 1971, 'The Separate Worlds of Operant and Non-Operant Psychology as a Case Study', *J. Appl. Behavioral Analysis* **4**, 61–70.

Kuhn, T.: 1962, *The Structure of Scientific Revolutions*, University of Chicago Press, Chicago (2nd ed., rev. 1970).

Law, J.: 1972, *Structural Change in Scientific Specialties: the Case of Protein Crystallography*, unpublished Ph. D. dissertation, University of Edinburgh, Edinburgh.

MacRae, D., Jr.: 1969, 'Growth and Decay Curves in Scientific Citations', *Amer. Sociol. Rev.* **34** (October), 631–635.

Menard, H. W.: 1971, *Science: Growth and Change*, Harvard University Press, Cambridge, Mass.

Merton, R. K.: 1957, 'Priorities in Scientific Discovery: A Chapter in the Sociology of Science', *Amer. Sociol. Rev.* 6 (December), 635–659.

Merton, R. K.: 1961, 'Singletons and Multiples in Scientific Discovery: A Chapter in the Sociology of Science', *Proc. Amer. Philosophical Soc.* **105**, 470–486.

Merton, R. K.: 1968, 'The Matthew Effect in Science', *Science* **159** (5 January), 56–63.

Milgram, S.: 1967, 'The Small World Problem', *Psychology Today* **22** (May), 61–67.

Mulkay, M. J.: 1972, *The Social Process of Innovation: A Study in the Sociology of Science*, Macmillan, London.

Mullins, N. C.: 1966, *Communication among Biological Scientists*, unpublished Ph.D. dissertation, Harvard University, Cambridge, Mass.

Mullins, N. C.: 1968, 'The Distribution of Social and Cultural Properties in Informal Communication Networks among Biological Scientists', *Amer. Sociol. Rev.* **33** (October), 786–797.

Mullins, N. C.: 1972, 'The Development of a Scientific Specialty: the Phage Group and the Origins of Molecular Biology', *Minerva* **10** (January), 51–82.

Mullins, N. C.: 1973a, 'The Development of Specialties in Social Science: the Case of Ethnomethodology', *Science Studies* **3** (July), 245–273.

Mullins, N. C.: 1973b, *Theories and Theory Groups in Contemporary American Sociology*, Harper and Row, New York.

National Science Foundation: 1971, *American Science Manpower 1970*, National Science Foundation, Washington, D. C.

Price, D. J. deS.: 1963, *Little Science, Big Science*, Columbia University Press, New York.

Price, D. J. deS.: 1965, 'Networks of Scientific Papers', *Science* **149** (30 July), 510–515.

Price, D. J. deS.: 1970, 'Citation Measures of Hard Science, Soft Science Technology and Non-Science', pp. 3–22 in C. F. Nelson and D. K. Pollock (eds.), *Communication Among Scientists and Engineers*, D. C. Heath, Lexington, Mass.

Price, D. J. deS. and Beaver, D.: 1966, 'Collaboration in an Invisible College', *Amer. Psychol.* **21** (October), 1011–1018.

Rapoport, A.: 1963, 'Mathematical Models of Social Interaction', pp. 493–580 in R. D. Luce, R. Bush, and E. Galanter (eds.), *Handbook of Mathematical Psychology*. Vol. 2, John Wiley, New York.

Rogers, E.: 1962, *The Diffusion of Innovation*, John Wiley, New York.

Ryder, N.: 1965, 'The Cohort as a Concept in the Study of Social Change', *Amer. Sociol. Rev.* **30** (December), 843–861.

Small, H. and Griffith, B.: 1974, 'The Structure of the Scientific Literature I', *Science Studies* **1** (4), 17–40.

Stoddart, D. R.: 1967, 'Growth and Structure of Geography', *Transactions and Papers of the Institute of British Geographers*, publication No. 46.

Travers, J. and Milgram, S.: 1969, 'An Experimental Study of the Small World Problem', *Sociometry* **32** (December), 425–443.

Westie, F.: 1972, 'Academic Expectations for Professional Immortality: a Study of Legitimation', *Sociol. Focus* **5** (Summer), 1–25.

Zuckerman, H. A.: 1967, 'Nobel Laureates in Science: Patterns of Productivity, Collaboration and Authorships', *Amer. Sociol. Rev.* **32** (June), 391–403.

Zuckerman, H. A.: 1968, 'Patterns of Name Ordering among Authors of Scientific Paper: a Study of Social Symbolism and its Ambiguity', *Amer. J. Sociol.* **74** (November), 276–291.

Zuckerman, H. A.: 1970, 'Stratification in American Science', *Sociological Inquiry* **40** (Spring), 235–257.

Zuckerman, H. A. and Merton, R. K.: 1968, 'Patterns of Evaluation in Science: Institutionalization, Structure and Functions of the Reference System', paper presented to the American Sociological Association Meeting, August (Boston).

GERNOT BÖHME

THE SOCIAL FUNCTION OF COGNITIVE STRUCTURES: A CONCEPT OF THE SCIENTIFIC COMMUNITY WITHIN A THEORY OF ACTION*

1. The Scientific Process as a System of Action

Our starting point is that science is a certain type of action. By this we intend to thematize what scientists actually do. We set off the cognitive structures of science as a behaviour by considering scientific action with respect to its material intention, namely the intention directed to truth.

Scientific action may be differentiated into three sections: scientific communication, research, and teaching. When we speak of communication we designate that part of scientific behaviour by which scientists interact as to their claims, projects, and results. Research here encompasses the total area of behaviour leading to scientific results, that is, data production and experimentation, but also theory construction and the formulation of hypotheses. Under the term of teaching we have comprehended all behaviour by which science is reproduced. In all three sections scientific behaviour does not proceed in an unpatterned way or merely individually, but follows certain rules, norms and patterns. These are in part obligatory for whole scientific branches, in part they characterize only particular disciplines, in some cases even only schools or specialities. We do not intend at this point to demonstrate the binding force of the rules and norms in particular – all that will be attempted in this section is to state what types of norms, rules, and schemes may in principle be thought to be operative here and to consider the relationship between the rigidity of norms on the one hand and of scientific innovation on the other.

1.1. *Argumentation for the Acceptability of Theses*

Within communication the individual scientist is primarily referred to the other scientists, within research primarily to his subject of inquiry.

It is true, each scientific communication is communication concerning a subject, and on the other hand theory construction (part of research) is

Knorr, Strasser, and Zilian (eds.), Determinants and Controls of Scientific Development, 205–225.

strongly determined by the reference group in the face of which the theory has afterwards to be justified – but the very terminus *ad quem* is yet different for both realms of behaviour.

In contrast to most other types of communication scientific communication, however, is argumentation; the coherence of communication here is the coherence of an argumentative context. This thesis may seem trivial, but that it is not so is shown by the fact that scientific communication is frequently understood to be an exchange of information (Hagstrom, 1965; Scherhorn, 1968; Luhmann, 1968). Moreover, the fact that each scientific utterance is an argument within a context of argumentation is by no means obvious, but rather tends to be concealed by the very circumstance that explicit controversies in science are rare. The reason for this is that in his publication every scientist anticipates as far as possible all objections so that he may be sure of the acceptance of his contribution by his fellow scientists. This anticipation is made possible by the scientist's socialization into his discipline and by his awareness of the state of affairs in his respective research area. This then can cause the impression that scientific publications only impart information. A more thorough examination of scientific publications would show, we suggest, that this is almost never the case.

Even the communication of pure measurement results usually is the adducing of empirical evidence for a hypothesis or even is itself an empirical hypothesis for which theoretical arguments have to be brought forward in the publication. The difference between information and argument is also not one of content throughout, it consists rather in the role which an utterance has in a certain context of communication. To be sure, the scientist needs information and he gets it from the publications of other scientists, although to a lesser degree from their published papers than from handbooks. Scientists, then, as it were, make technical use of the established knowledge. In this case it naturally becomes important whether their information is true or false. But the communication about scientific results is not concerned with the truth or falsity of information but with the validity or invalidity of claims.[1] This, however, has to be qualified. It might be supposed with some justification that the very concern of scientific argumentation was the truth of certain claims. This is not the case, or to put it better: the *truth* of what is claimed is not the immediate issue. Scientific argumentation is concerned with the justification of taking certain propositions *for true*. One reason for this may be adduced from the distinction we have made between the object level and the argumentation level of scientific behaviour and yet another from the logic of discovery. Without special reference to a corresponding theory of truth it may be asserted that truth is something that is concerned with the relation of the proposition to the object. Yet no judgment regarding this relation can be made on the ground of arguments alone, for with respect to decisions of the question of truth one is ultimately dependent on direct confrontation with the object.

The reason derived from the logic of discovery is the following: To restrict scientific communication to propositions the truth of which can be proven would to an intolerable degree impose limitations on the process of discovery. Besides the fact that in that event empirical universal statements would have to be excluded, the main point is a more general one, that is that the process of research is actually kept alive by the content of propositions which exceeds the empirical evidence. Scientists know that they have to work with unproved propositions, with conjectures, hypotheses, axioms – this is the essence of fallibilism – but propositions of this kind are neither proposed arbitrarily nor accepted without reasons. The adducement of reasons which serve to justify the acceptance of propositions in science: that is the concern of scientific argumentation. If someone asks what the reasons are, which lead to the acceptance of propositions within the system of scientific communication, the answer in the first place will be: A proposition is acceptable, i.e. one is entitled to take it for true, if it *is* true. This becomes tautological only if the difference between the object level and the argumentation level is not observed: On the level of argumentation the assertion that a proposition is proven (for example, by measurement or construction) is an argument justifying one's taking the proposition for true. For the rest our thesis gives a regulative for argumentation. In the long run just those propositions should be accepted which are true. Hence all arguments speaking for truth evidence in the long run are valid for the acceptability of a proposition. On the other hand, a minimum criterion for the acceptability of propositions, that is, for their establishment within the fund of science is their being scientific. The being scientific is made explicit by the demarcation criterion as elaborated by the philosophy of science. Thus, for instance, that a hypothesis is falsifiable is an argument for its acceptability. In addition, also arguments adducing the fruitfulness or simplicity of theories, and also those concerned with their adequacy in terms of certain more general ideational systems such as world images or those concerned with their legitimacy (on religious, political or social grounds) may be brought forward within the context of argumentation about the acceptability of scientific theses (cf. Böhme, 1974b).

1.2. *Consensus and Innovation*

If one asks whether there are certain norms and rules for argumentation which are shared by the participants, the straightforward answer will be: yes. Scientific argumentation, after all, consists precisely in the process of connecting one's own theses with what has already been accepted by the scientists adressed (one's audience, say, a specialty). In any argumentation a set of premises and possible arguments are presupposed as binding for all the participants. Nor is the connection between one's thesis with this already established consensus – say, within a discipline – made in an unpatterned way, rather, already the formulation of the thesis follows certain norms, the infringement of which entails the risk of not making oneself comprehensible to others. Further, there exist established schemes for the way an argumentation is to proceed: These have in part been analyzed by the philosophers of science. Thus, for instance, they have made explicit what an explanation, proof or refutation is, or by what kinds of arguments an induction can be justified. These investigations, to be sure, are largely of a general nature and provide rules of argumentation which are accepted by all scientists. In the present context it would be more interesting to discover whether there exist schemes of argumentation which are specific to certain research areas or disciplines. In this sense, Toulmin, in his *The Uses of Argument*, can be considered to have made a first important step. According to his investigations, it is the criteria but not the force of an argument which are region-dependent. But what the difference between a 'chemical' and a 'physical' argument really is, a difference also imputed by Ziman (1968: 71), has not been examined to date.

Argumentation is the process which in philosophy of science is termed the context of justification. While one is inclined to admit of every freedom and absence of rigid rules for the context of discovery, there exists unanimity regarding the rigorousness of procedure within the context of justification. Nonetheless, the process itself is not correctly described if one looks upon it as merely linking the thesis to an already existing consensus by already established means. This would correspond to a scholastic view of science. What characterizes modern science, however, is that each contribution is expected to be an innovation. What is of particular interest is, therefore, the enunciation and justification of such theses which are not merely supplemental to established knowledge, but which in some way deviate from the existing consensus or are not reconcilable with it by conventional means. Several scientists have on these grounds challenged the general validity of rules in science or questioned the supposition that scientific communication is argumentation: Thus Feyerabend (1970) suggests this is but propaganda by which a scientist attempts to gain public acceptance of his theses. This view, however, rests on a misconception of the essence of scientific argumentation. First, argumentation is not deduction from a completely explicit set of premises. What is actually the case is that a thesis which is only partially questioned will be justified on certain grounds, which are taken to be accepted by the partner. This signifies, that argumentation must be justifying for the proponent, but not coercive for the opponent. The latter always has the possibility of accepting the grounds given or to question them. This does not imply that, as far as the proponent is concerned, argumentation is an attempt to persuade the opponent. Secondly, one has to take into account that in actual argumentation a thesis is not only rendered problematic in an unspecified way, but that it is opposed by arguments. When we spoke of the logic of argumentation accepted by the participants it was understood that this logic must contain rules specifically defining what dissensus is, what situation arises from argument and counter-argument, and the like. Further, the fundamental consensus presupposed for scientific argumentation cannot be such as to categorically exclude dissensus on specific issues or questions. When Hagstrom takes consensus to be the supreme value in science and when Ziman takes it to be the defining principle of science, this rests upon a misunderstanding of the claim to universal validity associated with scientific theses. That universal validity, that is, validity for everybody, is claimed, does not mean that theses which have not yet achieved general validity[2], or the validity of which has not yet been proven, are not scientific.

Nor is it necessary for disciplines in which there is a definite dissensus that they must break apart (as Hagstrom says). It is rather a particular problem to find out when controversy must result in segmentation. While it is certainly possible to say that in the long run consensus is the aim of scientific argumentation, it is nonetheless false to attribute scientific value to the avoidance of controversies (as Ziman does, 1968: 78, 145): In the short run the function of scientific argumentation, rather, is to question theses by presenting counter-arguments. Consequently, the essence of scientific argumentation is misconceived if scientific consensus is seen as unchangeable and fixed. The essence of argumentation after all is to change a consensus by advancing new theses which, once accepted, can also be used as arguments; furthermore, to establish even new types of inference (for example, established scientific laws can be used as rules of inference[3]).

In addition one has to remember that argumentation connects a thesis with a more general cultural background (ideals of explanation, see Toulmin, 1961) so that what can be justified changes with that *from which* it is justified.

Finally, the consensus presupposed by argumentation is open in that in the process in which arguments are rendered problematical, the argumentation can always pass onto the meta-level: if it is impossible to agree on data or on theories, the rules of data production or of theory construction themselves can be questioned, which means that one is then engaged in a controversy on methods or on the foundations of science.

1.3. *Research and the Appertaining Communities*

We have opposed the sector of scientific behavior in which the individual scientist communicates with his colleagues through argumentation to that in which the scientist is primarily concerned with his object, this sector being termed 'research' or 'object-level'. This opposition corresponds to the traditional one of the context of discovery and the context of justification, although the view of the context of discovery we shall give is different from that stated by philosophy of science. One obvious difference between our distinction and the traditional one follows from the fact that at crucial points of the justification by arguments one again has to deal with the object. Thus in arguing about theses one again and again arrives at points where arguments do not get us any further, so that one has to measure or – for instance in mathematics – to construct.[4] Hence the domain in which scientific behavior consists in confrontation with the object is not only the area of discovery of the new, but also the area in which alone demonstrations are possible: proving a thesis to be true always has to be a demonstration on the object; by arguments one can only justify that one takes a certain thesis *for* true. In the tradition of the distinction made between the context of discovery and the context of justification the former was always viewed as an area of unpatterned and irrational behavior. This was only possible because attention was primarily directed to the process of the production of ideas. If the processes of selection and filtering and especially of experimentation are, however, included, then this area will be extended to cover the whole area of research and this changes the picture: it then becomes evident that in fact data and theories are produced in conformity with rules, so that in the context of justification they can be vindicated. It is the rules and norms of theory construction, of data production and experimentation operative here which make it possible to describe in general terms the practical research commitments of the scientist relative to his subject.

We suppose in particular that there exist not only rules for scientific practice which are binding for science in general but that it is possible to find out some rules which are specific for certain disciplines and that on the other hand it is sometimes possible to define scientific disciplines and specialities by their rules for research. To be sure, to what degree this expectation will be fulfilled depends on the definition of a discipline. It is not at all easy to say what a discipline is because in a scientific discipline the formal and the informal organization of science overlap. Hence disciplines are not throughout defined internally – that is by the coherence of research – but may be defined by external requirements and traditions (of the university for instance). To get rid of this ambivalence we shall confine ourselves to the informal organization and shall use the term 'communities of research'[5]. Scientists certainly form a community when they are concerned with one and the same subject. But this need not always be the case. For example, biophysicists are held together more by a common method than by a common subject (Certaines, 1973). Besides a common subject and a common method it has been taken into consideration that a common theory (Law, 1973) or a common goal may define a community of research.[6]

1.4. *Data Production, Experiment, Theory Construction*

The rules which regulate the practice of research in an obligatory way, are to be divided into rules for experimentation, rules for data production and rules for theory construction. Rules for data production are related to rules for experimentation but the sets are not identical, they only have a common subset. Rules for experimentation are the methodological prescriptions which must be observed if the experiment is to fulfill its function within the scientific method (a). The demand for control experiments or the demand for distinguishing object theory and measurement theory, both of which are to assure the function of testing hypotheses, may serve as examples. But the main function (b) of rules for experimentation is to present the object under investigation within the experiment precisely in the way it is intended by the special type of research: thus the social anthropologist will not investigate man by means of interviews and the psychologist not by means of a rule. This is perhaps a trivial example because the disciplines mentioned are too far from one another. We get to a higher level of sophistication, however, when the rules for experimentation divide only schools within a discipline, as is the case of experimental psychology examined by us (Böhme, 1974a). In this case the one school investigated the mind by a certain experimental procedure as an apparatus that is identical for all men, the other as an individually and typically different function.

Objects are only diffusively given within the ordinary, the life world experience. A regulation of the empirical approach is needed to transform the objects of the life world into the specific objects of the sciences to get well defined phenomena and effects. This way it is ensured that this and only this phenomenon is thematized within an experiment – for instance, in push-experiments the mass but not the weight or chemical attributes. Just as in geometry a figure is given by the rules for its construction so the rules for experimentation define the object or phenomenon one is empirically concerned with. Rules for experimentation of our set (b) are generally specific for particular types of research, rules for data production are generally not. Thus the rules, for example, regulating the assignment of numbers to observables which have been studied under the title of a general theory of scales within the philosophy of science, these rules – I say – do not have a validity which is only regional. In contrast to them the principles of quantification are different with respect to both region and phenomenon. The principles of quantification govern the procedure by means of which phenomena wich are not quantities in their own right become accessible to quantitative methods. Procedures of this type have been studied particularly in psychology where phenomena are almost never given immediately as quantities. Here as in the social sciences arises the problem of validation as well as the problem of reliability, in every case dealt with by determinate procedures. These too rank as rules for data production. Rules for experimentation and rules for data production together assure the reproducibility of experiments and thus their universal and intersubjective validity. It is here that the greatest rigidity with respect to the observance of rules is to be expected, for it should thereby be achieved that the experiences of the individual subject under contingent conditions come to be the specific and universal experience of an identifiable object.

The amount of norm rigidity within the area of theory construction will in any case not be comparable to this, for this is the very realm of 'creativity'. On the other hand, demands following from Popper's demarcation criterion – falsifiability, excess of empirical content, etc. – can hardly be disregarded. Further, there are region-specific demands to be made upon theory construction which are rather rigid, for instance, the demand of Lorentz-invariance within physical theories. These norms and their importance are not well examined yet. It may well be expected that by means of norms for theory construction it should be possible to make explicit what the difference is between a sociological and a psychological, between a physical and a biological theory.

1.5. *Teaching*

The section of scientific behavior termed 'teaching' here has been mentioned only for reasons of completeness. Teaching is not constitutive for what a scientific community is but makes possible its reproduction. It would, however, be worthwhile to search for region-specific patterns of socialization. Kuhn and Polanyi did some very fine work in this area, investigating for example the functions of text books, the meaning of data series within them, the significance of laboratory education etc. Here, it should be mentioned, the social relation (teacher-student relation) generally determines the relation of contents. This in part also holds good for the development of social networks. Therefore phenomena of member-recruitment and idea-diffusion should rather be conceived in analogy to teaching and be categorized under the third sector of scientific behavior.

2. Scientific Behavior within the Context of a General Theory of Action

2.1. *Elements and Basic Distinctions*

Little more will be achieved than *bon mots* in determining the relations between the content of science and its social forms if logical and propositional structures are simply juxtaposed with scientific institutions. In the last section we therefore conceived of science as a system of action and in this section we do not intend to investigate science as a social entity in its formal organization but will, to begin with, examine science in its informal organization. We hope to show that the informal organization rests in the organization of scientific work. Once this middle-ground between the propositional and institutional structures, that is science as a co-operative enterprise, has been explored, on the one side science as a propositional system will again be attained: for the product is essentially related to the rules according to which it was produced; on the other side the foundation will have been laid for understanding science as a reward system and as a formal organization.

The question is, therefore, how scientific behavior organizes itself, that is: how social entities are engendered by the (informal) organization of work within science. To begin with, we have to point out the particular characteristics of scientific action in contrast to action in general. According to Parsons, any action has a threefold orientation: cathectic, cognitive, and evaluative. Ego orientates his action according to the difference of gratification and deprivation (a), according to the conditions given with a determinate situation (b), and finally chooses between alternative possibilities of action regarding values to be realized. With a slight deviance from Parsons, scheme [7] we state that any action is oriented motivationally, cognitively, and intentionally. The characteristic of scientific action is the predominance of cognitive orientation. What generally is only of instrumental significance, namely "the location and characterization of objects" (Parsons *et al.*, 1967:5), is here the main issue. This could be stated in other words, that is by saying that the cognitive and the intentional orientation are identical: science is 'pure' action for orientation.[8]

Another characteristic of scientific action is related to the thematizing of cognitive orientation. Each scientific interaction is referred to an object. Each argumentation is concerned with an object, hence in their communicative relation scientists are also related by means of their common relation to the object. On the other hand, it is equally possible to say that within science each relation to an object implies a relation to the other scientist, for each reference to the object stands under the auspices of the demand for justification: Research always is research in face of the other scientist. As a basic pattern of scientific action a triangle formed by two scientists and one object may be drawn. The reference to the other scientist is performed by argumentation, the reference to the object by research (see Figure 1).

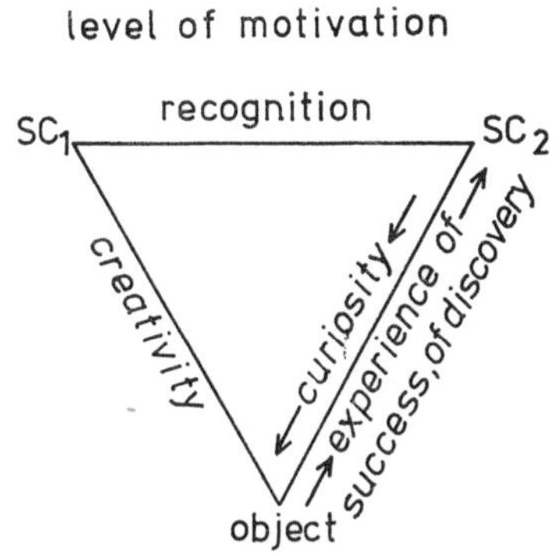

Fig. 1.

Thus formally the principle of the social organization of science is given. What sociology of science has generally failed to do is to account for the primacy of cognitive orientation with regard to the social organization of science. In contrast to this, the organization of science was constructed solely on the basis of motivational orientation: thus Storer took the demand for creativity and competent reaction to be fundamental, and Hagstrom the system of sanctions.[9] To be sure, the organization on the level of motivation is a reality, but it presupposes organization on the cognitive level. On the level of motivation another triangle corresponds to the basic pattern of scientific action, a triangle formed by two scientists who are immediately related by mutual recognition, and who are mediately related by means of their cathectic reference to the common object, that is, through curiosity and their experience of success within research (see Figure 2).

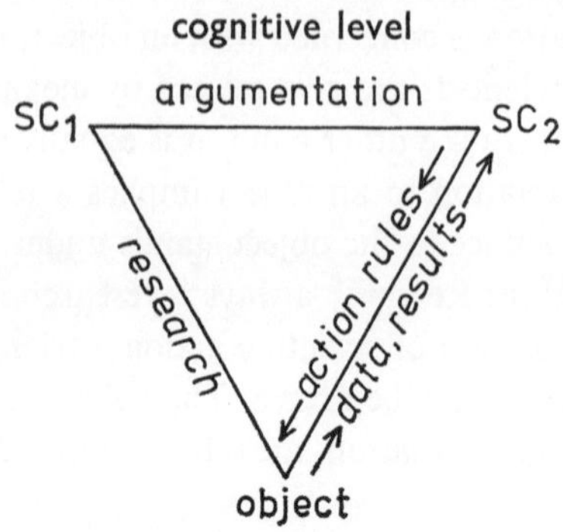

Fig. 2.

The two strata of scientific action here distinguished are actually found in each concrete action; they are only separable for analytic purposes. We will briefly make this clear for the case of argumentation: when someone asserts a claim he associates with the communication of a certain cognitive content the intention of getting attention for the point he has made and further that of it gaining acceptance. He can count on the former only if he enters into an existing context of argumentation; as to the latter, this may take the form of another scientist adopting his thesis and deducing further consequences from it. The cognitive use made of his thesis by another has for him the cathectic significance of a reward which he can, however, only be assured of (because of the impersonal character of the cognitive content) by way of certain norms, as, for instance, the duty of citation. If within the cognitive frame of reference his partner's reaction is that of refuting his claim, then on the cathectic level this will for him signify deprivation. Thus it becomes clear that the utterance of determinate arguments within the context of argumentation such as refutation, objection, corroboration, and so on, each have their cathectic value. But the context of argumentation determines their role in any specific case. Utterances of recognition, to be sure, may be found outside the context of argumentation. But then they are either not competent reactions in the sense of Storer, or they presuppose a context of argumentation and the performance of actual argumentations and are then the subsequent abstractions from these.

2.2. *Argumentation as the Organizing Principle of Scientific Communities*

The next question we must ask is what makes scientific action social.

Any set of people as such will not be considered to be a social entity, rather they will form a social entity by the manner of their behavior, that is, if their behavior is differentiated and conventionalized they will have left the level of individual behavior. On the basis of the same behavior as a common attribute they will form social categories or classes (in the logical sense). Their unity here is yet entirely abstract. To the extent, however, that they acknowledge common values they reach a higher level of social unity and are thus qualified for unified action, they form a collectivity. Further enhancement of social unity formation presupposes interaction or reciprocity of expectations, that is, the possibility of interaction. If there are more than two interacting partners there exist further distinctions. In the one case the interactions form only a network, so that the constellation of the whole is perceivable only if seen from the outside, in the other the different interactions, respectively expectations, are interrelated to a degree that the idea of the whole comes to regulate every particular interaction. The latter will be the case, for example, if the set of actors has a common goal, but even then when their interactions comprise a common orientation towards system maintenance, system stabilization or system demarcation. Let us designate the type of interaction constitutive for the latter kind of unity, the 'organizing principle'. We use the term 'organizing' because the social entity formed on this basis has some formal analogies to organisms: coherence of the particular elements with one another, the relation of this coherence to the idea of the whole, the demarcation against the environment.

Sociology, as far as I can see, has no established term for what is intended here. Sohn-Rethel (1972) uses the term *gesellschaftliche Synthese* (social synthesis). In contrast, the term 'organizing principle' has been used to designate the principle of exchange, which according to Hagström is the principle underlying the coherence of a scientific community, that is, the principle of the exchange of information for recognition.[10] Used in this sense, this term corresponds precisely to what we intend, except that we would contest that Hagstrom has actually specified what the organizing principle of scientific communities is. For, as we have previously emphasized, the possibility of recognizing information as gift-giving itself presupposes the existence of a group of interested consumers, organized by virtue of the coherence of their work. The type of interaction to be labelled the 'organizing principle' according to our definition was to be *constitutive* of the higher form of social unity. This does not mean that it must be the causal agent of unification – for this it is possible to adduce contingent facts – but that it is the principle which makes the social entity into what it is.

To elucidate this distinction by an example, we shall briefly deal with Storer's conception of the scientific community. We have stated that argumentation is the organizing principle of the scientific community. Storer, preferring causal-explanatory theories, would probably object that we have thereby not made explicit how the scientific community comes into being. This, because there must be some motivation on the basis of which people enter scientific communication. This we admit, but we would argue that their fundamental relation, once they have entered the scientific community, is given by argumentation. For only on the grounds of a context of argumentation can their achievements give rise to competent reaction and be evaluated as creative. Therefore, even Storer's exchange principle cannot be accepted as having the status of the organizing principle of the scientific community, although true enough, he has shown it to be sufficient for social system formation: his principle is thus but the organizing principle of the corresponding reward system.

Our claim is that argumentation is the organizing principle of scientific communities[11]: the type of interaction which holds a scientific community together and constitutes its unity is to be found in argumentation. The unity of a scientific community from the cognitive standpoint is hence a context of argumentation. A scientist who publishes a paper actually enters this context but by the definition of his work he is already located in it. His problems derive from the context of argumentation and taking account of the state of argumentation he must present his results if these are to be appraised as 'contributions'.[12]

What is called the state of research may be explained as the state of argumentation. For the former is not only defined by an aggregate of accumulated information: if this were the case the individual scientist would not continually find himself under pressure 'to keep up to date'. For otherwise he would only be running the risk of producing information which already exists. The greater risk, however, is to produce information which is not relevant. Yet what is relevant, is in fact defined by the context of argumentation. For example, some facts may be known which do not accord with the generally accepted views: they require an explanation. Or a supposition has been formulated, a hypothesis stated: these must be supported by empirical evidence or examined as to their further theoretical consequences. Or again, it may be necessary to develop theoretical approaches into a research program, or research programs should be extended to new areas of inquiry. Or certain methods of data production or types of inference are being questioned from the viewpoint of more general methodological concepts. But it is beyond the scope of the present essay to analyze the possible situations which may define a state of research. This is in part a matter of the logic of argumentation, in part of case studies in the history of science. What should be stated here is simply that the state of argumentation defines the problem situation for the individual scientist, that it defines for him 'the status' of the theory he refers to, it prescribes the way in which he can present his contribution as an argument, the premisses which he can build upon, and what counter-arguments he has to take into account.

The context of argumentation, as it were, is something objectively real: it is to be found in the publications, although it has diffuse ends in unpublished papers and in laboratory discussions. The crucial advantage of this principle, in contrast to the concept of social networks as the sociological principle of scientific communities, lies in this objectivity: it is thereby possible to take into account that scientists may live with other scientists in the *same* research community without actually having relations of communication. By the very fact that, even prior to publication, scientists connect their work with a state of argumentation, they are integrated into the corresponding research community. Further, persons who are members of the social network in merely a 'social' sense (i.e. citations for reasons of courtesy or amity, acknowledgements for support not definable in terms of any content, and so on) are not considered a part of the scientific community.

Further, the principle of argumentation as the unifying principle can very well account for the openness of scientific communities. This openness has been adduced frequently as an argument against the existence of paradigm groups in the sense of 'belief circles'. Thus Ziman (1968:4) conceives the criticism from 'outside' to be the most important corrective against possible tribualization of science. Science is not – or at least not yet – only *one* context of argumentation. Contexts of argumentation are regional, usually defining smaller unities than the (formally defined) discipline. But they are frequently inter-connected, forming hierarchies, overlapping or containing each other. They also have distinct boundary lines. Scientists know very well what their contributions are contributions to, and on the other hand the main part of the production of others is excluded as irrelevant to their own context. But this does not mean that the boundary lines are impermeable, or that they could not suddenly open to incorporate new areas of research. Any scientist may enter the scientific community which is defined as a community of argumentation (irrespective of his particular socialization!) provided that he succeeds in referring to the state of argumentation, or eventually even due to being an 'outsider' to problematize certain elements of the argumentation as it proceeded so far.

To conclude, argumentation as the organizing principle of scientific communities is particularly suited to conceive the latter as innovative systems. To be sure, certain norms and rules are binding for determinate contexts of argumentation, certain patterns of argumentation and types of arguments are established. But in as much as they are more specific than the norms of a general logic of argumentation, they are perpetually altered along with the development of the state of argumentation. Their validity is always provisional. But being provisional they have indeed a binding force: if a scientist intends effectively to enter an argumentation context he either has to accept the *coinai ennoiai* or to question them explicitly.

2.3. *The Solidarity of Scientific Communities*

Solidarity, formally understood, in sociology usually is conceived of as that property of an aggregate of persons which enables them to become a whole and to act as a whole. We expect to find the solidarity of scientific communities to be based upon the acknowledgement of certain technical-methodological rules and paradigms (in the narrower Kuhnian sense). Durkheim in his analysis of traditional and modern societies distinguished two types of solidarity: the mechanical and the organic one. In the case of mechanical solidarity the possibility of a social entity achieving cohesion is given in the fact that each individual is immediately related to this whole by shared values and on that basis accepts his position as ascribed to him. In the case of organic solidarity the possibility of cohesion is based on the functional differentiation of individuals and on the division of labor. As was to be expected, attempts have been made to conceive of the solidarity of scientific communities within this framework. Downey (1969) has pointed to the remarkable phenomenon that the solidarity of scientific communities is a more mechanical one – remarkable in that in other respects the scientific community is regarded as the forerunner of modern societies. The crucial point in Downey's argument is the fact that hitherto there has almost never existed any division of labor in science. In fact, the only case of institutionalized functional differentiation mentioned by Hagstrom (1965, Ch. V) is the differentiation of experimental and theoretical physics. Downey, to be sure, is right when he states that the most general solidarity which makes all scientists feel that they belong to *one* scientific community, is mechanical. This solidarity is based on their commitment to such values of a particular professional ethos as Merton has made explicit, but on the other hand also on the acceptance of such norms as analysed by Popper under the catchword of 'demarcation criterion': that is, methodological norms which demarcate science from non-science.

But there are, in contrast to this, solidarities in science which are more specific. These differ from each other depending on whether a social entity in science identifies itself primarily in terms of a common subject matter, shared methods and theories, or through a common goal. The last case implying the possibility of division of labor, and hence organic solidarity, has so far, in practice, hardly been realized. Only within the perspective of 'finalization of science' [13] it could become a more universal reality. The 'fit' of one scientific result to the other is so far generally not the outcome of any definite division of labor but appears rather to be a felicitous coincidence. This may sometimes be the case; but as a rule the fit of scientific results can be explained as the product of a particular kind of solidarity. This type, in contrast to the well known other types, may be labelled 'pre-established' or 'concerted' solidarity. The astonishing harmony of scientific results produced individually, without arrangement, and at different places, is grounded in what Kuhn has called the acceptance of a paradigm. In contrast to mechanical solidarity where the whole the individual persons are referring to, as it were, lies in front of them, here the whole is behind them: the work of the individual scientist is, by acceptance of a theory, regulated in such a way that the results fit together.

A further type of solidarity in scientific communities is given by the reference of scientists to a common object. We have noticed above that the scientific object is specifically given so that within a research community its identity has to be ensured by the agreement upon determinate rules for experimentation. By experimentally investigating one and the same object these scientists form a community, even if they do not have a common (paradigmatic) theory as yet. Thus we have mentioned four types of solidarity within scientific communities: (1) a general solidarity in methods (guidance by method), (2) a particular solidarity in a common goal (guidance by purpose), (3) a solidarity founded in a common theory or model (guidance by theory), (4) a solidarity founded in a common object (guidance by object).[14]

It should not cause surprise that we have made no decision for one or the other type. What is rather to be expected is that social entities in science will show, from case to case, historically different types of solidarity. On the other hand it should be kept in mind that solidarity is only a very weak social unity, no more than the precondition for the emergence of a social system. Thus argumentation associates scientists into one research community, scientists who are already solidary with one another somehow, through their object, methods, theories or goals: each argumentation presupposes a basic consensus which, for instance, may be assured by their referring to a common object [15]. To be sure, the solidarity area may be larger than the area held together by argumentation.

Thus, I think, we have again turned Kuhn's approach on its head, as is appropriate in this case. Contents, methods and technical norms are what defines the primary communities in science. It is true, scientific development is describable as the formation and reorganization of scientific communities. But the reasons why the coherences of argumentation break apart and why new solidarities arise, have to be determined on cognitive grounds.[16]

Max-Planck-Institut zur Erforschung der Lebensbedingungen der wissenschaftlich-technischen Welt, Starnberg, West Germany

NOTES

* In the following essay a framework is given for investigating the relations between social and cognitive structures in science. We look for such relations as a consequence of T. Kuhn's theses that scientific development is connected with the reorganization of scientific communities respectively the formation of new ones. For this reason the original paper presented at the Graz conference contained a typology of the different reception of Kuhn's works within sociology of science which had to be left out of this publication for lack of room. For a more complete treatment of this, see Böhme, 1974c.

[1] Here we are in a sense abandoning Habermas' consensus theory of truth, to which we are otherwise in many respects indebted. Habermas (1973; 1974) identifies the truth of a proposition with the justification of its validity claim, something we would deny for reasons grounded in the logic of discovery to be specified below.

[2] It is therefore nonsensical to claim that the science of history or the social sciences are 'unscientific', as does Ziman (1968: 10).

[3] Cf. Toulmin (1969); the conception of laws as rules for inference can be traced to Schlick and Scriven.

[4] Such situations, when one is at the end of one's wits can be highly dramatic in the history of science: take for instance Pasteur's challenge to Liebig to carry out certain fermentation experiments which were a point of controversy between them before a commission of the Académie Royale.

[5] Whitley (1974) calls the smallest unit to be considered in this connection a 'research area', a notion that in itself does not have a sociological meaning. When we use the term 'research communities' we intend to mark out those groups of scientists in which scientists determine their identity on the basis of the research they are concerned with: 'I am doing ribosome research', 'I am an embryologist', and therefore not on the basis of their disciplinary training at the university nor on the basis of their formal appertainance to associations, departments etc. In this sense the sentence 'I am a physicist' is ambivalent.

[6] For physical research: Janich (1973). Hagstrom too always refers to the 'shared goals' of a scientific community without, however, indicating what these consists of.

[7] For the philosopher this deviation is a minor one: for every intention is grounded in a final 'for the sake of what' and every concrete goal-determination is orientated to a value. In the same way the distinction we make between motivation and intention (as against the Parsonian scheme) is familiar to the philosopher. To justify it in the sociological frame of reference is something that remains to be done in the future.

[8] That there is no particular intention underlying science is a claim asserted, although in another way, by the thesis of value-neutrality. We are neither satisfied with this formulation, nor with that given in the Parsonian scheme above, but intend to leave them as they are, in order to be able to start from familiar notions. In our view a multidimensional structure must be premissed for any intention: Any action is rooted in an objective, a personal and a social intention. The distinguishing feature of science is in these terms that its objective intention is cognitive orientation.

[9] Luhmann, in his respect, is to be regarded as an exception in that he conceives of the reputation system as a mechanism of regulation operative in a science, the organizing principle of which is a cognitive one (the flow of information).

[10] The expression is not used by Hagstrom himself, rather Barnes in his Reader (1972) has included an extract from Hagstrom's book entitled 'Gift-Giving as an Organizing

Principle of Science.' Whitley utilizes the expression (1974) to characterize the principle of organization of problem areas, that is, not in a sociological sense.

[11] Here we come close to what K. O. Apel has tried to do in giving, following Peirce and Morris, a reinterpretation of the Kantian foundation of natural science within the frame of universal pragmatics. Here the community of science conceived as a community of interpretation (1973) is assigned specific functions of the transcendental subject.

[12] The context of argumentation is hence also an organizing principle in Whitley's (1974) sense: It organizes problem areas. As against the principles enounced by Whitley's argumentation, however it has the advantage of at the same time having a cognitive and a social function: For the context of argumentation is a cognitive constellation which is represented by real partners in a process of communication. Whoever makes his scientific *work* enter the cognitive context of argumentation, at the same time *himself* enters a certain context of interaction.

[13] So far this refers in particular to the projects in Big Science. Compare the concept of finalization in Böhme *et al.* (1973).

[14] For the connection between the typology of scientific communities and the types of solidarity within them, see also Law (1973), especially pp. 278, 279, 302. The reasons why my results – notwithstanding a general agreement – differ in some points are in short the following: (1) There is no reason to take the Durkheimean alternative to be complete. (2) Law's general concept of solidarity, which may be abstracted from his definitions (1973: 278, 279) as being "the ground on which relationships can develop", is a too wide one. (3) There are research communities which are held together by a common goal but without having a shared subject matter (project-oriented research).

[15] Seen from the level of argumentation the forms of solidarity are constituted by types of the consensus which must be presupposed for any argumentation, that is of the basis from which a discussion can take place.

[16] I am indebted to Mrs. Suzanna Libich, MPIL Starnberg, for her assistance on the preparation of the English version.

BIBLIOGRAPHY

Apel, K. O.: 1973, 'Szientismus oder transzendentale Hermeneutik', *Transformation der Philosophie* **2**, Suhrkamp, Frankfurt.

Barnes, B. (ed.): 1972, *Sociology of Science*, Penguin, Harmondsworth.

Böhme, G.: 1974a, 'Die Bedeutung von Experimentalregeln für die Wissenschaft', *Zeitschrift für Soziologie* **3** (1), 5–17.

Böhme, G.: 1974b, 'Die Bedeutung praktischer Argumente für die Entwicklung der Wissenschaft', *Philosophia naturalis* **15** (1), 133–151.

Böhme, G.: 1974c, 'Die soziale Bedeutung kognitiver Strukturen. Ein handlungstheoretisches Konzept der scientific community', *Soziale Welt* **25**, 188–208.

Böhme, G., Daele, W. v. d., and Krohn, W.: 1973, 'Die Finalisierung der Wissenschaft', *Zeitschrift für Soziologie* **2**, 128–144.

Certaines, J. D.: 1973, 'La mort des disciplines closes: naissances de la biophysique?' Maison des Sciences de l'Homme, Parex Réunion de 18 et 19 décembre, Paris.

Downey, K. J.: 1969, 'The Scientific Community: Organic or Mechanical?', *Sociological Quarterly* **4**, 438–454.

Feyerabend, P. K.: 1970, 'Against Method', *Minnesota Studies in the Philosophy of Science*, Vol. 4, University of Minnesota Press, Minneapolis.

Habermas, J.: 1973, 'A Postscript to Knowledge and Human Interest', *Philosophy of Social Sciences* **3**, 157–159.

Habermas, J.: 1974, 'Wahrheitstheorien', *Wirklichkeit und Reflexion. W. Schulz zum 60. Geburtstag*, Neske, Pfullingen.

Hagstrom, W. O.: 1965, *The Scientific Community*, Basic Books, New York.

Janich, P.: 1973, *Zweck und Methode der Physik aus philosophischer Sicht*, Universitäts-Verlag, Konstanz.

Law, J.: 1973, 'The Development of Specialities in Science: the Case of X-ray Protein Crystallography', *Science Studies* **3**, 275–302.

Luhmann, N.: 1968, 'Selbststeuerung der Wissenschaft', *Jahrbuch für Sozialwissenschaft* **19**, 147–170.

Parsons, T. and Shils, E. A. (eds.): 1967, *Toward a General Theory of Action*, Harvard University Press, Cambridge, Mass.

Scherhorn, G.: 1968, 'Der Wettbewerb in der Erfahrungswissenschaft', *Hamburger Jahrbuch für Wirtschafts-und Gesellschaftspolitik* **14**, 63–86.

Sohn-Rethel, A.: 1972, *Geistige und Körperliche Arbeit*, Suhrkamp, Frankfurt.

Toulmin, S.: 1961, *Foresight and Understanding. An Inquiry into the Aims of Science*, Hutchinson, London.

Toulmin, S.: 1969, *The Uses of Argument*, Cambridge University Press.

Whitley, R. D.: 1974, 'Cognitive and Social Institutionalization of Scientific Specialities and Research Areas', in R. D. Whitley (ed.), *Social Processes of Scientific Development*, Routledge and Kegan Paul, London.

Ziman, J.: 1968, *Public Knowledge: The Social Dimension of Science*, Cambridge University Press, London.

KARIN D. KNORR

THE NATURE OF SCIENTIFIC CONSENSUS AND THE CASE OF THE SOCIAL SCIENCES*

The data presented here derive from a comprehensive study probing into the characteristics of 'external' and 'internal' (Leinfellner, 1974) control in social science development. The concepts used in the study are tied up with Kuhn's theory of scientific revolutions (1962, 1970), according to which the systems of rules and theses in mature sciences encounter a more or less world wide acceptance on the part of the scientists involved with them. During the pre-paradigmatic stage of a science, however, there is a series of schools that are ensnared in reciprocal controversies about their specific metaphysical basic assumptions, which ultimately constitutes their only raison d'être. The scientific efforts are dialogues between the members of different schools rather than with nature. The transition to mature science is identical with the triumph of one of these schools and with the general acceptance of the first paradigm. Simultaneously the nature of the paradigm changes. Only after the change is normal puzzle-solving research possible.

In his postscript of 1969 Kuhn points out that evidently also the schools of 'pre-paradigmatic' science show paradigms. Hence, it is not the existence of paradigms, but their acceptance beyond the boundaries of the individual schools as well as their applicability as a tool for scientific problem solving that mark the transition to mature science. Various critics have reproached Kuhn that he drastically overstates the unanimity of scientists in their allegiance to a paradigm. This reproach is at least in part due to a misunderstanding, since no dividing line is drawn between the various levels on which a paradigm is operating. Obviously individual paradigm components (or entries in a 'disciplinary matrix') can at the same time furnish the consolidating basis for an entire series of areas of research as well as they can themselves be made the subject of controversial analyses in other speciality fields. Therefore, the concept of professional consensus must always be understood in the sense of a *preliminary* consensus. In our view such a preliminary consensus is a logical correlative of the growth of scientific knowledge, since scientific results can only be converted into

Knorr, Strasser, and Zilian (eds.), Determinants and Controls of Scientific Development, 227–256.

cumulative 'knowledge' by inter-subjective consentaneous acceptance. The actual dissensus, which Kuhn and in the following others have attested to the social sciences, is therefore rightly regarded as an indicator for the lack of progress of knowledge (see Kuhn, 1962; Storer, 1967; Zuckerman and Merton, 1971; Lohdal and Gordon, 1972; Lammers, 1974).

The introduction of professional consensus as a demarcation criterion between pre-paradigmatic (social) and paradigmatic (natural) sciences evokes a series of questions some of them we would like to illustrate in the present paper by reference to empirical data. The most fundamental question is undoubtedly under which conditions a general agreement can be expected to arise in scientific communities. Furthermore there is the question of whether the commonly held judgement on the actual dissensus in the social sciences can at all be sustained. Another question which comes up is as to which effects the differing 'consensus potential' in the natural and social sciences will engender within the science-internal control systems; and finally there is the question as to whether the observed dissensus in the social sciences is in accordance with the cognitive expectation structures of the scientists themselves.

1. Consensus as a Function of the Triviality of Decisions

The first thing we are interested to explore is the question of the conditions which enhance or hinder the development of consentaneous agreement transcending the boundaries of 'schools' in science. The answer given by those adhering to a materialist theory of knowledge may well be that consensus can only be achieved under the condition of identical interests and the hence resulting convergence of opinions (e.g. Sandkühler, 1975). Analytical philosophy of science, on the other hand, will refer to the inherent rationality of the scientific operation, which, being backed by the scientists system of norms, implies that a scientist would be obliged (and competent enough) to decide in favour of the scientifically 'better' alternative. Conflicting career interests and differential access to resources make it appear impossible to reconstruct consensus in terms of identical interests without reference to a system of norms. However, the assumption of the inherent rationality of scientific action is also met by objections which impose themselves immediately: firstly, the possible *incom-*

mensurability of scientific theories or paradigms, in itself the object of unresolved controversies in the system of science; and secondly, the *context-dependence* of scientific action which – irrespective of incommensurability – makes the possibility of aggregating the individual adherences to norms appear unlikely. Since the individual scientific decision is taken in a cognitive field, which must be relatively restricted with respect to the consideration of factors relevant for decision making, this decision is of necessity only partially rational, even if no external non-cognitive factors are introduced. In addition, the individual rationality of the scientist is permeated by particularistic considerations, which by no means can all be rendered functional for the scientific enterprize with the help of the reward system.[1]

In the light of these difficulties we consider it useful to substitute the optimistic inference of science-internal consensus from cumulative rational individual decisions by a more pessimistic alternative, whose advantage mainly consists in having to make much weaker assumptions on the aggregating potential of individual interest-linked rationality. Accordingly, consensus would not be a matter of convergence of opinions due to identical interests and/or equivalent normative orientations, but rather a *function of the triviality of decisions*[2] involved. We call a decision trivial if one cannot avoid its outcome without calling into question one's own rationality as a competent subject. We now maintain that the degree to which decisions on acceptability of knowledge-claims are trivial depends on the structure of the chains of argumentation[3] and justification which they refer to. According to Toulmin's (1958) structure chart[4] an argument would have to include a *conclusion* (C) or a knowledge-claim; in order to establish the merits of the conclusion we appeal to facts as the foundation for the claim – what Toulmin refers to as *data* (D). Being required to indicate the bearing on our conclusion of the data already produced we would have to propose a *warrant* which shows that the step to the original claim or conclusion is an appropriate and legitimate one. Warrants, in contrast to data, are appealed to implicitly; furthermore they are general, certifying the soundness of all arguments of the appropriate type. Standing behind our warrants there will normally be other assurances ('*backings*', B) which vary from one field of argumentation to another. Finally, the conclusion is subject to certain *conditions of exception* or rebuttal (R) which have to be specified. The argument thus assumes the following simplified form (Figures 1 and 2):

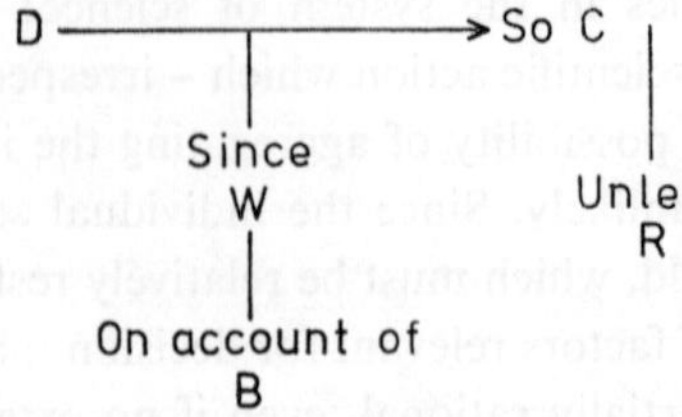

Fig. 1.

The following example Toulmin's should serve as an illustration:

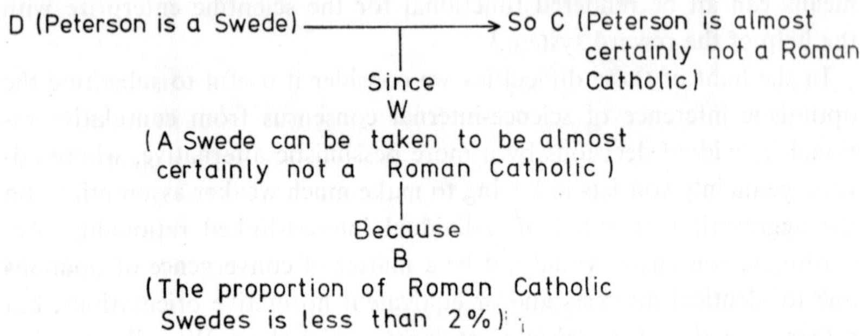

Fig. 2.

According to the type of backing put forward a tentative distinction of the following structures of argumentation can be made:

(a) *Analytical argumentations* according to Toulmin refer to those arguments whose warrants are backed not by experience, but by entailment, or, in other words, in which checking the backing of the warrant involves ipso facto checking the truth or falsity of the conclusion (Type 1).

(b) *Empirical argumentations* accordingly refer to those arguments, whose backings revert to empirical data, which can supposedly be attained through the operation of rules of correspondence (Type 2)

(c) *Dialectical argumentations* finally refer to arguments where the attempt to justify a generalization W_1 involves a new argument containing a further generalization W_2, etc. (Type 3).

The above distinctions allow us to differentiate between structurally *open* and structurally *closed* chains of argumentation in terms of the consensus which can be reached about the 'truth' or 'falsity' of the arguments involved. With structurally open arguments it should in principle be difficult to reach consensus about the truth of their knowledge-claims solely on cognitive grounds – a situation which might enhance the operation of factors such as power or fashion in the process of establishing the number of followers of a thesis. On the other hand, structural closure of an argumentation depends on the standards of truth and falsity presupposed or accepted and on what constitutes a 'proper' calling into question of knowledge claims. Standards of truth presupposed are those required by the process of symbolization itself such as principles of logic and those required by the process of communication such as acceptance of what we can observe as 'facts' (cf. Feigl, 1963). By standards of truth accepted we mean those methodological rules and procedures in science adherence to which guarantees that truth is (held to be) conserved over or brought out by a whole series of steps of interaction with 'reality' based on scientific instruments of some kind. In a clearly analytic argument, because of the presupposition of the rules of symbolization to any argument failure to accept the truth of the message is a sign not of potential counter-evidence, but rather of a 'defect of reason' on the part of the person raising the question. Hence, with *analytic arguments*, closure in terms of consensus is implied by the very structure of what is put forward. On the other hand, *dialectical arguments* are potentially open arguments, since they constantly have to rely on new, unproven non-methodological warrants. With dialectical arguments as with empirical arguments the range of proper callings into question is much wider than with analytic arguments. We refer to 'proper' callings into question because of the fact that the potential of critique and objection embedded in every argument is not fully exploitet by fellow scientists. Thus, a correctly established empirical proof of a thesis will in the normal practice of natural science hardly raise objections as far as the factual result is concerned in spite of the fact that empirical proofs because of all the background knowledge and ceteris paribus clauses involved will never be ultimately conclusive. Consequently, closure of the argument by consensus is reached because actual areas of *proper* and *effective* (i.e. taken serious by fellow scientists) callings into question are much more limited than potential ones.

A legitimate area of objections in the case of empirical arguments involving *theoretical terms* (cf. Carnap, 1959) is that of correspondence between knowledge claims and the data presented. We held that in the social sciences consent on and hence closure of type 2 argumentations based on theoretical terms meets with fundamental difficulties because of the lack of a *theory of meaning* establishing connections between those terms and measurement results.[5] Furthermore, measurement results themselves as far as they transcend clearly observable factual information have to be made sense of in the light of such a theory of meaning. If, however, structural consent in empirical chains of justification remains restricted to arguments based on *observational terms*, growth of knowledge in the social sciences is reduced to the level of empirical generalizations. This implies that the social sciences are banned from an effective reduction of complexity which goes together with cumulative theoretical development (Luhmann). One of the symptoms of such a situation is the 'gap' between theory and research observed by many social scientists, for which empirical evidence is now being accumulated. Cole (1975) has proven for a sociological specialty field that only 3% of the empirical studies which explicitly rely on a certain theory do in fact attempt to test the respective theoretical assumptions, whereas the rest only vaguely relates to them thus continuing the gap.

That a theory of meaning or a theory of symbolical communication is fundamental for the further development of the social sciences has been demonstrated in particular by Cicourel (1964, 1970) in his methodological and linguistic studies. Numerous other authors have also pointed to the necessity of developing a theory of the social and cognitive constitution of meaning and its relationship to knowledge claims (e.g., Schütz, 1932; Winch, 1958; Fodor and Katz, 1964; Habermas, 1967; Luhmann, 1974). We consider it extremely important that distinctions like that between theoretical and observational terms be included into such a theory. In our view it is no coincidence that the main explanatory variables of empirical sociology once compiled by Baldamus (1973) are concepts of mainly factual nature: income, profession, education, sex, age, status, religion, race, residence (urban-rural origin), political attitude and social class, the latter being the most strongly theory-oriented concept in the enumeration[6]. Obviously this repertory of explanatory concepts has evolved pragmatically in the course of time from the fact that these variables are indeed effective in a typical sociological survey study. That they were at all considered is probably due to their every-day relevance rather than to any embedding in scientific theories. We do not want to deny that these concepts yield as a matter of fact high empirical correlations. We only think that the 'paradigmatic traits' shown by many areas of empirical sociology and indicated by a potential consensus on explanatory variables (as well as on a certain methodological approach, and the like) could not be imagined without the objective or objectifiable character of the dimensions mentioned. These paradigmatic traits are also not impaired if the empirical sociology in question has been severely criticized in the meantime. For it was not the falsity of the results that was critized, but their substantial limitation or inadequateness in dealing with the problems involved.

In the field of sociological theory, however, no approximately equivalent consensus on relevant concepts can be found, not to speak of the consensus on theoretical generalizations. Freese (1972) rightly indicates that in this context it is not only the problem of the 'gap' between theory and research that is relevant for the assessment of the situation, but above all also the mutual interplay of many theories or of the theoretical implications of many empirical studies. Yet the problem does not seem to primarily reside in the fact that we allow our theoretical knowledge to evolve haphazardly; in the natural sciences there also is no a priori integration of research results. What there is, however, is a sort of 'pre-established harmony' on the basis of the fact that concepts can be well defined (which does not necessarily mean that they are always well defined) in the framework of their theoretical and factual references. The latter guarantees that a positive or negative (i.e. falsifying) integration of the research results is in principle possible, just as well as an appropriate linking of the different theory fragments. The possibility to formalize empirical and theoretical generalizations in wide areas of the natural sciences is generally considered as an indication of the assumed well-definition of meanings. As opposed to this in sociology, for instance, as well as in most other social sciences, it is the *haphazard relation of meaning in our knowledge claims* due to unknown mechanisms of the social and cognitive constitution of symbolic content that we held to be responsible for the lack of cumulativity whenever theoretical terms are involved.[7]

We are now in the position to present a more detailed version of our initial hypothesis: accordingly, the amount of consensus in a certain field is a function of the *degree of triviality of decisions*, and this again is a function of the *structure of the chains of argumentation and justification* involved. The amount of consensus will be greatest when structures of argumentation and justification of *Type 1* dominate. The potential closure of empirical arguments in the social sciences depends on the type of terms used: on the level of argumentation operating with observational terms the possibility to render decisions trivial and hence to promote selections leading to consensus seems to be guaranteed. In the field of what has been called 'theoretical' terms, on the other hand, there will remain difficulties in achieving consensus as long as no theory of meaning brings us closer to the solution of two main interlinked difficulties: the relation between theoretical and empirical generalizations (*the measurement problem*) and the relation between different theoretical contents (*the problem of meaning*). Since a clear-cut distinction of areas operating with observational terms from those operating with theoretical terms hardly seems possible in most social sciences and since scientists themselves show a tendency to create theory-based interpretations for their observational findings, in sum total we expect rather little consensus for the *empirical structures of argumentation* and justification. One can assume that, for the latter reasons, the extent of agreement in the area of empirical argumentation will hardly differ from the expected low degree of consensus in the area of dialectical argumentations.

2. Consensus and Cumulativity of Scientific Knowledge

The asserted correlation between consensus and structures of argumentation and justification has an important consequence for the relation between consensus and cumulativity of scientific knowledge postulated by Kuhn: it implies that it is the predominance of the respective chains of argumentation and justification or their codified methodological versions [8] and *not* necessarily the *level of development* of a discipline that is essential for the explanation of a given amount of consensus or dissensus. Therefore, our second main assumption states that consensus is though a *necessary* correlative to cumulative progress of knowledge, yet does not represent a *sufficient* condition for achieving substantially cumulative results. Hence – depending on the methodology – the potential gap between consensus and cumulativity makes it appear possible that high rates of consensus go together with a granted lack of cumulativity, for instance when analytical structures of argumentation are linked with substantial problems in the realm of theoretical terms. This is the case, because complex decision making processes (such as for instance within the research process) according to our hypothesis enhance an orientation towards those decision criteria which are easily decided and evaluated. Within the context of decision making in scientific research practice such an orientation brings about a *cumulative selection* in favour of those aspects and topics which involve *trivial* (i.e. easily decided) *decisions*, both in the framework of the decision making problems of the individual scientist as well as in the framework of the diachronic development of individual scientific fields[9]. Such a selection, however, implies a shift of perspective in a direction which potentially neglects the original substantial problem involved.

Selection processes in favor of trivial decisions not only derive from an orientation towards easily decided and evaluated criteria, they are also encouraged by other related factors. One is that non-controversial criteria are apt to unambiguously signalize the *professional competence* of the author; another is that rewards are allocated on the basis of those aspects of a scientific effort which can easily (trivially) be controlled by those in charge for the allocation of rewards; a third is the fact that only consentient decisions allow for 'definite' *error identification* as well as for a (cumulative) selection of 'correct' results basing on it. Obvious examples for such selections are the model platonism in economy as well as in general the much deplored methodism of social science disciplines. In both cases one can speak of a universally recognized deepening of knowledge with respect to the development of mathematical and statistical procedures and other codifiable problem aspects at the cost of the substantial problems involved. The most common explanation for this phenomenon points to the artificial imitation of natural science interests and procedures in order to account for this development (e.g., Habermas, 1968). Our hypotheses do not deny this interrelation; they rather attempt to specify those mechanisms that were responsible for converting general value orientations into the actual flourishing of certain subjects.

3. Scientific orthodoxy as factual falsificationism

Our last assumption is a specification of Kuhn's thesis on the relative orthodoxy of normal scientific orientations in the light of the statements so far made. It maintains that the predominance of potentially trivial decisions within a certain research practice leads to the formation of *convergent*[10] expectation structures whereas the respective contrary practice ought to lead to divergent orientations. Here the dichotomy convergent-divergent denotes the difference between expectation structures which are oriented towards consentaneous, problem-solving results easily integrated into the original frame of reference on the one hand, as opposed to *divergent*, disagreeing results, on the other hand. The assumption becomes understandable when it is pointed out that trivial decidability also means *trivial controllability* in the dual sense of the word: firstly, if the appropriate decision criteria are present, the scientist is well adviced to control his own results before submitting them to the scientific community (primary control), because of the very ease of control on the part if his fellow scientists; and secondly, the further use of the results by other scientists implies an indirect and mostly also unintentional secondary control in that possible problems are identified as errors and can be fed back to the original results. In our view this last possibility presents a particularly decisive difference to those fields which do not dispose of easily closed chains of argumentation and justification.

In order to illustrate this we would like to begin with an ideal type of example. Let us imagine a hypothetical natural scientist, whose empirical results do not correspond with his (paradigmatic) expectations. Such a scientist will first of all repeat his experiments in order to be sure that no procedural error has crept into his results. If again the results do not coincide with his expectations he will embark upon a detailed examination of the method used (re-calibration, etc.). If no flaws can be found here either he will consider whether he had not made some general conceptual mistake which ran through all his checks and therefore prevented him from finding the 'correct' results. If his results again withstand this examination he might – maybe after another thorough control of the individual steps – do the following:

(1) The results disappear in a drawer, because our scientist or his professor do not want to run the risk of being reproached with bad working methods, or, worse, scientific incompetence;

(2) The results are published, but with the specific comment that they cannot be generalized, and with an emphasis on the special ceteris paribus clauses under which the experiments were carried out;

(3) Our scientist goes on checking his results, which implies that he

submits them to a still more scrutinized and profound investigation.

The time and money consuming severe tests of his results are not undertaken because our scientist is interested in the *falsification* of the underlying theory. Rather, scrutinity was provoked by the ease of primary and secondary control of results and by the fact that easily identified errors in the controllable sphere of his work might make people doubt the scientific competence of their author, thus bringing about serious existential consequences for the natural scientist. This is quite contrary to the situation of a typical social scientist, say a sociologist, whose empirical examination has led to results which contradict some of the results generally accepted in the literature. Due to the problems of measurement and meaning a sound control of his survey responses is in principle only possible when the survey has been restricted to concepts based on observational terms. The same applies to the possibility of validating the respective method. Since pragmatic difficulties usually prevent the two control steps from being taken, checkups of possible conceptual mistakes do not exactly impose themselves, with the effect that there are no obstacles impairing the immediate publication of the results including an emphasis on their being new and different. The best controllable and most controlled aspect of a survey is that of the proper use and correct application of statistical methods and techniques. A step, which does not tie up with the items of data selection, data quality and data meaning that are strategical for obtaining valid results.

We summarize our examples as follows: the production of research results under the condition of open structures of argumentation and justification in wide areas of the social sciences operating with theoretical terms is characterized by a *verification circle*, in which the lack of adequate control possibilities as well as of rigid expectation guidance by generally accepted knowledge (the entries in a 'disciplinary matrix') leads to a situation in which the accumulation of different and contradicting results imposes itself. The primary and secondary control possibilities in the natural sciences, on the other hand, represent a *negative control circle*, which stabilizes the system by orienting it towards the target value 'truth'. As in the social sciences the steering mechanism is not a direct one, but one which operates via the system of norms and rewards serving as a guide-line for the members of scientific communities. Their relative orthodoxy and their orientation towards the production of 'consonant' results conforming with the generally accepted knowledge turns out to be a directed *falsificationism* that submits *those* results to a specially rigid investigation that question the ruling and effective patterns of explanation.[11]

If orthodox patterns of expectation and the assumption of severe (directed!) testing of results can be reconciled with each other within the negative control circle, the verification circle described by us should not be confused with the lack of any expectations or any constraints (e.g., Stehr, 1974). However in the case of verification, expectation structures are different in the following respects: firstly, these expectations are not oriented towards the elaboration of *convergent* results, since the lack of a generally recognized body of knowledge and the difficulties of a rigid endogeneous control do not promote such an orientation. Therefore, the norm of novelty and innovation characterizing science works in a direction where an almost unrestricted variety of isolated, diverse and contradicting results can be produced. Secondly, it is true that work within a rigid scientific tradition (a school) will yield accordingly orthodox expectations – and a critical control will primarily attack those results that contradict these expectations – yet the expectations themselves are not grounded in *generally accepted* knowledge, but on particular doctrines. And finally it is also true that the complexity and diversity of the results produced by a faulty endogeneous selection is reduced by the expectation structures, but, contrary to the natural sciences, this reduction is governed by influences *external* to science (cf. Leinfellner, forthcoming).[12] Since generally accepted theoretical interpretation calculi are practically nonexistent and since an effective internal elimination of errors and erroneous interpretations is largely lacking because of the mal-functioning methodological control circle, said reduction of complexity is carried out in the social sciences by ruling *societal patterns of interpretation.* We are not in a position here to present a detailed definition of the concept of societal patterns of interpretation. All that should be mentioned is that value judgements as well as class interest and metaphysical assumptions flow into the 'every-day knowledge' which constitutes the basis of theoretical hypotheses as well as of empirical variables in the social sciences. This knowledge is conveyed by the concepts and conceptions rooted in everyday language; it is imparted by the structures of expectation of the individual scientists who determine the selection and interpretation of the data.

The arguments presented so far lead to a qualification of our assumption of the potential gap between cumulativity of knowledge and consensus: it imposes itself to make the coincidence of substantial cumulativity as a criterion for the maturity of a science, on the one hand, and of consensus, on the other hand, depend on the *adequate functioning* of the control circle described. For only the functioning of this control circle is tantamount to a primary science-internal error reduction and selection (obtained by an effective further utilization) of 'true' results meeting the requirements of correct inference and tieing up with accepted bodies of knowledge.

In this paper only two of the above assumptions can be substantiated empirically. Firstly the assumption of the interrelation between consensus in a specific field and easily closed 'trivial' decisions in the research process; and second the hypothesis that the expectations scientists place in their results reflect their methodological orientations, the latter being linked with certain structures of argumentation and justification. For a more detailed presentation of the relevant empirical data, see Knorr (1975).

4. Design and Implementation of the Study

The above mentioned hypotheses were operationalized in a study that was carried out in Austria in 1973–74. It is a survey of all research units working in the social sciences, our definition of social sciences comprising the following disciplines and fields: psychology, paedagogics, sociology, political science, economy, business administration, ethnology, modern history, regional and urban research. In the respective research units the head of the unit was personally interviewed, while his research staff was asked to fill in written questionnaires. Since the survey addressed institutionally defined 'units' and not individual scientists, individual scientists without an institutional backing fitting our definition of social sciences were not included. Not considered were also those few units exclusively engaged in teaching social sciences. Members of a unit were regarded as research staff when they either had an academic degree or at least held the position of a research assistent.

In total responses from 624 scientists were obtained. The rate of response differed slightly according to the various survey regions as well as according to whether the data were collected at universities or in extramural sectors. An average of 58% for questionnaire responses was achieved. (65% at universities). The rate of response for the unit heads interviewed orally amounted to an average of 86%. Since in many questions the questionnaire presupposed research experience as well as adequate training in one of the social sciences and/or experience in one specialty field, a major portion of the loss of respondents is due to the fact that the questionnaire could not be applied to those 'researchers' in a unit who had no research experience of their own or whose functions were primarily administrative. In these cases the worse rates of response do not imply a distortion with respect to the questions investigated in this study in that these persons do not belong to the target population of the communities of social scientists.

5. Operationalization of the Concept of Consensus

For controlling our assumptions it was doubtlessly important to be able to operationalize the concept of consensus. The following aspects appeared to be relevant: on the one hand, the respondents had to be offered response categories which took into account the possible context dependence of the respondent's consensus perception. In other words, it was to be prevented that a sociologist claimed a consensus in a certain field, which was as 'strong' as a physicist's would be only because the implicit frame of reference 'sociology' made degrees of consensus appear high, which would receive a completely different rank if an interdisciplinary comparison were made. Secondly, the sociology of science concept of a 'paradigm' had to be disassembled into those components known to the respondent with which he would supposedly be able to associate convergence or divergence of opinions. Thirdly, if we wanted to check out on the assumption of an interplay between consensus and actual procedure in the research process independent of and irrespective of the discipline involved, the concept of consensus was not a priori to be linked with any discipline[13]. And finally a possibility had to be found to distinguish between the forefront of scientific events, which presumably is controversial in every field, and the potentially consentient decision making process on subjects which have been under discussion for a longer period of time.[14] The definite step towards operationalizing the concept comprises a series of questions on the extent of consensus as to different paradigm components, which once refer to the most competent speciality field of the respondent (supposedly in a state of potential consensus), the second time to the concrete subject the respondent has been studying lately (supposedly characterizing the forefront of events). The selected paradigm components range from 'epistemological presuppositions' to 'generally accepted results'. In order to counteract the possible context dependence of the consensus perception mentioned above the question was not on the amount of consensus, but on the *number* of *different positions* existing on the subject within the relevant group of scientists.[15] In the following we shall primarily refer to those consensus items concerning the specialty field of the respondent.

6. Consensus and Methodology

The evidence of an interrelation between consensus and concrete methodological procedures in the research process is based on the question which sources of data collection, which methods of data analysis and what level of theoretical presentation of the results is used by the respondent, and to which extent[16]. In order to examine the role of time the question was posed for a period three years ago as well as for the last year. The individual items were combined into different indices, which all yield the same results as shown in detail in Knorr (1975). Here we shall restrict ourselves to a summary presentation of the results of three indices which classified the respondents according to the methodology used into scientists working '*dialectically*', '*empirically*' and '*analytically*'. All indices filtered out those respondents who had used the *same* procedures 'three years ago' as well as 'during the last year'. As a criterion for distinguishing the first two orientations mentioned Index 1 mainly used the question as to whether the social scientist's argumentational basis consisted of arguments stated in texts and historical sources, or of primary data collected by observation, experiment or interview.[17] Index 2 differentiated between dialectical, empirical and analytical orientations exclusively by drawing from the methods of data analysis used by the respondents during the last three years[18]. Index 3 was produced on the basis of Index 2 and was combined with the variable discipline in that the individual methodological orientations were aggregated only for those disciplines which were typical for them[19].

The indices elaborated were tabulated with the various consensus-variables while controlling for the existence or non-existence of practical research experience[20] on the part of the respondent, and the level of significance as well as the degree of association were calculated[21]. Thereby the hypothesis was confirmed that the respective methodological procedure correlated with the consensus perceived; in addition, we found interaction effects such that the significant relations mentioned were only observed, with some exceptions, when the respondent already had some practical research experience. The following figure illustrates for Index 1 the interrelation between methodological orientation and consensus perceived in the respective specialty field:

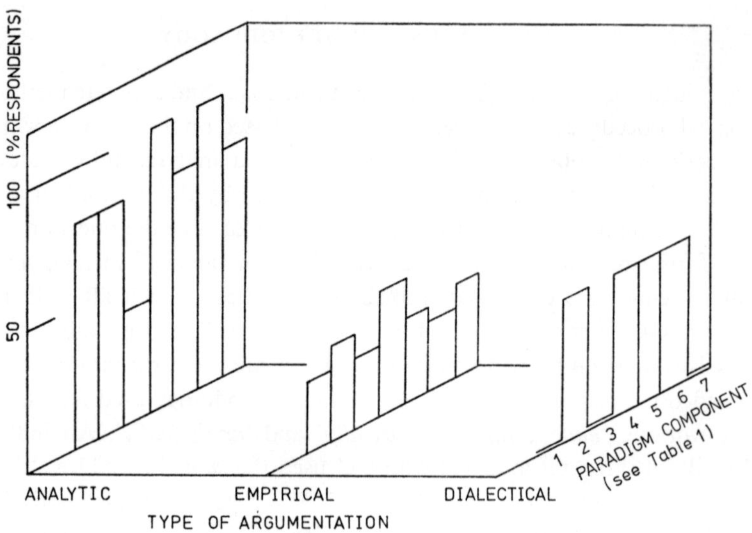

Fig. 3. Percentage of representatives of a specific methodological orientation (or type of argumentation) and consensus (Index 1).

Figure 3 shows for all items on consensus collected that the percentage of social scientists who for their specialty field claimed a high degree of consensus on specific fields of study[22] increases significantly from the adherents of 'dialectical' or 'empirical' methodologies to the scientists proceeding in a formal-mathematical ('analytical') manner. Thus not a single social scientist proceeding 'dialectically', whereas 80% of the formal type claimed consensus on the central problem of their field of study. Or: only 33% of the supporters of an 'empirical' methodology, but 80% of the adherents of analytical procedures claimed consensus on the epistemologically 'proper' pre-suppositions of their specialty field or on the acceptance of methods and techniques respectively. Even for the theoretical approaches, the component of paradigms of social science with the evidently worst results, the percentage of social scientists proceeding in a formal – mathematical manner who perceived consensus on relevant theories is still three times higher than that of the scientists proceeding qualitatively (namely 40% as compared to 0%). The last mentioned share hardly increases for social scientists proceeding empirically and drastically confirms – in the self-evaluation of the social scientists – the well-known theoretical disorder of the disciplines in question. The percentage of respondents who claimed consensus on theoretical approaches is also the only one remaining below the 50% limit. There follows a summary of the associations between methodological orientation and consensus as yielded by all 3 indices (see Table I).

TABLE I

Degree of interrelation (Cramér's *V*) between perceived consensus and research methodology

Consensus in the specialty field with regard to:	Index 1		Index 2	Index 3
	N=31 Cramér's *V*		*N*=51 Cramér's *V*	*N*=17 Cramér's *V*
1. Central problem of the field	.47**	(.04)	.31*	.43
2. Epistemological presuppositions	.35*	(.30)	.29	.43
3. Theoretical approaches	.20	(.06)	.22	.39
4. Priority research problems	.41*	(.31)	.34**	.70**
5. Accepted techniques and methods	.39*	(.07)	.29	.32
6. Most promising techniques and methods	.57**	(.12)	.43**	.59*
7. Accepted results	.41*	(.08)	.32*	.65**

* significance <.10
** significance <.05

Table I presents the coefficients under the condition of practical research experience. The figures in brackets for Index 1 show the equivalent coefficients for respondents without practical research experience. A comparison of the two kinds of coefficients for index 1 shows (except for variables 2 and 4) that significant associations between perceived consensus and research methodology obviously are obtained only for those respondents who already have of practical research experience.

In other words, whether or not opinions are perceived as crystallizing into consensus or dissensus depends on whether practical decisions in the context of *comprehensive* research tasks have been *coped with.* Since practical research experience has been operationalized by means of the research 'projects' a scientist has completed, the wholeness and task character implied by the concept of 'project' stresses the importance of the role of entire series of interrelated research steps.[23]

7. The Role of the Discipline

It was obvious that the interrelations obtained should first of all be checked with respect to the variable 'discipline'. Within the framework of this study the analysis of the role of the discipline confronts the difficulty of obtaining significant results in spite of the drastic reduction of the cell sizes linked with index construction and multivariate fragmentation. In the following nevertheless we summarize some of the results obtained:

(a) No significant relations were observed between consensus claimed and the field originally *studied* by the respondents; a partly significant, but not very strong, interrelation between perceived consensus and the discipline *practiced* at present, and a slightly stronger interrelation when these two fields *coincided*[24].

(b) If the responses to the consensus items are aggregated to an overall consensus index the following classification of disciplines[25] with a high degree of consensus is obtained (see Table II).

This order with slight variations is also confirmed for the other measures of consensus. It surprises on the one hand by the unexpectedly consolidated picture of history (or rather: modern history) and on the other hand by its agreement with the results of a survey carried out in 80 U.S. university institutes, a survey which, albeit being operationalized in a completely different manner, resulted in the same ranking of economy

TABLE II
Classification of the disciplines with respect to the percentage of respondents claiming consensus

Discipline	Respondents claiming consensus, in %	N
History	43.5	23
Economy	33.7	101
Business administration	28.8	66
Psychology	24.5	49
Sociology	17.1	70
Paedagogics	14.9	47
Political science	10.5	19
Urban and regional research	7.4	27
Significance	.01	
Cramér's *V*:	.21	

before psychology, sociology and political science and placed these social sciences all together behind physics, chemistry and biology (Lohdal and Gordon, 1972: 60)

(c) The interrelation between consensus and methodological procedure is generally stronger than between consensus and discipline[26]. Since in general the contingencies between methodology and consensus are maintained *within* the individual disciplines too, they can*not* be explained by the effect of the discipline. However, deviations between the contingency coefficients in the various disciplines indicate an interaction effect.

(d) It is important that no significant interrelation exists between the evaluation of the cognitive capacity of their discipline by the respondents themselves and the degree of consensus claimed, not even then when the effect of practical research experience is controlled for[27] (see Table III).

If the opinions of the respondents are accepted as opinions of experts and hence as the best possible measurement of the *level of development* of the respective discipline, the low or missing contingencies imply that the fundamental thesis of Kuhn's theory on the interrelation between consensus and cognitive capacity in a discipline is not confirmed here. This would result in another, though negative, evidence in favour of our assertion of the *primary* importance of the structure of argumentation and the corresponding decisions in the research process. However, apart from

TABLE III

Interrelation between consensus and the evaluation of the state of development of the respective discipline under the condition of practical research experience made by the respondents

	Cramér's V: (N=112)	(A)
1. Central problem of the field	.05	33%
2. Epistemological presuppositions	.18	21%
3. Theoretical approaches	.11	13%
4. Priority research problems	.09	38%
5. Accepted techniques and methods	.10	31%
6. Most promising techniques and methods	.18	31%
7. Accepted results	.16	19%

(A): The percentage of those scientists who assert that their discipline has either 'not yet found a fruitful approach for coping with their specialty field' or that 'it is at present going through a fundamental crisis', but who nevertheless claimed a high degree of consensus.

the problem whether our question was an adequate operational instrument it is not at all clear whether the state of development of a discipline can adequately be measured by the opinions of the respective representatives.

Taken together, the results presented with respect to the discipline confirm our initial assumption that it is primarily the *methodological procedure* and the corresponding structures of argumentation and justification and not the discipline that we have to turn to if we want to explain the amount of consensus observed. However, it is likely that the discipline plays a not unimportant role as a context-variable influencing, through the type of its research object, the possibility of instrumentalizing methodological procedures for achieving cumulative knowledge.

8. The Role of Practical Research Experience

As a complement to the role of the discipline we considered it necessary to carry out some further analysis on the unexpected importance of practical research experience. With regard to our original result, namely that the interrelation between methodological procedure and consensus depends on the existence of practical research experience of the respondent, we obtained the following additional results:

(1) There is possibly a *direct* influence between practical research experience and consensus in that the perceived consensus increases with the growth of experience.[28]

(2) The interrelation mentioned can*not* be explained by the effect of the discipline. If the factor discipline is controlled, the coefficients in many cases increase, which is an indication for an original suppressor-effect of the discipline (see Table IV).

TABLE IV

Interrelation between practical research experience and consensus[29] within the individual disciplines

Discipline	Cramér's *V*:	*N*
Business administration	.32	44
Economy	.47	34
Psychology	.35	26
Sociology	.25	10
Modern history	.59	14
Political science	(.55)	7

Due to low class frequencies the results of some disciplines can only be interpreted as a tendency.

(3) Furthermore the interrelation mentioned can*not* be explained by the *age* of the respondent. Whereas a tabulation of the respective consensus items with age did not yield any significant interrelations[30] the association between research experience and consensus was maintained within the individual age groups.[31]

Taken together, the results confirm that it is the actual involvement in chains of practical research decisions and argumentations and not the mere knowledge of certain types of procedures that is decisive for the development of an interrelation between consensus and methodology. We can throw some light on this result by pointing to the fact that structures of argumentation and justification can obviously only be effective if they are incorporated into actual practical decisions. On the other hand, it is not perfectly clear to us why a several year's use of a certain methodological procedure corresponding to a specific type of argumentation does not result in the same interrelationship with consensus as research experience in terms of projects does.[32] Ultimate clarification of this aspect will have to wait for some more analysis.[33]

9. Expectation Structure and Methodology

Our hypothesis with regard to the structure of expectations was based on the following considerations: if the degree of consensus constitutes a function of the possibility of trivializing methodological decisions and this again ties up with the structure of argumentation processes, it is obvious that the expectations of the individual scientist underlying his results reflect these structural conditions – mediated and enhanced by the scientific control and reward system. Accordingly, representatives of a structurally 'open' methodology should be significantly more oriented towards argumentative dialectics than those scientists who apply structurally closed procedures of justification. In order to operationalize our assumption the scientists were asked which type of (positive) criticism they would welcome most; as response categories we offered the dimensions 'problem solving', exploration of 'fresh lines' of research, 'critical' discussion and 'insight' as well as methodical 'implementation' and 'argumentative' evidence.[34]

For the analysis the above mentioned indices of methodological procedures (see Table I) were applied. The results are summarized in the following table; for every dimension of the expectation structure the methodological orientation with the highest class frequency is presented. It is important that the interrelations are again only observed under the condition of practical research experience (see Table V).

As can be seen from the following table the adherents of a '*dialectical*' qualitative methodology show the expected orientation towards *critical disproof*, doubt or exposure as well as towards the dimension of '*insight*'. On the other hand, the adherents of an '*empirical*' quantitative procedure primarily go for the items encompassed by the concepts '*fresh lines*' and '*implementation*', and the social scientists proceeding '*analytically*' consistently prefer the dimension '*problem solving*'. Inconsistencies are only observed with respect to the dimension 'argumentation', which the re-

TABLE V

Methodological procedure with highest class frequency per dimension of expectation

Dimension of expectation	Index 1	Index 2	Index 3
1. 'Criticism' and 'insight'	D	D	D
2. 'Argumentation'	D	A	D
3. 'Problem solving'	A	A	A
4. 'Fresh lines' and 'implementation'	E	E	E
Cramér's *V*	.30	.44	.49
Significance	–	.009	.12
N	(26)	(45)	(21)

'D': Representatives of a dialectical qualitative methodology account for the highest percentage of this dimension of expectation

'A': Representatives of an analytical methodology account for the highest percentage

'E': Representatives of an empirical quantitative methodology account for the highest percentage.

spondents possibly interpreted in different ways[36]. The following block diagram gives an impression of the respective percentage of respondents adhering to a certain methodological orientation who decided in favour of items of the first dimension (Figure 4):

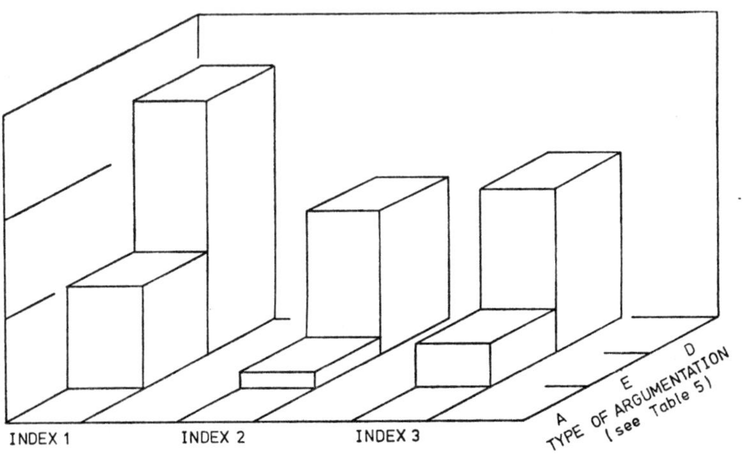

Fig. 4. Percentage of Respondents adhering to a certain methodological orientation (type of argumentation) and choosing critical discussion and 'insight' items.

Finally one more result is important for illustrating the role of the discipline with respect to structures of expectation: The degree of association between the latter and methodological procedures *within* the individual disciplines in general increases slightly, which clearly proves that it is again the *practical research procedures* and not the discipline that play the key role in determining expectation structures (see Table VI).

However, due to the low class frequencies the results can again only be interpreted as a tendency.

TABLE VI
Interrelation between expectation structure and research methodology within various disciplines (Index 2)

Discipline	*V*-Coefficient	*N*
Business administration	.58	15
Economy	.49	31
Psychology	.54	16
Sociology	.55	22
Political science	(.58)	(4)
Modern history	(.66)	(9)

10. Summary and Conclusion

The present paper has tried to explain the amount of consensus in a given field of social science by the *structures of argumentation and justification* entrenched in the methodological procedures used. It has been shown that the assumption of cognitive disunity among social scientists must be modified in that – depending on the methodological procedure used – extreme consolidation and extreme disagreement coexist side by side. The high degree of consensus can have two meanings: either the relatively extensive consolidation of areas proceeding formally is taken as an indication for the '*mature*' level of development of these areas in contrast to all the other fields of social science; or one accepts the thesis of the *potential gap* between consensus and substantial growth of knowledge. We asserted that the proved consolidation does though speak in favour of a cumulative selection leading to easily closed 'trivial' structures of argumentation and justification, yet does not guarantee substantial cumulativity. The latter depends on the functioning of *endogenous methodological control*, which

– in addition to the unresolved problem of measurement and meaning – does not seem to operate effectively in those areas of social science which ground their results extensively in theoretical terms. Instead of the oriented and directed falsificationism of orthodox normal science we are confronted in these areas with verification circles which pave the way for the production of a variety of divergent and dissonant results. The complexity and diversity produced is reduced by *current societal patterns of interpretation*: by means of the language of social science that is rooted in everyday language, and by means of social science theory formation that is rooted in everyday knowledge, they guide and govern the expectations of the scientists. Nevertheless, this influencing process is not unilateral. Both the possible accumulation of empirical generalizations in the realm of observational terms as well as the systematic articulation and elaboration of theoretical concepts and systems of knowledge in the area of individual disciplines guarantee feedback effects. It will be the task of further studies to examine these.

Institute for Advanced Studies,
Vienna

NOTES

* Parts of the results presented in this paper have been published in the special issue 'Wissenschaftssoziologie' of the *Kölner Zeitschrift für Soziologie und Sozialpsychologie*, edited by Nico Stehr and Erwin K. Scheuch (1975).

[1] A system of norms integrated with the gratification system should imply that by allegiance to the norms a simultaneous maximization or at least satisfaction of one's personal interests can be achieved. However, the cooperative model of society of the functionalist approach underlying the assumption of successfully functioning norms in the above mentioned sense has been strongly criticized if not completely abandoned. Therefore, today, a normative foundation of scientific progress where the latter is based on socialization, operated by motivation and backed by prompt gratification is today less plausible than ever before.

[2] The triviality of decisions is a matter of degree. The concept is to indicate that the border-case of perfect triviality of a decision actually constitutes a reduction to a non-decision.

[3] The concept of 'argumentation' has been recently emphasized by Böhme (see his contribution to the same volume) as characteristic of scientific activity. But while Böhme analyzes the existence and function of interrelated arguments we are, by leaning heavily on Toulmin (1958), concerned with analyzing the *structure* of argumentation processes.

[4] As compared to the Hempel-Oppenheim reconstruction of the concept of explanation Toulmin's layout of arguments has the advantage of greater generality and therefore receives priority in our discussion.

[5] For a more detailed investigation into the meaning of different terms in methodological procedures see also Knorr (1974).
[6] The concept of social class belongs to the set of directly accessible concepts insofar as it is based on a conceptual combination of profession, income and education. The problems of classification and categorization, which appear regularly when measuring variables like that of 'profession' do not in principle rule out the possibility of measurement. The question is whether there is a sufficiently unambiguous everyday classification such that there are clearly identifiable links between 'facts' and 'labels'.
[7] Before returning to our initial assumption we would like to indicate that the distinction between studies based on observational terms and those based on theoretical terms is not equivalent with the distinction between 'descriptive' and 'analytical' (or 'theoretical', 'systematic') procedures. This, for instance, is shown by the 'purely descriptive' ethnographical tribe-monographs, which usually apply inadequate theoretical concepts, i.e. concepts borrowed from a different culture, such as 'economy', 'religion' or 'totemism'.
[8] We maintain that the structures of argumentation and justification are rooted in the methodological procedures of the individual fields of science insofar as these procedures circumscribe the codifiable and standardizable aspects of scientific activity to which the structures we describe doubtlessly belong.

We are using the term 'methodology' instead of 'methods' in order to stress the importance of the general approach or the structure of the procedure as opposed to the concretely elaborated versions.
[9] This assumption does, of course, not hold true in an unqualified manner: For instance, the use of analytical procedures (i.e. quantitative or formal procedures) must present a real alternative within the reach of the individual scientist if selective choice is to work. Besides, one can imagine a series of constraints which would still prevent such a selection. Here further analysis is urgently required.
[10] We borrow the terms convergent-divergent from the work of Hudson (1966, 1968).
[11] Compare Lakatos (1971) for an epistemological analysis of protective attitudes towards scientific results.
[12] Leinfellner (forthcoming) develops the argument that once a science has reached the theoretical state it becomes nearly immune to external influences while internal determinants begin to regulate the development of this science.
[13] For the reasons mentioned we could not demand from the respondent, like Lohdal and Gordon (1972) did, a rankordering of disciplines according to the degree of consensus on paradigms; nor could we use simple Likert-scales in order to determine the extent of consensus.
[14] Accordingly, the questions on consensus concerning the problems studied most recently generally ought to yield less consensus than the questions on the specialty field – which indeed turned out to be the case.
[15] The exact wording of the questions on consensus concerning the specialty field of the respondent is: "If you consider the relevant literature on this (meant is: your most competent) *specialty field*, how much *agreement* is there with respect to: – the *problem* which today ought to be considered as central for the field; – the *epistemological* presuppositions of the field; – the *theoretical* approaches which ought to be applied for exploring the field; – those *problems* which ought to be granted top priority; – those techniques and methods that can be considered as generally *accepted* – the most *promising* and useful methods for exploring the field; – the *results* which so far can be

considered as generally accepted." The respondent was asked to indicate whether the topic in question was 'practically not discussed', whether there was 'farreaching unanimity of opinion' on it, whether there were 'two very different basic positions', or 'several different positions'.

[16] The exact wording of the question was: "Please state to which extent you have used the following research approaches and methods (a) during the last year and (b) three years ago"; the response categories 'never', 'in some studies' and 'in practically all studies' were offered. The sources of data collection were: texts and historical sources, existing statistics, existing data from projects, observational procedures, questionnaires, interviews, experiments. The methods of data analysis offered ranged from the 'interpretation of texts and qualitative analyses' to 'purely mathematical procedures', the categories for the degree of theoretical presentation of the results from 'description' to 'axiomatic theories'.

[17] Or rather: Those respondents were classified as scientists proceeding 'dialectically' who had used 'texts and historical sources' in practically all their studies, possibly had also referred to existing data from projects and existing statistics, but had never themselves collected data by means of observations, experiments or interviews. On the other hand, those were considered 'empirical' scientists who in practically all their studies had made use of the latter methods of data collection.

The methods of data collection present the difficulty that they do allow for a distinction between data-oriented and literature-oriented procedures, but none between the retrieval of 'facts' and that of 'opinion'.

[18] Those respondents were classified as scientists proceeding 'qualitatively' or 'dialectically' who in 'practically all their studies' had used interpretations of texts and qualitative analyses, but never statistical or mathematical procedures. Those respondents who in practically all their studies had used contingency analyses, mean values, significance tests, scaling procedures, regression statistics or multivariate procedures were placed in the category 'empirical-quantitative' whereas those scientists were considered as proceeding on a formal-mathematical ('analytical') basis, who in practically all their studies had used purely mathematical procedures such as differential equations, matrix algebra, and the like.

[19] Or rather: The qualitative methodology of index 2 (see Table I) was attributed to the historians and political scientists, the empirical-quantitative one to the sociologists and psychologists and the formal-mathematical procedures to the economists and business managers. This classification is based on the appropriate correlations between these procedures and the disciplines mentioned.

[20] The practical research experience was operationalized by the question as to how many projects the respondent had concluded in his research unit. Thereby the concept of project did not refer to one distinct methodological procedure. The restriction to the last three years was made in order to establish a correlation between the respondent's specialty field of today and the consensus perception. Projects carried out jointly with others were not included in the analysis because of the danger of contamination by opinion formation processes in the project group.

[21] Apart from the X^2 test of significance Cramér's V was generally used as a measure of the degree of association, since compared to the Pearson's contingency coefficient and Tschuprow's T it has the advantage of an upper limit 1. The formula for V^2 is:

$$V^2 = \frac{X^2}{N \min(r-1, c-1)}$$

r meaning the number of rows, c the number of columns and min $(r-1, c-1)$ the smaller of the two numbers $r-1$ or $c-1$. See Cramér (1961).

The X^2 test was carried out in spite of the fact that we surveyed the complete universe of Austrian social science units. Since our questions aim at generalizations which apply to the social sciences as a whole we must consider our respondents as a sample from a fictitious totality, which reintroduces the necessity of performing significance tests. (see Pfanzagl, 1967).

22 The response categories of the consensus questions were dychotomized in order to prevent low class frequencies in the cells. The categories 'practically not discussed' and 'farreaching unanimity of opinion' were qualified as 'consensus', whereas 'two very different basic positions' and 'several different positions' were qualified as 'dissensus'.

23 The interrelations between methodology and consensus with respect to the epistemological pre-suppositions as well as with respect to the problems which require top priority obviously exist irrespective of the practical research experience. Since in the case of the epistemological foundation we have to cope with a meta-problem that can be easily ignored in substantial scientific work, the irrelevance of practical experience does not surprise. The additional irrelevance with respect to the 'priority problems' might make it seem advisable to draw a dividing line between paradigm components 'external' to substantial science (meta-aspects, aspects induced by society, such as for instance the priority problems) and science-internal components, i.e. components which depend on research practice.

24 The distinction between the discipline studied, the discipline practiced and the same discipline for both cases was made in order to take account of the different educational preconditions for social scientists, who are often working in fields that have been institutionalized and professionalized only recently, such as political science. The assumption was made that the correlation between discipline and consensus would be generally higher when the discipline studied and the discipline practiced coincided; this proved to be true. For the individual coefficients see Knorr (1975).

25 If not specified differently the discipline practiced was used.

26 The V-coefficients between the discipline practiced and the consensus-variables 1 to 7 in the case of practical research experience are: .30, .29, .19, .18, .22, .26 and .22; a significance of .05 or .10 respectively was achieved in the first two cases only.

27 The 'cognitive capacity' of a discipline was operationalized by the question whether in the respondent's view his discipline 'has not yet found a fruitful approach for coping with its field', is 'at present going through a fundamental crisis', 'is progressing as a whole, but showing disruptions in some fields', is developing 'continuously' or is at present in a 'period of strong progress'.

28 As the following contingency coefficients and significances (in brackets) presented in the order of the consensus-variables indicate, the interrelations are in general not very high: .16 (.01); .18 (.01); .16 (.01); .24 (.001); .08 (--); .14 (.05); .13 (.05).

29 The ratings of consensus-items were aggregated to an overall measure for all mentioned consensus-variables; the table shows those respondents for whom the discipline practiced today coincided with the discipline studied.

30 V-coefficients between the age of the respondents and the consensus-variables in the order of these variables: .13, .10, .09, .09, .07, .07.

31 The V-coefficients in the various age groups (=years of birth) for consensus-variables 1–7 run as follows:

Age group up to 1915:	.20,	.35,	.18,	.19,	.29,	.31,	.32
Age group 1916–1925:	.18,	.54,	.19,	.23,	.14,	.31,	.23

Age group 1926–1935: .45, .47, .30, .34, .23, .36, .15
Age group 1936–1945: .19, .13, .17, .25, .13, .10, .17
Age group 1946–1955: .19, .13, .25, .26, .15, .18, .08

Significances of .05 or less were mainly obtained in the strongest age groups 3 and 4. The differences between the age groups indicate interaction effects which call for further investigation but can be neglected for the present.

[32] This is due to the fact that the *V*-coefficients frequently tend to increase when the diachronic indices reaching over three years are used as opposed to the synchronous indices, yet the formation of significant coefficients on the whole depends on whether the respondent has at all conducted research projects.

[33] However, the importance of finished research tasks can be linked with the fact that it is only within such tasks that the chains of argumentation and justification can be concluded with a chance of achieving consensus.

[34] The exact wording of the question was: "Assuming you receive a (rather positive) criticism in a respectable journal, which comments on your scientific work would you be most pleased about?"; the respondent was supposed to choose up to three of the following possibilities of response: "Brilliant argumentation..." (1), "Excellent counter-evidence...," (2), "Extraordinary approach..." (3), "Problem brilliantly solved..." (4), "Minute details..." (5), "Ingenious analysis..." (6), "Aggressive exposure..." (7), "Fresh line of research..." (8), "Attractive insight..." (9), "Unassailable implementation..." (10), "Pitiless doubt..." (11) and "Finally precise clarification..." (12). As operationalizations of the same aspect the items 4 and 12 were drawn together to the dimension 'problem solving', 3 and 8 to the dimension 'fresh line', 2, 7 and 11 to the dimension 'criticism', 6 and 9 to the dimension 'insight'.

[35] Possible divergent interpretations are: 'correct evidence and deduction' as well as 'good presentation'.

BIBLIOGRAPHY

Baldamus, W.: 1973, 'Relevanz und Trivialität in der soziologischen Forschung', *Zeitschrift für Soziologie* **2** (1), 2–20.

Carnap, R.: 1959, 'Beobachtungssprache und theoretische Sprache', in *Logica Studia Paul Bernays dedicta*, Neuchatel, Suisse, pp. 32–44.

Cole, St.: 1975, 'The Growth of Scientific Knowledge: Theories of Deviance as a Case Study', in L. A. Coser (ed.), *The Idea of Social Structure*, Harcourt, Brace and Jovanovich, New York.

Cramér, H.: 1961, *Mathematical Methods of Statistics*, Princeton University Press, Princeton, N. J.

Feigl, H.: 1963, 'De Principiis Non Disputandum...?', in M. Black (ed.), *Philosophical Analysis. A Collection of Essays*, Prentice-Hall, Englewood Cliffs, N.J.

Fodor, J. A. and Katz, J. J. (eds.): 1964, *The Structure of Language: Readings in the Philosophy of Language*, Prentice-Hall, Englewood Cliffs, N.J.

Freese, L.: 1972, 'Cumulative Sociological Knowledge', *American Sociological Review* **37** (August), 472–482.

Habermas, J.: 1968, *Technik und Wissenschaft als Ideologie*, Suhrkamp Verlag, Frankfurt am Main.

Habermas, J.: 1971, 'Vorbereitende Bemerkungen zu einer Theorie der kommunikativen Kompetenz', in J. Habermas and N. Luhmann (eds.), *Theorie der Gesellschaft oder Sozialtechnologie*, Suhrkamp Verlag, Frankfurt am Main, pp. 101–141.

Hudson, L.: 1966, *Contrary Imaginations*, Methuen, London.
Hudson, L.: 1968, *Frames of Mind*, Methuen, London.
Knorr, K. D.: 1974, *Toward a Positive Epistemology*, Paper presented at the VIII World Congress of Sociology, Toronto (Mimeo).
Knorr, K. D.: 1975, 'Konsensus und Kumulativität in den Sozialwissenschaften', *Kölner Zeitschrift für Soziologie und Sozialpsychologie*, special issue 'Wissenschaftssoziologie', Westdeutscher Verlag, Opladen.
Kuhn, Th. S.: 1962, *The Structure of Scientific Revolutions*, University of Chicago Press, Chicago; second edition, enlarged 1970.
Lakatos, I.: 1970, 'Methodology of Scientific Research Programs', in I. Lakatos and A. Musgrave (eds.), *Criticism and the Growth of Knowledge*, Cambridge University Press, Cambridge.
Lammers, C. J.: 1974, 'Mono- and Polyparadigmatic Developments in Natural and Social Sciences', in R. D. Whitley (ed.), *Social Process of Scientific Development*, Routledge and Kegan Paul, London.
Leinfellner, W.: 1974, 'Wissenschaftstheorie und Begründung der Wissenschaft', in W. Leinfellner *et al.* (eds.), *Forschungslogik der Sozialwissenschaften*, Bertelsmann, Düsseldorf.
Leinfellner, W.: forthcoming, 'Interne und externe Kriterien der Wissenschaften und der kybernetische Charakter des wissenschaftlichen Fortschrittes', in H. Strasser, K. D. Knorr und H. G. Zilian (eds.), *Beiträge zur Theorie der Wissenschaftssteuerung*.
Lodahl, J. B. and Gordon, Gerald: 1972, 'The Structure of Scientific Fields and the Functioning of University Graduate Departments', *American Sociological Review* **37**, 57–72.
Luhmann, N.: 1974, 'Einführende Bemerkungen zu einer Theorie symbolisch generalisierter Kommunikationsmedien', *Zeitschrift für Soziologie* **3** (3), 236–255.
Pfanzagl, J.: 1967, *Allgemeine Methodenlehre der Statistik*, Vol. 1, Walter de Gruyter & Co., Berlin.
Sandkühler, H. J. (ed.): 1975, *Marxistische Wissenschaftstheorie. Studien zur Einführung in ihren Forschungsbereich*, Fischer Athenäum, Frankfurt am Main.
Schütz, A.: 1932, *Der sinnhafte Aufbau der sozialen Welt*, Wien.
Stehr, N.: 1974, 'Paradigmatic Cristallization: Patterns of Interrelations among Areas of Competence in Sociology', *Social Science Information* **13** (1), 119–137.
Storer, N. W.: 1967, 'The Hard Sciences and the Soft: Some Sociological Observations', *Bulletin of the Medical Library Association* **55**, 75–84.
Toulmin, St. E.: 1958, *The Uses of Argument*, Cambridge University Press, London.
Winch, P.: 1958, *The Idea of a Social Science*, Routledge and Kegan Paul, London.
Zuckermann, H. and Merton, R. K.: 1971, 'Patterns of Evaluation in Science: Institutionalization, Structure and Functions of the Referee System', *Minerva* **9**, 66–100.

H.ROM HARRÉ

IMAGES OF THE WORLD AND SOCIETAL ICONS

The problems to which this conference is addressing itself of scientific progress, of conceptual development, of 'external' influences on science, of the relationship between science and the social order, cannot be properly enquired into unless an adequate theory of scientific thinking and its relation to its subject matter, as well as an adequate theory of society, has been developed. I hope to show in this paper how theories of science and theories of society which might be adequate to each, must develop together. The inter-linked theories I shall be proposing involve a number of new and unfamiliar ideas, both about the nature of scientific thinking and about the world to the understanding of which that thinking is directed, as well as the re-emphasis of some well known ones.

Many of the difficulties that beset the philosophies of science and society derive from the continued acceptance of unthinking assumptions about the nature of science and the nature of the social world. It is still often assumed that the world to which science is directed is a flux of atomistic events independent of the conceptual structures which we bring to its understanding. At the worst, this point of view generates a corresponding atomism in language and concepts in that the atoms of the world are thought to be matched by linguistic or conceptual atoms, the notorious independent empirical predicates of logical empiricism. There is a sibling theory of the social world supposed to exist in some sense 'out there' beyond the experience of individual human beings whose task in subjugating that world to concepts is conceived of on precisely the same model as the positivistic theory of a science of nature. I have sufficiently examined and criticized logical empiricism (positivism) as a philosophy of science in other places, and I shall not detain the conference by examining it further. I proceed, then, to the positive construction of a more adequate theory of science linking it to the most reasonable theory of the social world I know of, namely that of ethnomethodology. By this means I hope to provide a schema for handling the problem to which this conference is directed.

Knorr, Strasser, and Zilian (eds.), Determinants and Controls of Scientific Development, 257–283.

1. A THEORY OF SCIENCE

At the back of any theory of science lies a theory of concepts – the means of scientific thinking – and a theory of individuals and properties – the subject-matter of that thinking. Recent theories of science in the empiricist tradition have assumed that a theory of concepts adequate to science can be elaborated in terms of a language model derived from elementary predicate logic and that a theory of properties need proceed no further than the elements of sensory experience. Neither theory can stand close examination and I shall proceed to construct new theories.

1.1. *A Theory of Concepts*

Not only is it not at all clear that all the concepts essential to a scientific theory can be carried by simple, independent predicates and logical constructions out of them, I shall show that, at least, it is essential to invoke concepts in scientific thinking that are not naturally expressed verbally. These will prove to be particularly prominent when thinking is directed towards structures. I shall be concerned, unashamedly, with a theory in which the concepts and properties I shall be suggesting invoke a world of active agents, though I do not wish to be understood to be promoting any sort of Pan-psychism or *anima mundi* theory.

Our problem, then, is to identify in scientific thinking some vehicles for thought, bearers of concepts, which are capable of forming the ground of thinking in structural and dynamic terms. For the bearers of such concepts I shall introduce the term 'icon'. I prefer this, now, to the much-abused notion of model, which has recently fallen into dissuetude since it has been used carelessly both for sentential structures and for the objects which are described in such structures. Furthermore, the use of models in the latter sense has been seriously misunderstood in that some critics have supposed that the introduction of real or imagined model-objects was in contrast to a conceptual theory of thinking. Nothing could be further from the intention of those who have directed their attention to such models since they are intended to express the manner of real scientific thinking when it is concerned with structures and dynamic or power concepts (Harré, 1970). They are, thus, no less the bearers of concepts than are words. However, not wishing to embroil myself in what is largely a resolution of a mere misunderstanding, I shall speak in this paper only of icons.

An icon is a non-verbal image (often a pictorial or quasi-pictorial symbol) which is a vehicle for thought or a bearer of a concept, but, unlike words and such constructions, is projectively related to reality. The concept of an icon will become clear only when I have explained what it is for an entity to be projectively related to reality.

When a projective relation exists, two objects are related by structural isomorphism in that the elements of each correspond according to some rule and the invariants by which the integrity of the structures of each are maintained also correspond according to rule. An icon is a member of a pair of objects related by such an isomorphism. In scientific thinking the object to which the icon is related need not be a concrete entity. It may be an abstract entity, or entities equally related by systematic relationship. This can be made clear by comparing the iconic character of a wiring diagram for, say, a television receiver, which is an object related to another object of the same ontological character, with the system of military insignia, with which for example in the British Army, three chevrons are related to two chevrons as the rank of sergeant is related to the rank of corporal. The relation of the ranks is an abstract relationship exemplified in a wide range of relations between those individuals who have that rank. The icons which are sewn on the sleeves of the appropriate uniforms show in a non-verbal projective relation the structure of the abstract relation.

An adequate theory of concepts must, at least initially, assume that conceptual thinking is carried by both verbal and iconic means. I shall not assume lightly that there is necessarily any verbal equivalent to a specific icon. We shall see instead that the central role that the conceiving of icons performs in scientific thinking, suggests rather that the icon is an indispensable element in that thinking.[1]

1.2. *A Theory of Properties*

The theory of properties which I shall sketch considers a property to be assigned to an individual. From the scientific point of view a property is treated as the cause of a certain manifestation as a quality or pattern in experience. In the language of philosophical logic, such a property will be a power or a grounded disposition (Fisk, 1974). For example, considered naively, the property of redness as assigned to an object, is nothing but the quality that object shows when viewed by a human being, but considered scientifically, the property of redness is the power or grounded disposition of an object to cause such an experience. Thus, when the property of redness enters a scientific context, it is no longer a simple quality but a complex property. It is clear that, for a scientist, redness is a grounded dispositional property or power of many, but not all, red things. I think it is indisputable that this is how properties are understood in science. Are we obliged to accept this way of understanding? Philosophers have sometimes disputed the propriety of the scientific understanding of properties in favour of a theory of properties which identifies them with the elements of experience, particularly qualitative elements. This is the theory of phenomenalism. Despite tremendous efforts, it has not proved possible to construct a satisfactory phenomenalism, one which does not collapse into unconnected, individual, experiential worlds of atomistic elementary experiences. I have no wish to add to the mountain of objections which has been built up to that view. I shall assume that the only theory of properties worth advancing is that which is current in the sciences.

A property, then, is to be considered as a power or grounded disposition: that is, to say that an object is red, speaking scientifically, is to say that it has such an internal structure, and, if the quality is used to identify the object, such an intrinsic nature, that it looks red to well-placed observers. What is the nature of that internal property in which its disposition to look red, or power to produce such experience, is grounded? In scientific thinking there is a remarkable uniformity in the answer to that question. The grounds of dispositions or the natures of things upon which their powers depend are almost always thought of as structures, usually spatial, though not necessarily,[2] of elementary agents. Such elementary agents are themselves the possessors of properties which can again be treated as grounded dispositions or powers leading to further and more refined structural hypotheses. It is my general contention that in actual scientific thinking the property of an object which is thought of as a structure of elementary agents is thought of by means of an icon. So the concept of that property is carried by the appropriate icon. Sometimes that icon will be a physically realizable diagram or picture; sometimes, when the structure and elementary agents are abstract, it will not. And iconic representation may extend to the power of the elementary agents, as well (c.f. Maxwell's fluid images for thinking about potentials).

I can now bring together the theory of concepts and the theory of properties in a brief description of the nature of the predicates required for the assignment or attribution of such properties and can say a word or two about their truth conditions. Such predicates as 'red', conceived as incorporated in a science, will have identical logical form to such predicates as 'charge' or 'potential'; that is, they will ascribe to an individual, either concrete such as a thing, or abstract such as a field, a certain nature which is the cause of the manifestation of the power or disposition as the given quality in experience, so that the logical form of a scientific predicate will be:

If certain unspecified conditions obtain, then a certain effect will be produced by virtue of the nature of the individual involved. Thus, the disposition to present a certain quality or pattern to experience is grounded in an unspecific hypothesis about the nature of the things and substances involved. This logical structure allows us to link the predicates by which properties are ascribed to real things with a sensory language, since, of course, it is required that the properties of real things appear either directly in personal experience as qualities and patterns, or indirectly in the responses of instruments. But this linkage is loose, since of course, the qualities may appear for other than their usual causes. Something may appear red, or hot, or a galvanometer may give a twitch, for reasons other than the standard reasons which are captured in the structure of scientific predicates.

It is possible now to see why one wishes to say that science is the attempt to conceive and describe (or depict) icons of reality under certain constraints, in terms of which dispositions to produce experience can be grounded, and thus the powers of things to act on other things can be explained. It is clear, then, that there is a logical reason for the necessity for thinking with such icons. Since experiental predicates *must* be transformed into grounded dispositional predicates, or power predicates, to have a place in science, thought is forced on to the task of conceiving an unexperienced world in terms of which those dispositions can be grounded and those powers explained.

There is, then, no substance whatever in the idea that science must be in some way or other inductive, growing by accumulation of little bits of knowledge. Generality is achieved in science in the manner conceived by

William Whewell – that is, by the addition of what he called 'an idea' to a field of experience. Whewell's ideas are not concepts corresponding to elements of that experience, nor are they merely logical orderings of it. They are such ideas as that of chemical atom. Clearly, they enable us to identify genuinely similar experiences – that is, similar beneath their skin of qualities – and to identify within experience those patterns which are genuinely general. For example, Darwin's theory of evolution is a great scientific theory because it describes an icon of reality, an icon of biological history and its dynamic process. Empirical evidence enters Darwin's work, not as particulars from which generalizations are drawn, but as accompanying anecdotes which illustrate the capacity of the icon to represent, and thus its description to explain, diverse matters. The world of animals and plants is made intelligible by Darwin's theory, because Darwin's theory describes an icon of the mechanism by which recognised patterns are generated, and previously inchoate experiences come to be seen as intelligible and orderly. It is to this notion of making intelligible that the iconic theory of science is directed. Notice that in special circumstances a mark or sign of such intelligibility is the fact that a description of the pattern generated can be deduced from a description of the generating mechanism, together with a description of the conditions under which it is active. But, according to this view, this is a consequence of the satisfaction of the more fundamental condition rather than being itself the necessary and sufficient condition for the existence of an explanation. On this view, the paradoxes of the deductive conception of explanation are very easily resolved in that there may well be logical forms which allow predictions to be made which our intuition tells us could not count as explanations, and the iconic theory makes it perfectly clear what the ground of such an intuition might be; namely, the absence, in the content of the propositions satisfying that form, of any reference to the structure of elementary agents which generates the pattern upon whose generality the prediction depends.

2. Consequences for Science

2.1. *The Form of a Scientific Problem and Its Solution*

The specific form that a problem and its solution takes, depends upon very broad, general, almost ideological differences in conceptions of knowledge and experience. I have hinted at such differences already, but

I would like at this point to make them a little more precise. In a great many fields of scientific investigation one seems to come across a dichotomy of approaches, which I shall call for convenience, atomism and structurism. (One version of structurism being structuralism).

It is typical of the atomistic ideology to see the world of experience as a flux of independent, non-overlapping, unconnected and passive elementary parts and the dynamics of such a world to be structured in terms of an efficient cause externally imposed. Thus, the explanation of the appearance of a particular elementary unit of experience is to be achieved by reference to a specific antecedent element which is concomitant with it. The only relation, it is assumed, which we can legitimately suppose to obtain between those elements, is one whose ultimate explanation lies in the psychology of the people who experience such concomitances. Such an approach is exemplified in psychology, for example, by the theory of language which is based upon the ideas of Skinner. In that theory, the sentence produced by a person is a linguistic atom, which is under the control of features of the environment which call it forth.

The Skinnerian theory is in strong contrast to the theory of language based upon the ideas of Chomsky and others, in which the sentence which is produced by a speaker is thought of as a structure which has evolved from another, more fundamental structure – the underlying form of semantic representation. Such an idea, when generalized, forms the basis of the competing ideology-structurism. The world process is conceived in terms of the operation of formal causes, that is, of the imposition of structure or pattern on an unstructured, unpatterned base. Such imposition is conceived by structurists as the evolving of an antecedently given structure, under the constraint of certain invariants into the structure or pattern which is manifested, sometimes under a very different appearance, in experience. The nature of this evolution and the manner in which the evolved structure imposes its pattern on the unpatterned base, varies from science to science.

It is within the structurist ideology that the theory of grounded dispositions or powers is most naturally fitted, since the pattern of experience in which the disposition or power of an object is manifested, is derived from structural properties of that object, those very structural properties which are the grounds of the disposition.

It should now be clear that the fundamental form of a scientific problem is 'What is the source of this or that pattern in experience?' And the fundamental form of the answer to any such question will be that the icon of the real world, by which the grounds of the disposition or power to produce those patterns are conceived, contains a plausible template for the production of that pattern in experience. That, of course, is a very abstract way of putting the matter. I can dispel uncertainties best by citing some examples. The pattern of a crystal of diamond is to be explained in terms of the structure of carbon atoms. Indeed, the pattern of crystal of diamond can be shown to be a derivative of the pre-existing structure of the carbon atom. A synchronic pattern such as the anatomy of an organism is the product of a diachronic process by which the elementary parts of that organism are assembled in the pattern by reference to a pre-existing template. This is the form that hereditary explanations now take; so that one synchronic pattern is the source of another. Such patterns are the patterns of things, but there may equally be patterns amongst events, diachronic patterns, the structure of processes. And symmetrically with the explanation of synchronic patterns, so diachronic patterns will be explained by the existence of synchronic templates such, for example, as is exemplified in a punched tape which determines and controls the production of a process. This is how, currently, we understand the patterns to be observed in physiology or in chemical interactions. It seems, then, as if the structural properties of our icons of the real world are central to the form that specific scientific explanations take.

In pursuit of the preoccupation of this conference we must ask whether there is any general feature of icons of reality which mark off those which are likely to prove satisfactory as the bases of scientific thinking. We might look then for the sources of these features in certain constraints upon conceptions of reality. It has been a commonplace since the later middle ages that the construction of scientific theories, whether conceived of logico-verbally as instruments of prediction, or realistically as icons of reality, must come under some constraint since there are certainly infinitely many instrumental theories possible in any field of the phenomena and a finite though large number of icons of reality are certainly possible. One of the most serious objections to the use of purely logical concepts in analyzing scientific thinking is that by deliberately eliminating attention to content from the philosophy of science, this point of view leaves theories constrained only by logical possibility which is hardly a constraint at all, or by some formal criterion such as simplicity, whose application is uncertain. The shift to the iconic theory makes the problem of the identification of constraints more tractible, though it does not, by itself, solve it.

If we regard theories as descriptions of icons of reality produced by the human imagination, it is clear that there must be some account of the constraints upon that imagination, for the human imaginative faculty is well-known for its capacity to generate mere fantasy: and yet, it is plain that the conceptions of reality which scientists have drawn upon from time to time, are not fantasies, though in the end some have been abandoned as unrealistic.

To understand the complex intellectual process involved in the constraints on products of the imagination, I will proceed by first looking at the structure of the thinking that is involved in the creation of particular icons of reality for the solution of particular problems. One must remember that the reality represented is to be causally responsible, usually in the formal cause sense, for the structure of the behaviour or anatomy of some thing or class of things. So the reality conceived must come under the primary constraint of behaving in such a way as to match the behaviour of the real world. We know the way the real world behaves since it is those patterns we have observed. What we do not know is how the real world generates those patterns. Thus, the first constraint which we could call 'behaviour-matching', can be put like this:

> The icon of reality must represent a structure which could produce behaviour similar to that we know the world produce.

But in order for the form and elementary parts of the icon to be satisfactory as a source of content for the theory there must be thought to be a possible reality. We might describe this as 'content-matching'. This leads us to a statement of the second and more constrictive constraint:

> The structure of elementary agents which constitutes the specific content of an icon must be a possible existent.

Thus, the behaviour of the icon is matched to experience and its content is matched to existing conceptions of reality.

The constraints under which the content structure and elements of the icons must come are much more rigorous constraints than the instrumentalists demand for matching behaviour. The constraint can be considered under two headings:

(1) The behaviour of the structure and the elements of the icon should, in general (though there are exceptions) be according to the principles of already accepted science, for example, gas molecules must obey Newtonian laws (Bunge, 1974).

(2) The most pervasive but the least easily examined source of constraint is what is usually referred to dismissively by philosophers as the 'background'.

The background consists of what Toulmin called the Principles of Natural Order (Toulmin, 1961) or what others have called the General Conception of Reality, etc. Since this background is so vitally important to the kind of icons of reality that the minds of a given generation will generate, we must not simply dismiss it in these rather general terms and return to logic-chopping.

Before proceeding to look into the structure of the background, I would like to draw the attention of the conference to the correspondence between the pair of constraints, behaviour and background, and the elements in our theory of properties, namely the disposition and its ground. The behavioural component corresponds to the disposition; the constraint upon the content of the icon corresponds to the ground of that disposition. In this way, a theory constructed in accordance with the iconic paradigm serves as a scientific explanation, in that it satisfies both parts of the scientific attribution of a property to a logical subject. It has long been my contention that no other theory of theories has this capacity and no other theory of properties is adequate to an understanding of science.

2.2. *Sources of Sources*

The background to icon-conceiving, which serves to place rather sharp constraints upon what is possible, can be thought of in terms of choices among certain options with respect to the main Aristotelian categories. A background will then consist of a certain path chosen amongst alternative views as to the nature of substance, individual, quality, relation and so on. For example, substances may be regarded as continuous or granulary, bounded or unbounded, complex or simple. Individuals may be regarded as atomic or inter-related, qualities may be regarded as the manifestation of powers or as the units of sensory experience, and so on. Only certain choices amongst the background options form consistent wholes. It is the traditional task of metaphysics to examine the consistency of various choice patterns amongst background options.

Recently, a quite new feature has entered our discussions of such matters. Namely, the idea that the icon of reality generated in accordance with a certain set of choices amongst background options, has certain structural features, that is requires that certain invariants be preserved during the history of the universal entity conceived in accordance with the choices. This approach has become particularly prominent in physics, in linguistic studies, and in anthropology. Indeed, in anthropology we have had a movement, the main burden of which has been to draw attention to the value of the search for structural invariants in a wide range of what, in this paper, I am calling Icons of Reality, e.g. Lévi-Strauss's alleged demonstration of the structural isomorphism between the totemisms of various Australian Aboriginal tribes (Lévi-Strauss, 1962: 127). I do not myself attach much importance to the fact that this approach has come under strong criticism since the critics have, it seems to me, contented themselves merely with a critique of details. I do not think any successful general attack has been made upon the approach. However, the undoubted rightness of many detailed criticisms has led me to use the word 'structurism' rather than 'structuralism' for the approach I wish to advocate; thus standing firm with the structuralists only on the most general issue, namely that the world and the thought matching it are both to be conceived of as structures which evolve in time, preserving certain invariants.

The problem then, of sources of sources, resolves itself into two distinct areas:

(1) What are the invariants, and what are the elements of the various structured icons which form the most general background to thought? This problem is a matter for an informed historical investigation, informed that is, by the iconic structurist theory.

(2) But as one can see from the preoccupation of this conference, the dynamics of change in such structures is a much more interesting since much more obscure area to investigate. I shall, in the end, in this paper, devote some attention to the sketching of some preliminary points about it.

At present there is little more that we can do but remark upon the enormous influence that something we might call 'fashion' has upon the background. One can perhaps put the matter in a preliminary way in a quotation from Henry Fielding. 'Fashion is the governor of this world. It presides not only in matters of dress and amusement, but in law, physics, politics and religion, and in all other things of the gravest kind. Indeed, the wisest of men would be puzzled to give any better reason why particular forms in all these have been at times universally received and at other times universally rejected, than that they were in or out of fashion'. I am sure that we can do better than Fielding's wisest of men. But our problem is part philosophical (what are the invariants?) and part social-psychological (what are the processes by which the evolving structures are passed from mind to mind?) Our business will be with a general account of the nature of such structures and of the possible dynamics of their evolution. But we are not likely to get far in speculative psychology until the psychologists can offer us at least the glimmerings of a theory of fashion, even in that most obvious world fashions in clothes. However, there is, even at this early stage, something that can be said and I shall return to the matter in the last section.

3. The Direction of Science

Participants in the congress were asked to consider among other things the problem of identifying a direction in science, an essential preliminary to any talk of progress. I think it abundantly established by the difficulties that Popper and others have run into in attempting to formulate their intuitions that science does develop towards something that might be called 'truth', that any attempt to discuss this matter in terms of a logic coextensive with the logic of deductive reasoning runs into enormous difficulties. I think there can be no doubt that Popper has established once and for all the limitations of the deductive approach to rational thinking, in his distinction between the logical structure of science and the methodological rules, where it seems to me, all we want to say about science as a unique activity lies in the latter (Popper, 1958). I do not propose, therefore, to consider progress in any quasi inductive fashion in terms of universality or anything of that sort. However, I do believe that we can make considerable progress in terms of the iconic theory which I have so long been advocating.

3.1. *Stratification*

The idea that science is pursued according to the grounded disposition or powers theory of properties and predicates, leads us to a stratified theory of knowledge which is reinforced by a further, very pervasive feature of scientific thinking.

(1) The theory of powers or grounded dispositions (Madden and Hare, 1971) leads us to say that the behaviour of some class of substances and their anatomical structure derives from an inner and, in general, other set of properties, which we have found frequently to be expressible in terms of a structure of elementary agents; thus the valency of a chemical element is explained in terms of the structure of the elementary charges in terms of which we conceive its atomic constitution. So the dispositions of things and substances are grounded in hypotheses as to their natures. In expressing the results of a power-natures investigation, we are naturally led to lay out our knowledge in strata, each stratum consisting of two levels. Thus in one stratum there would be, in level one, a description of the pattern of behaviour of a chemical element, and in level two of that stratum, a description of the structure of the elementary agents that constituted the nature of the atoms of that element.

Now this is a *stratum* because the expression of our hypotheses in this form leads immediately to the question of the explanation of the behaviour of the elementary agents referred to in level two. This explanation will take exactly the same form, that is the behaviour of electrons will be conceived as the manifestation of certain dispositions to behave and, in theory, those dispositions will be grounded in structural analyses of the nature of electrons, which must on this paradigm, be conceived of as some structural form of yet more elementary entities which themselves have powers. At present physicists are unable to make this step, but it is not clear to me that there is any prohibition in principle upon it being made. In other sciences, the stratified form is even more clearly exemplified. It has been one of my contentions in recent years that this stratificatory form should also be aimed at in the social and psychological sciences, in preference to the one-dimensional instrumentalist positivism by which they have been recently plagued (Harré, 1971).

(2) I have taken for granted in talking of the setting up of a grounded disposition account of behaviour that the powers of things and substances are grounded in structural hypotheses as to the natures of those things and substances. It has certainly been a very pervasive feature of scientific thinking to prefer structure to quality. I can perhaps illustrate this pre-

ference in an example. There are at least several hundred thousand different substances to be found on the surface of the earth. Let us say that there is a million-fold diversity among the substances which we can find or synthesize. Chemical thinking involves the substitution of the million-fold diversity by a structural hypothesis that there are a million different structures involving, roughly, one hundred different kinds of atoms. Thus a million-fold qualitative differentiation is reduced to a hundred-fold difference of substance and the disparity is taken up by structural theories of chemical compounding. Chemists have not rested at that. The one-hundred-fold diversity of chemical elements has itself been transformed by a structural theory in which the one-hundred-fold qualitative diversity of chemical elements is explained in terms of a hundred different structures of three fundamental substances – protons, neutrons and electrons. Unfortunately, nature has not exactly sustained the original beauty of this conception, but nuclear and particle physics is at a very early stage of development as yet and I am prepared to hazard the guess that any further development will consist in the reduction of diversity amongst so-called elementary particles to a structural diversification of some yet more and yet fewer elementary substances.

It is clear that the structural explanation of qualitative diversity matches almost exactly the powers-natures explanation of the behaviour-dispositions of natural substances, in that the same structure is responsible both for their disposition to manifest certain qualities and their disposition to behave in other chemically relevant ways. So these factors are in essence the same and they define the strata of a science, each stratum overlapping with the next and consisting, as always, of two levels; in one a substance or individual is described in terms of dispositions or powers, and in the other those dispositions are grounded in structural hypotheses about its nature.

3.2. *A Tentative Definition of Progress*

Assuming a fairly stable background to iconic thinking (an assumption which we will examine in the last section of this paper), a limited conception of progress can be defined in terms of the stratified theory of knowledge. Thus progress would consist in supplying an icon of reality such that three conditions are satisfied:

(1) The icon represents a reality that exists in a stratum further from

experience than all previous icons, so that, for example, the invention and promulgation of the double helix theory of genetic material is an example of scientific progress in that in the strata of knowledge we have about heredity, the double helix is the structure of an icon of a reality which is further from experience than the simple atomistic gene of the Mendelian theory: that is, it is further from simple experience than the count of the proportions of green and yellow peas, and further still than the chromosomes which were the individuals at the second level of the first stratum of hereditary theory.

(2) An icon of reality represents progress if it maximises structural diversity at the expense of qualitative difference; thus the electron theory of the chemical atom is progressive over the Dalton theory which differentiated chemical atoms wholly in terms of their qualities and powers.

(3) The appearance of an icon of reality is progressive if the same icon reappears at the deeper levels of many stratified fields of knowledge. Something like this was proffered many years ago by Ernest Nagel as a criterion for the reality of a substance, process or individual, postulated in a theory; thus, the physico-chemical atom could be taken to be real insofar as it appeared in the explanation of a very wide range of very different phenomena (Nagel, 1961: 147–149). Nagel, I believe, had perceived something important, but had misunderstood it. The idea of there being a criterion of reality which does not involve any kind of demonstrative reference, nor in strata remote from experience, transcendental arguments, seems to me a serious error and yet there is something important about the reappearance of a certain icon and its associated concepts all over the place. I believe that this marks not the reality of the structures and entities imagined, for that question is preempted by the metaphysical choices amongst background options long before (logically speaking) any specific icon is imagined. Rather the pervasiveness of a certain conception of nature is a measure of the progress which has been achieved by its introduction.

4. Society as a product of theorizing

In recent years a number of trends in the social sciences have led to a critique of traditional views about the nature of society and the mode of its being as society, that has been at its sharpest among ethnomethodologists (Filmer *et al.*, 1972) and at its most articulate in the work of Scott and Lyman (1970). I must confess myself wholly convinced by their line of argument and their criticism of positivist sociologists, who suppose that society has a nontheoretical mode of being and that truths about it constitute social facts. I shall call this the 'naive' theory.[3] It is in accordance with the naive theory that much contemporary empirical work in sociology is still carried on as if there were a reality 'out there' which, by various techniques, we could sample and investigate and ultimately describe. I am now very much of the opinion that we should begin by insisting that there is no such reality and regard the very existence of patterns of conformative behaviour as the problematic element in human life. It follows then, that in order to give an adequate account of the relation between social factors and the sciences we must attempt a new departure, taking the ethnomethodological conception of society as the groundwork of our argument. Thus, the social sciences will be in a very different position from the natural sciences with respect to their ontology and metaphysics, though not, I believe, with respect to the intellectual skills and theoretical methods which are brought to bear upon their problems.

According to the view which I wish to advocate, society and the institutions within a society are not to be conceived as independent existents, *of* which we conceive icons. Rather, they *are* icons which are described in explanations of certain problematic situations. Thus, the *concept* of a Trade Union, or a University, is to be treated as a theoretical concept judged by its explanatory power, rather than a descriptive concept judged by its conformity to an independent reality. The form of a social and a scientific explanation are identical, but their ontological commitments and metaphysical structures are quite different. Beyond the icons of reality which are conceived for purposes of explanation in the natural sciences, lies a real world of active things; but beyond the icons conceived for the explanation of social interactions by social actors lies nothing but those very actors, their conformative behaviour and their ideas.

For what purpose, then, are social explanations attempted and icons conceived? Social thought, whether expressed in ordinary language or in one of the official rhetorics of sociology, is aimed at the injection of meaning into a flurry of individual, face-to-face, encounters, in the course of which messages are sent from one person to another, which we *assume* reach their intended recipients and are understood. Without the active intervention of sense-giving explanatory activity, this flurry is without meaning though, as we shall see, not wholly without form. I can illustrate the point with a simple and restricted example. Physical actions enter social reality when they become embedded in semantic fields and syntactic structures. For example, a gripping of hands comes to be meaningful when it is embedded within an introductory ceremony, and acquires within that fundamentally syntactical structure a certain distinct meaning in relation to other actions and words, and certain structural relations with other parts of the ceremony. Thereby a hand-grasp becomes part of social reality, but only as the bearer of a recognised meaning, conformative to its intended meaning. The gripping of hands has no social meaning taken outside all possible embedments. Within a karate encounter, of course, its meaning is quite different from that which it has within an introduction ritual.

So an analysis of the social world must begin with the fundamental thought that the components, or elements, of that world, have no meaning in themselves and therefore do not exist as social objects. Once we pass from the simple and restricted, and immediately apprehendable, structures, such as are exemplified in introduction ceremonials, the joint procedure by which meaning is given to an element or set of elements, and structure is imposed upon them, has to be achieved by reference to some icon, for example, an introduction ceremony makes sense only if people are perceived according to the social dichotomy, intimates and strangers.

I will illustrate the role of such icons in two examples of increasing relative generality.

(1) *A microsocial icon.* Sometimes, perhaps even within a single building, the flurry of personal encounters is made intelligible for those within it by theoretical explanations, based upon descriptions of a shared icon in terms of which some of the flurry can be treated as consisting of activities within an institution, for instance, a police station. Such an icon frequently appears in pictorial form as a chart showing, for example, the hierarchical organization of the management structure. Of what is that diagram a picture? Certainly not any structure existing independently of the icon itself which the icon mirrors. Rather it is a conception of the institution in terms of which actions are performed and made meaningful, and in terms of which the flurry of such actions has structure.

(2) *Macrosocial icons.* It happens from time to time that we become aware of differences in the dress, speech, habits and tastes, jobs, etc. of other people. Along with this perception may go certain feelings of respect, resentment or contempt, and we may provide an explanation of such feelings by reference to the differences we perceive between us and them. In real life this phenomenon occurs in an unordered flurry of momentary, personal interactions, some few of which may persist. Sociologists provided us some time in the nineteenth century with a societal icon in terms of which this flurry of isolated experiences could be made intelligible. I refer to the class theory. It describes a well known image or icon. It is not a theory which describes an independent reality, but it is the basis of a wide variety of explanations in terms of which various differences between people can be made intelligible. It replaced an icon of reality which paid attention to other features of the pragmatic boundaries of concepts like respect, namely, the situational or place icon which allowed us to conceive of society rather as a network of transformable relations than as a layered structure, within each layer there being more or less random movement. Now it is my contention that classes exist only insofar as they are thought to exist and the function of such a notion is to provide a standard, ready-made, easily acceptable explanation for understanding what has happened on some occasion.[4]

The point of view which I am sketching here is certainly not new but has rarely been defended in a pure form, since the temptation to suppose that there are 'social facts' and structures in the social world independent of thought, is perennial, a temptation based upon the correct perception that there is a problem of how an individual's forms of thought are related to the shared thought of his community, which community is created by the sharing of the thought. I should like to describe the pure form of the theory as the Borodino theory, in acknowledgement of the exquisite exposition of the theory by Tolstoy. Tolstoy uses the Battle of Borodino both as an example and as a model, for understanding society. If I may remind you of the account he gives, the battle is seen as a flurry of relatively isolated, inter-personal encounters occurring both on the battlefield and amongst the officers of the High Command. Messages are sent from Napoleon to the various parts of the battlefield but of course they rarely arrive, and if they do are either garbled or not understood. And if rarely understood, are not acted upon. At various levels in the army hierarchy an icon physically exemplified in the badges of rank, generates the illusions of command that exist throughout the day, though smoke and the geographical structure of the battlefield prevent any empirical verification of individual hypotheses of command. Yet, Tolstoy reminds us, the battle was written up by historians in a rhetoric which conceives of that flurry of personal interactions as a structured entity controlled from one side by Napoleon and from the other by Kutuzov. Historians' accounts include such items as 'brilliant tactical decisions' and so on, and *assume* a structure of command with reliable interrelations. All of this accounting, Tolstoy makes clear to us, is essential since it makes sense of an affair which though concentrated in space and time was without overall structure, but being human we persist in pursuing the theoretical activity of endowing it with one. When reading *War and Peace*, were we momentarily in the presence of the greatest of all social analysts? I am inclined to think so.

I see myself then, as accepting and advocating a Tolstoyan or Borodino theory of the social. That is, that we cannot take for granted that without an implicit social theory describing assumed structural icons we are left with a mess or flurry of individual encounters which are then given form and meaning by accounting procedures which draw upon various rhetorics. For example, we cannot assume that any people in society are endowed with *actual* power, since messages are rarely understood, seldom reach their targets, and are usually ignored, but in accordance with some picture of the way things are, people are endowed with the property of 'having power' and so occupy a certain niche in the icon, generated by and at the back of the accounting rhetoric. Such a person can then become the focus of certain important social feelings, not in himself, in which he is frequently a pathetic figure, but rather in terms of the imaginary capacities which, by our act of supposing him to have them, serves to make our experience intelligible. It is in terms of a Borodino theory, then, that I propose to examine the relation between science and society.

But before I proceed to the final section of my argument, there are two qualifications to the pure Borodino theory that must be entered.

(1) There are patterns in social action which are not represented in any individual consciousness – for example, the patterns of unintended consequences – which we can *come to know*, and which must, therefore, exist prior to our knowledge. It is the existence of these patterns which has encouraged the reification of 'social facts'. I contend that though there are

such patterns as, for example, the socially divisive migration of people inside cities which has been associated with changes in the structure of the school system, the activity of representing such patterns in thought is a bad model for the activity of representing institutions, ceremonies, role-presentations and the like; but when we come to attempt to explain the patterns of unintended consequences, we must return to the individual consciousness and the theories which individual citizens form concerning their situation.[5]

Thus the patterns of unintended consequences are not produced by independent societal causal processes, but are either,

(a) the accidental products of myriads of stages modulated through individual consciousnesses, as, for example, messages are passed and habits are acquired; or

(b) the intended products of myriads of individual understandings of messages and acquisitions of habits according to some promulgated scheme, for example, the successful promulgation of the idea that we should drive on one or other side of the road in any particular country.

In the case of (a) the patterns are imposed by us in the search for intelligibility and appear in our accounts. Their objective status is like that of the castles in the embers or the cherubs in the spring leaves.

Even in the case of (b) the promulgation of a pattern through society is an uncertain matter and dependent upon matters of which we are quite ignorant. Compare our understanding of someone poisoning everybody through the water supply and our lack of understanding of someone ordering everybody through the media.

(2) The icons which make our actions intelligible as social actions are shared conceptions, and the conception of social order held by any individual may fit it more or less well. I am not the first to suggest the analogy of this problem with Saussure's distinction between language and speech. But is an individual who only gradually comes to understand the structure of a university like someone who only gradually comes to understand the structure of a protein? Or like someone who only gradually comes to understand the current theory of electrical conductivity? It is my contention that the person adapting himself to society is like the latter as far as the form of his intellectual activity is concerned. But whereas for a scientist there is a three-fold system of structures – that of his own icon, that of a common icon in the shared theory of the scientific community, and

that of the bit of the real world represented; for the social actor there are in many crucial cases only two.

In terms of *his* icon, the individual constitutes himself *a* mayor, and in terms of the *shared* icon he is constituted by his co-societals as *the* mayor; and they can verify their attribution in his way of going on because his icon sufficiently matches each of theirs. Only if the other recognise him as such does his constitution of himself leave the realm of fantasy. And it is well worth noticing that his fantasy does have a measure of latitude even with respect to the shared icons of those who recognise him as mayor in that he has considerable latitude in the style in which he performs those duties which devolve upon him to his place in the picture of the village that the villagers have.

5. The true relation of science and society

Taking the theory of science together with the theory of society, we reach the inevitable conclusion that the relation between science and society is an internal one. The icons of reality conceived by scientists, and society itself, since it is just another icon – though of enormous importance for human life – are generated by the self-same process and according to the self-same constraints. However, one should notice that while it is necessary that society be generated as an icon by the same process as scientists conceive of the structure of the world, it is contingent that scientists fall under the very same set of constraints at any given time as do those who conceive the social world. There may well be historical times in which science and society are out of joint, but in general we would expect, in the absence of a solution to the problem of the generation and promulgation of fashion, that contingently the icons of science will be constrained by something of the same set of background options as are their contemporary conceptions of society.

It follows from this that there are no causal connections *from* society (or even patterns of conformative behaviour existing pre-theoretically to any particular consciousness) *to* thought patterns. Even so sensitive a student of symbolic forms as Mary Douglas slips into an unexamined assumption that such connections exist. "The body", she says, "is a symbolic medium which is used to express particular patterns of social relations" (Douglas, 1970: 13–14). Not so, patterns of social relations *and* bodily symbolism could equally be regarded, as far as the ethnographic evidence goes, as both representations of a conceptual pattern, syntactically structured and intersecting with semantic fields. Again this assumption shows up in her remark that each set of metaphysical assumptions is *derived* from a type of society. Let us say instead, each type of society is a representation of a set of metaphysical assumptions. By taking this turn one can quite readily understand Douglas's true perception "that the ideas about the human body, its potential and its weaknesses, which are found in particular social types, correspond uncannily well with ideas current in the same social types about the potentials and weaknesses of society."

In Bernstein's well known terminology (Bernstein, 1965), I would claim there is no elaborated code, there are only different restricted codes, since there is no code which is independent of a social form, in that some social patterns may develop to be another representation of its syntactic and semantic structure.

Thus the communality of 'backgrounds' and structural principles in 'social reality' *construction* and natural reality *matching*, explains how social factors are actually involved in scientific thinking and its products. That is, societies are conceived according to the same set of options that are present in natural science in terms of which icons of physical reality are conceived. For example, two well-known and well-articulated sets of background options are:

(1) Numerous passive atoms in random motion interacting by collision.

(2) A structure of agents transforming itself under the constraint of some invariant.

According to the hypothesis I am proposing in this paper, it should be possible to show that when nature is conceived according to one or other of these sets of options, so society is conceived and thus created in a similar mould.[6] Now as I have pointed out, it may be historically the case that there are timelags and disconnections in thought. For example, it may be that in some epochs the societal icons in terms of which social actions are made meaningful remain fairly fixed, while the ways of conceiving of the structure and nature of the real world through iconic representation are changing. And the reverse may also be the case in other epochs. There are, I believe, two very striking cases where the background and structural invariants across the whole range of human thinking had a marked universality. These are the Romantic movement – culminating in the science of Michael Faraday and the social forms of Victorian Europe, and the Positivist movement culminating in the psychology of Skinner and the conception of society as bureaucracy. Skinner has proposed an explicit theory of society which is intended to promote an icon of the social world which is precisely adjusted to the set of background options that appear in classical positivism – namely a theory of human beings as passive automata and a conception of moral virtue centred, one might say, from outside the theory, around the notion of perfect obedience, i.e. all actions as produced by the environment. The internal relation between the scientific and societal icons is very clearly evident in Skinner's recent work, *Beyond Freedom and Dignity* (1972).

I offer, therefore, as a tentative hypothesis, yet to be fully developed in the context of certain specific examples, the idea that societal factors are not causally efficacious in the thinking of scientists. The occasions on which they seem to be so are, according to this theory, an illusion. There is no mysterious causal process which Mannheim and others looked for in vain.[7] Students of the sociology of knowledge have found no plausible causal relations, since there are none to find. However, it is my firm belief that the connection between science and society is very strong and in the theory I have proposed, the connection has certain features which are no less than necessary in that the icons of the inner forms of things which form the subject matter of science, are generated by the very same intellectual processes by which all of us, scientists and laymen alike, generate societal icons in terms of which our experience can be made meaningful, and are likely to share the same set of metaphysical options and be subject to the same constraints. Thus the social order is one set of realizations of the basic structure of our icons, scientific theories are another.

6. The Dynamics of Icon Change

While the facts of social and conceptual change are treated as wholly separate problem fields, it is unlikely that any progress will be made in understanding either, but the sociology of knowledge, as conceived by Mannheim, was a false start, since it assumed an implausible causal theory. Equally, conceptual change in science has been, in my view, wrongly studied by paying exclusive attention to its products, verbal structures and the meaning of words, as opposed to its subject matter – Conceptions of

Reality. The dynamics of change in the logico-verbal part of science is, according to the theory of this paper, derivative from a more fundamental dynamics – that in which the background options are revised, and in accordance with certain, as yet not clearly identified invariants in the structure of icons of reality, control the transformation of one icon into another plausible picture of the world. Meaning change is then derivative from this underlying process. On the other hand, if society is conceived of as an 'out there' real entity, we may be tempted to look for an underlying dynamics – for a process which simply does not exist. If societies are theoretical constructs for the explanations of our situations, then they will be subject to the same dynamics as are the conceptions of nature which lie behind the sciences, but as I have emphasized, whereas there is an independent reality to the natural world, society has no being except as an icon. Thus, I can offer to this conference no answers to the problems of change, but instead what I hope is a well-articulated theory as to the problem-fields in which such answers may be sought. Only when we direct our attention to the right field of study are we likely to come up with some useful answers.

I would, however, like to offer two more specific ideas, which could be applied both to science and its social analogue, political controversy – say that surrounding the behaviourist-Skinnerian paradigm – discussed in Chomsky's political works (Chomsky, 1973). If I am right in a general way, then, if for the sake of reducing the problem to manageable dimensions, we ignore – though of course only temporarily – the whole moral dimension, then we seem to have two quite distinct methods by which we justify a background. And, of course, it may well be that in the means of justification, particularly in the failure and success of justification, that we may get the first hint of a dynamics of the structure of icons.

The two methods which seem in fact to be involved in the justification of backgrounds, are as might be expected, one theoretical and the other empirical. Theoretically, one can take up the hint that the study of backgrounds is a form of metaphysics and look for transcendental arguments by means of which items and structures present in the background and conceived as the products of choice among various category options, might be justified. By a transcendental argument I mean a demonstration that, preferably, the only conditions under which our experience is possible, are realised by a particular set of options. Of course, in any transcendental argument, a very great deal depends upon the prior analysis of experience. The particular form the necessary conditions for experience take will be dependent upon how that experience is understood. Thus, if experience is held to be inherently atomistic, analyzable into independent elementary parts, then this will influence rather strongly the form of demonstration by which the set of background options are shown to make it so. But at this point the ethnomethodology of everyday life, our common-sense sociology and psychology, begins to diverge from the ordinary natural sciences, since there is no external criteria by which societal thinking can be brought under some kind of existential constraint. But it is clear that background options and structural properties assigned to the physical world do generate existential hypotheses which can, under certain restrictive conditions, be subject to test. Thus, if the background is 'randomly moving atoms', then there are certain situations in which it may be possible to ask whether such entities exist, and even to find an answer empirically. Or if a more sophisticated theory of the background is in play, it may be possible, for instance, to provide an existential test for the existence of a potential or a charge, or something of that sort, at a place for a time. Now such testing proceeds by a progressive passage through the hierarchy of strata that I have identified earlier as an essential feature of an adequate theory of scientific knowledge. For example, we pass from diseases to bacteria, to viruses. We pass from the anatomical features which differentiate species and individuals one from another, to genes and chromosomes through to molecules. Existential tests are relative to strata and what is shown to exist in one stratum may be further analysed and indeed re-categorized from the point of view of a deeper stratum. Thus the products of the background, even if established in being, are liable to continuous categorical transformation as that background changes in essential ways.

But for institutions and other societal entities, to believe in them is to bring them into being. So the existential test for an element or a structure in a societal icon is liable to be self-fulfilling; though not, therefore, invalid.

There remains the problem of transformation. Here again I can only propose some problems. Transformation of the structure of icons occurs in two dimensions in this theory. We can ask how does the underlying synchronic structure, say, a conception of a system of elementary agents, transform into an icon of the world, natural, social or even artistic, in each of which the mode of representation may be very different. This is the problem of the mechanisms of contemporary corepresentation. But that underlying structure may itself suffer diachronic change, and we can pose the problem of the manner and means of such transformations too.[8]

Linacre College,
Oxford, England

NOTES

[1] A theory on this view must describe an icon of reality to be truly scientific. Thus purely formal theories, consisting only of signs or other free linguistic constructions, must be regarded as primitive. Thus constraints on, or *a priori* universal invariants in, icon conception constrain theories.

[2] A non-spatial structure would be exemplified by the old theory that substances were made up of different proportions of the Four Elements.

[3] Durkheim's difficulties with this crux are worth noticing. In the *Rules of Sociological Method*, and *Suicide*, his general view that society is a system of ideas, the animating theory of the *Elementary Forms of the Religious Life* seem to be negated by a reified conception of 'social fact', forced upon him by his acceptance of three criteria of objectivity: (i) our social ideas are not selfcreated, (ii) an individual can come to know more about his social situation, (iii) shared social conceptions are coercive over an individual's desires. While proposing that social facts consist of 'representations and actions', an opinion from which I would not dissent, he nevertheless argues that 'their substratum can be no other than society'; the same kind of error as the Cartesian reification of minds on the grounds of the incompatibility of physical and mental properties, so that he, Durkheim, proposes that we should look for the sources of social facts in other social facts.

[4] E.g. 'He did so and so *because* he was middle class' which would be interpreted on the iconic theory as 'In doing so and so he represented an aspect of sociologist S's layered icon of reality.'

[5] There is a corresponding error in transferring the causal mode of the pattern of unintended consequences to the arena of political action. One should notice here the folly of the socialist radical in attacking 'institutions' while leaving untouched the system of meanings that sustain them, and of which they are part representations. The hyperradical would aim first at semantic change.

[6] So far as I can understand, this is the meaning of Foucault's proposal for an archeology of knowledge, identifying the structures common to forms of thought, modes of practice, and explanations of conformative behaviour, which each, in its own way expresses, and then seeks a theory of change in their transformation (Foucault, 1972).

[7] Mannheim's analogy between artistic style and forms of thought as socially conditioned, and indeed any of the examples he cites (Mannheim, 1936: 264–290), is susceptible of explanation in structurist theory, without *assuming* any direction of influence *from* social position (situatedness) to intellectual product, since in the structurist view 'social position' is as much an intellectual product as anything else.

[8] I am grateful to Dr. J. McGeochie for some useful criticism of an earlier draft.

BIBLIOGRAPHY

Bernstein, B.: 1965, 'A Socio-Linguistic Approach to Social Learning', in J. Gould (ed.), *Penguin Survey of the Social Sciences*, Penguin, London.

Bunge, M.: 1973, *Method, Model and Matter*, Reidel, Dordrecht and Boston.

Chomsky, N.: 1973, *For Reasons of State*, Collins.

Douglas, M.: 1970, *Natural Symbols*, Cresset Press, London.

Filmer, P., Phillipson, M., Silverman, D., and Walsh, D.: 1972, *New Directions in Sociological Theory*, Collier-Macmillan, London.

Fisk, M.: 1974, *Nature and Necessity*, Indiana University Press, Bloomington and London.

Foucault, M.: 1972, *The Archaeology of Knowledge* (transl. by A. M. S. Smith), Tavistock, London.

Harré, R.: 1970, *The Principles of Scientific Thinking*, Macmillan, London.

Harré, R.: 1971, 'Johnson's Dilemma', *Bull. Br. Psychol. Soc.* **24**, 115–119.

Lévi-Strauss, C.: 1966, *The Savage Mind*, Weidenfeld and Nicolson, London.

Madden, E. H. and Hare, P.: 1971, 'The Powers that Be', *Dialogue* **10**, 12–30.

Mannheim, K.: 1936, *Ideology and Utopia* (transl. by L. Wirth and E. Shils), Harcourt Brace, New York, pp. 264–90.

Nagel, E.: 1961, *The Structure of Science*, Routledge and Kegan Paul, London.

Popper, K. R.: 1958, *The Logic of Scientific Discovery*, Hutchinson, London.

Scott, M. B. and Lyman, S. M.: 1970, *A Sociology of the Absurd*, Appleton-Century-Croft, New York.

Skinner, B. F.: 1972, *Beyond Freedom and Dignity*, Cape, London.

Toulmin., S. E.: 1961, *Foresight and Understanding*, Hutchinson, London.

SECTION C

SOCIETAL COMPONENTS IN SCIENTIFIC DEVELOPMENT

INTRODUCTION

In highly industrialized societies no one can deny that three fundamental facts characterize the world of science: (1) Science *has grown* enormously in terms of personnel, publications, scientific disciplines, and costs. (2) The *position* of science in society has changed as its results became increasingly relevant – socially and politically. (3) Its contribution to *economic growth* increased markedly as better and more qualified manpower in general and management, organization, and planning made possible the systematic use of productive forces in particular.

Research into science development has evolved in the wake of the growing importance of science in society. Society as a whole and science as one of its subsystems have grown in complexity as a consequence of generalization of expectations (mainly through institutionalization) and functional differentiation into subsystems. Since greater system complexity implies an increasing number of possible events, which, in turn, means elimination of more possible choices in actual human conduct. Thus, the need for selective power of social processes also increases, that is, for structural arrangements that order processes as consequences of events and hence improve the selective capacity of the processes themselves (Luhmann, 1972: 125f.). One way to strengthen the selectivity of these processes is to provide them with a reflexive structure, that is, to apply them to themselves. Thus, in pedagogy one may go about teaching how to teach or in a bureaucratic setting one decides about decisions.

In science, under the given circumstances, it became increasingly necessary to reflect upon scientific thinking, to investigate the causes of scientific development in general and of scientific disciplines, specialties and research areas in particular. These new reflexive mechanisms are designed to improve the selective potential of the system that produces science as well as the systems that use science. Selection that takes place in various institutions and on all levels of society including the science system itself, presupposes criteria for scientific choice. Research into science development, as indicated in the Preface to the volume, is there-

Knorr, Strasser, and Zilian (eds.), Determinants and Controls of Scientific Development, 287–293.

fore confronted with the task to provide answers as to the potentials and limitations, the problems and consequences of science control with regard to internal and external goals.

Therefore, science research, properly conceived, is to be defined *not* by its objects of investigation; rather, some relevant topic, question or problem should be studied in its relation to the fundamental questions concerning the conditions that determine science development and circumscribe the controllability of science directed toward certain goals. The analysis of the factual and potential contexts of determinants and controls of science development is the central concern of science research. The relationships between determinants and controls point to questions and analytical foci, which refer as much to the system of scientific knowledge production as to science using systems such as politics, the economy, education, international organizations, etc.

The papers presented in Sections A and B seem to have elaborated upon the following assumptions, which at the same time constituted the philosophical and sociological points of reference in the explanation of scientific development: (1) Any time and everywhere scientists operate within a *minimal set of methodological and logical norms*, which can claim validity for all scientific production. This assumption, as demonstrated primarily in Section A, led to further questions regarding the succession and cumulativeness of theories, the definition of scientific progress, etc. (2) Science always takes place within a *framework of communication* of a more or less comprehensive community of specialists. The contributions to Section B dealt with the question of how the communicative networks of particular scientific communities effect cognitive changes in the respective scientific fields. This is not to say that these studies were not concerned with problems of science control or external interests in using science.

The red thread, however, that goes through the papers in the present Section C is the assumption (3) that scientists are embedded in a *world of political, economic, and cultural experiences* that go beyond the scientists' sphere of communication while at the same time exerting considerable influence on contents and forms of communication among scientists. Analyses come into focus, which deal with these societal components in scientific development, their controlling effect as well as the extent to which science contributes to the realization of objectives set by different science using systems such as the state, economic corporations, and foundations. The main thrust of the papers to be presented in this section will be along these lines.

The papers by Maria Nowakowska, Jerry Gaston, and Walter Hirsch deal, at least in large parts, with career and role characteristics of scientists as they relate to the organizational setting, the scientific community, and the societal context. These authors' contentions testify to the well established thesis that a scientific community is structured in terms of communicative networks, reputational and power differences, research styles, and theoretical schools (cf. Griffith and Miller, 1970; Zuckerman, 1970; Radnitzky, 1970). These structural elements not only determine its internal policy of recruitment, socialization, and distribution of resources, but also the scientific community's external policy of legitimizing its interest vis-à-vis funding agencies and the wider public. In her paper on 'Measurable Aspects of the Concept of Scientific Career', Maria Nowakowska develops a model, in which scientific career is defined as an interplay between three functions representing the changes of scientific authority, social power, and recognition as a scientist. One of the main assumptions of the model concerns the mechanisms that underly the choice of the level of difficulty of research problems attacked by the scientist. A taxonomy of scientists into sixteen categories permits making differential predictions of the type of career as described by the three functions above. Such a model of career patterns of scientists is of relevance for science research insofar as scientific career patterns may be analyzed under the aspect of their functionality or disfunctionality for scientific development (cf., e.g., Mulkay, 1972a, b).

The factors that are used by Nowakowska to characterize scientific careers do not, of course, exhaust all aspects of the concept under consideration. In his comparative study on 'Scientists from Rich and Poor Countries', Jerry Gaston adds the dimensions of cultural history, economic development, and institutional support. He attempts to describe the actual career patterns of scientists in poor countries and finds that, relatively speaking, they are not as different from those in rich countries as one would perhaps expect. He also points to a few interesting paradoxes. On the one hand, there is the widely held view that science in economically advanced countries is more widespread and also better. It is probably also true that the more scientists there are in a country, the more this society approves of, and supports (must support?), science. On the other hand, data presented by the author suggest "that the more scientists there are, the less the rate of scientific productivity" (see Table IX). Gaston's tentative explanation is "that when a country has few scientists, they are left to do science, and they do it well, but when a country increases its stock of qualified manpower the best scientists are drafted into administrative roles." This scientific brain drain has long been noted by students of complex organizations (cf. Thompson, 1969).

Walter Hirsch ('The Autonomy of Science in Totalitarian Societies: The Case of Nazi Germany') brings into focus the dimensions of the value structure of scientific communities, the cultural tradition, and the political system. In his case study on the development of (natural) science(s) during the Third Reich, the author claims to come up with some evidence against Robert K. Merton's hypothesis regarding the special affinity between a democratic social structure and the ethos and practice of science. Hirsch arrives at the conclusion that, given a long established scientific tradition and its attendant institutions, scientific autonomy could be maintained in the face of totalitarian controls to a greater degree than widespread belief suggested. In fact, after a short period of 'German physics', these natural scientists managed to maintain social recognition and cognitive autonomy and to continue basic research to an extent that can well be compared with other (democratic) societies at war at that time. Hirsch did limit his study to the physics community though. Apparently, also in a totalitarian society it is important to know the criteria, according to which the scientific community distributes social rewards when resources to promote certain goals are to be allocated (cf. Spiegel-Rösing, 1973: 18–19). Finally, the question could be raised to what extent the cognitive structure of science (as opposed to the institutional structure) has changed under external political pressures.

David E. Morrison ('Philanthropic Foundations and the Production of Knowledge') and Lewis A. Coser (Comment) deal with an aspect of societal conditions of science development, which often shows most explicit influence on the way sciences, or better: particular scientific fields, develop. These authors address themselves to the thesis that foundations are particularly 'open' to social pressure, and that this pressure shapes their granting policies and hence the extent of promotion of scientific activities. To determine the reasons for this openness to social pressures, the foundations' position within American society is examined. The conclusion drawn is that a major reason consists in the failure of foundations to legitimate themselves. Interestingly enough, foundations, propelled by their own principle of "improving man's estate," were compelled to enter the social arena, which involved the social sciences and consequently controversy. Thus, Coser suggests that "What foundations need... are new role partners with a measure of real power so that the dialogue with them would be more than polite prattle that now characterizes its exchanges with the public at large." Of general relevance is the question of how various funding and control institutions affect the development of those institutions and sciences they support (cf. MacLeod and Andrews, 1968). Morrison, for example, observes that there is a reciprocal relationship between the organization of empirical research and the financial support by the foundations, since just as the former requires the latter, so the foundations need organized research to help dispense their funds.

Finally, a third group of papers attempts to explain the development and present structure of two different scientific fields by relating them primarily to conditions existing in the respective societies. By drawing on case studies on the development of selected disciplines, these papers aim at formulating discipline-bound theories of scientific continuity and change that also include the disciplinary units actually subject to change as well as some information concerning the directionality of scientific change. Hilary Rose ('The Social Determinants of Reproduction Science and Technology') discusses some of the problems of examining science in society and society in science and fears that in our enthusiasm to escape the sociological paradigms of the past we are now in danger of moving towards the aridity of metatheory. Within the context of the long standing ideological debate on regulating human reproduction, the author first presents some of the new technologies and then considers how a theoretical analysis of the position of women and the role of science under capitalism may be used to cast some light on the interpenetration of science into the woman's position, and the patriarchal and class bound society into the science and technology of reproduction. Rose makes us aware of the *static* aspect imputed to an analysis of the *determinants* of science development as contrasted with the *dynamic* aspect of science *control*. We are encouraged to look at the effects of science and the technologies made possible by science in terms of a feedback mechanism on the development of the scientific field under consideration. In that the author takes into consideration the intricate relationships between science and other social subsystems, on the one hand, and the societal presuppositions and the consequences of science, on the other, she views science as a social control agent that, for instance, undertakes measures to bring about adjustment to socially recognized needs of reproduction.

In his paper on 'Social Technologists and Social Emancipists: Factors in the Development of Sociology', Hermann Strasser makes an attempt to separate the factual, the hypothetical and the normative elements involved in the development of social thought. The author concentrates on those variables, most particularly on the fact that social groupings compete for public interpretation of social existence, that contributed to the emergence of major types of sociological theories. He also demonstrates that the persistence of competition and succession of generations among (social) scientists does show an important influence on the development of the sociological discipline. Finally, some evidence is presented supporting the thesis that sociology is poly- or part-paradigmatic, whose (part-) paradigms lack individual self-sufficiency and are, to some extent, complementary.

In that these papers focus on the structure of the scientific community, on institutions that provide funds for doing science, and on structural characteristics of different societies, they raise questions and attempt answers as to the causes of scientific continuity and change as well as those aspects of science subject to change and the direction in which change takes place. Often implicitly, the assumption is made that the *organization* of scientific work at the various institutions of scientific research is the critical determinant; that is to say, in what way the organizational set-up translates the social and cultural conditions into cognitive-scientific processes.

As the case may be, it seems at times that research into the development of social (and some natural) sciences has not really gone beyond the thesis professed by Scheler (1926) half a century ago that the relationship between 'ideal factors' and 'real factors' in science development is merely a regulative one. In other words, only the socio-historical *selection* of cognitive-scientific productions is amenable to sociological analysis, but not their socio-historical *causation*. 'Real factors', it is suggested, may only regulate the conditions under which certain 'ideal factors' appear in the history of science; the former, however, could not affect the content of the latter. Nevertheless, research into determinants and controls of scientific development does represent a reflexive mechanism that increases the selective potential of systems of science production as well as of systems that use or support science.

BIBLIOGRAPHY

Griffith, B. C. and Miller, A. J.: 1970, 'Networks of Informal Communication Among Scientifically Productive Scientists', in C. E. Nelson and D. K. Pollack (eds.), *Communication Among Scientists and Engineers*, Heath, Lexington, Mass., pp. 125–140.

Luhmann, N.: 1972, 'Soziologie als Theorie sozialer Systeme', *Soziologische Aufklärung* (second ed.), Westdeutscher Verlag, Opladen, pp. 113–136.

MacLeod, R. M. and Andrews, E. K.: 1968, 'Scientific Careers of 1851 Exhibition Scholars', *Nature* **218**, 1011–1016.

Mulkay, M. J.: 1972a, 'Conformity and Innovation in Science', in P. Halmos and M. Albrow (eds.), *The Sociology of Science*, The Sociological Review Monograph, No. 18, Universiy of Keele, Keele, St., pp. 5–23.

Mulkay, M. J.: 1972b, *The Social Process of Innovation: A Study in the Sociology of Science*, Macmillan, London.

Radnitzky, G.: 1970, *Contemporary Schools of Metascience*, Akademiförlaget, Göteborg.

Scheler, M.: 1926, *Die Wissensformen und die Gesellschaft*, Der Neue-Geist Verlag, Leipzig.

Spiegel-Rösing, I. S.: 1973, *Wissenschaftsentwicklung und Wissenschaftssteuerung*, Athenäum Verlag, Frankfurt a. M.

Thomspon, V. A.: 1969, *Bureaucracy and Innovation*, University of Alabama Press, University, Alabama.

Zuckerman, H.: 1970, 'Stratification in American Science', *Sociological Inquiry* **40**, 235–257.

MARIA NOWAKOWSKA

MEASURABLE ASPECTS OF THE CONCEPT OF SCIENTIFIC CAREER

The present paper concerns a theoretical model of a scientific career. It is an extension and modification of a model suggested earlier (Nowakowska, 1973).

The model considered here is based on some quantitative measures, and may therefore be treated as a basis for empirical studies on scientific careers.

The latter concept is, of course, rather complex, and one could easily list a number of its relevant aspects. In this paper, however, the scientific career will be interpreted as an interplay between three functions, describing changes in scientific authority, social power, and social recognition.

Needless to say, these three factors do not exhaust all aspects of the concept of scientific career (to mention only few, the other factors are economic conditions, social context, availability of funds, etc.). One could argue, however, that the distinguished three aspects constitute the main dimensions of scientific career.

The present paper may, in some sense, be considered as complementary to the analyses of the development of science (see, for instance, Whitley, 1974a, b; Weingart, 1974; Klima, 1973; Mullins, 1974; Nowakowska, 1975), to the analysis of cognitive factors underlying the individual research (Klima, 1974; Frost and Whitley, 1971), and to the impressive analysis of success in research (Lemaine *et al.*, 1973).

The assumptions of the model, together with their brief discussions, are presented in Sections 1–9, while the remaining paragraphs deal with consequences of the model.

1. Scientific events

We consider a class, say Z, of scientific events. The typical examples of elements of Z may be: solving a research problem, obtaining a given scientific degree, nomination to a given position, receiving a scientific

Knorr, Strasser, and Zilian (eds.), Determinants and Controls of Scientific Development, 295–322.

prize, becoming a member of the editorial board of a journal, or a member of a scientific committee, refereeing someone's Ph.D. dissertation, obtaining an invitation to a scientific congress, etc.

The class Z will be partitioned into three subclasses, Z_1, Z_2 and Z_3, described as follows.

The elements of Z_1 are events consisting of solving research problems.

To describe classes Z_2 and Z_3, let us distinguish a class of certain scientific positions, characterized by the property that a scientist may get promoted (nominated, elected) to a given position, and he may also cease to occupy it. Typical examples here would be: professorship at a given university, membership in a committee of experts or editorial committee, etc. (as opposed, say, to professorship understood as a scientific degree, which is in effect permanent).

Accordingly, class Z_2 will consist of events which are either entering or leaving the positions as described above.

Finally, class Z_3 comprises the remaining scientific events. The typical examples here might be: refereeing someone's Ph.D. dissertation, obtaining a given degree, election to the Academy of Sciences, etc.

2. The concept of scientific career

The scientific career of a scientist will be identified with the process of occurrence of events from the class Z for him, and we shall associate with this process three functions, to be denoted by $A(t)$, $M(t)$ and $R(t)$.

In the sequel, we shall denote by $E_1, E_2, \ldots$ the consecutive events from the class Z which occur for the given scientist, and by $t_1, t_2, \ldots$ – the respective moments of their occurrence. Furthermore, $E_1^{(1)}, E_2^{(1)}, \ldots$ will denote the process of events from the subclass Z_1, which occur at the moments $t_1^{(1)}, t_2^{(1)}, \ldots$ (so that $\{E_n^{(1)}\}$ and $\{t_n^{(1)}\}$ are subsequences of $\{E_n\}$ and $\{t_n\}$ respectively). Similar notations will be used for subprocesses of events from classes Z_2 and Z_3.

3. The principle of choice

In order to describe the first function, $A(t)$, representing the scientist's scientific authority at the time t, the following assumptions will be made.

The research problems which are solved by a given scientist are taken out of a certain universe of research problems. This universe is, of course, changing in time, and forms a rather complex structure, subdivided according to disciplines, and interrelated by logical connections (e.g. a solution of one problem may automatically yield solutions to the others, or at least facilitate them, etc.). Each problem is assumed to have its own intrinsic difficulty (which may change as other problems become solved, new techniques are invented, etc.).

Needless to say, when a scientist chooses a problem for research from the universe of research problems available at the time, his choice is heavily constrained: obviously, he is not likely to choose problems from outside of his own domain of interest, and within it, his choice may be further constrained by his personal preference towards particular types of problems, the necessity of conforming to the general research plans of his institution, available funds, etc.

Next, we make the following assumption about the choice of the problem by a scientist, and the probabilities of success in solving the selected problems.

Each scientist has a certain level of creative talent[1], to be denoted by c. Given that he attempts to solve the problem with difficulty d, and devotes all his time to it, the probability that he will succeed in solving it before time t, equals $p(c, d, t)$. We assume that the function $p(c, d, t)$

(a) increases in c for all d, t;
(b) decreases in d for all c, t;
(c) increases in t for all c, d.

Thus, it remains to specify which problems will be attacked by a given scientist. Since the only factor distinguishing problems in the suggested model is their difficulty, the crucial question is reduced to specifying how difficult problems will be selected by a scientist as his research goals in the nearest future. Here we shall rely on a simplified version of the achievement motivation model of Atkinson (1964). Thus, it will be assumed that each scientist, in selecting his task difficulty by the appropriate choice of the level of difficulty of the problem, will select that (or those) research problem(s) among those feasible for him, for which the expression

$$(1) \qquad Cp(1-p)$$

attains its maximum; here p is the probability of success in solving the problem, and C is a certain constant[2], positive for those for whom the need of achievement exceeds the fear of failure, and negative for those for whom it is opposite.

It follows from (1) that a scientist for whom the need of achievement exceeds the fear of failure will choose problems for which the probability of success in solving them is as close to $\frac{1}{2}$ as possible (since for $C>0$ the parabola (1) attains maximum at $p=\frac{1}{2}$). On the other hand, scientists for whom the fear of failure dominates over the need of achievement will select problems for which the probability of success is as close as possible either to 0 or to 1 (i.e. they will choose either the most difficult or the easiest problem available; this follows from the fact that for $C<0$ the parabola (1) attains the maxima, in the range of p between 0 and 1, at the extreme points $p=0$ and $p=1$).

4. Some Consequences of the Principle of Choice

The above postulates concerning the dependence of the probability of success p on ability c, difficulty d, and time horizon[3] t, and the assumed model of choice, lead to the following conclusions.

THEOREM 1. *Assume that $C>0$, i.e. the need of achievement exceeds the fear of failure. Then*

(i) *for the same time horizon t, the more able is the scientist, the more difficult problem he will select;*

(ii) *the longer time horizon t is allowed (given the same creativity level c), the more ambitious project will be undertaken (i.e. the more difficult problem will be chosen).*

To prove this theorem, consider first the functions $p(c, d, t)$ for a fixed t. Qualitatively, for different values of c we have here a family of decreasing functions of d (assumption b), as presented on Figure 1. The arrow on this figure indicates the growth of c for different curves (i.e. the higher curve corresponds to a higher value of c; this follows from assumption a).

Consequently, the solutions of the equation $p(c, d, t)=\frac{1}{2}$ (marked by circles on Figure 1) have the property that to higher values of c there correspond higher values of d, which yields the first assertion of the theorem.

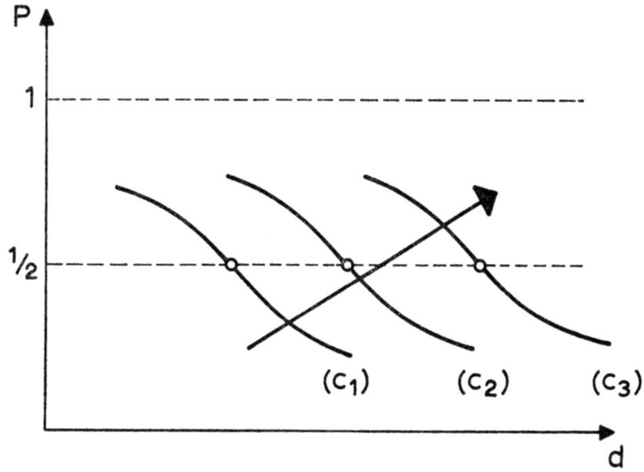

Fig. 1.

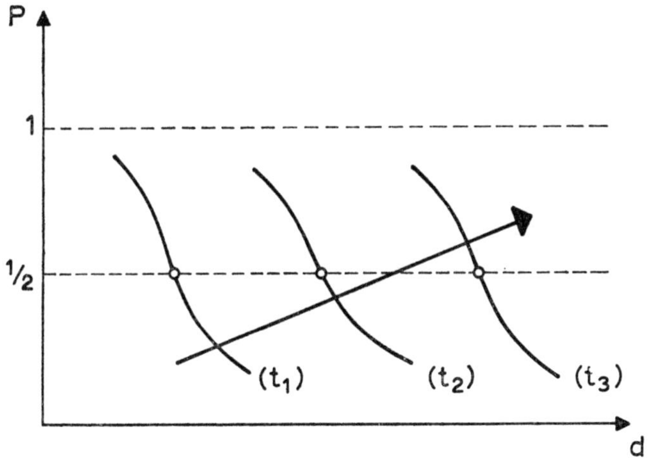

Fig. 2.

Consider now the analogous family of curves for a fixed value *c*, and varying values of *t* (Figure 2). This time the arrow indicates the growth of *t* for various curves (assumption c). By the same reasoning as before, we see that the solutions of the equation $p(c, d, t) = \frac{1}{2}$ have the property that the higher is the value of *t*, the higher will be the corresponding value of *d*, which gives the assertion (ii).

The situation for the scientists with $C<0$, i.e. for those whose fear of failure exceeds the need of success, is somewhat more complex, and at the same time, more unexpected. We have, then,

THEOREM 2. *Assume that $C<0$, and that the choice is to be made from a fixed set of problems with varying degrees of difficulty.*

(i) *Given two scientists, and the same time horizon for both of them, if the less able one chooses the easiest problem, so will the more able (and equivalently, if the more able scientist chooses the most difficult problem, so will the less able one).*

(ii) *If for a fixed time for solution, a scientist chooses the easiest problem, then given more time, he will also choose the easiest problem (and equivalently, if given more time he chooses the most difficult problem, then given less time, he will also choose the most difficult problem).*

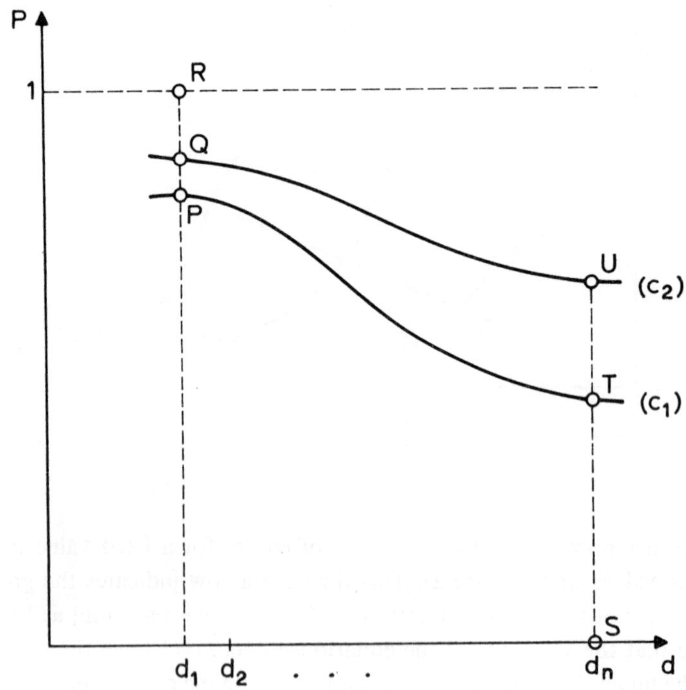

Fig. 3a.

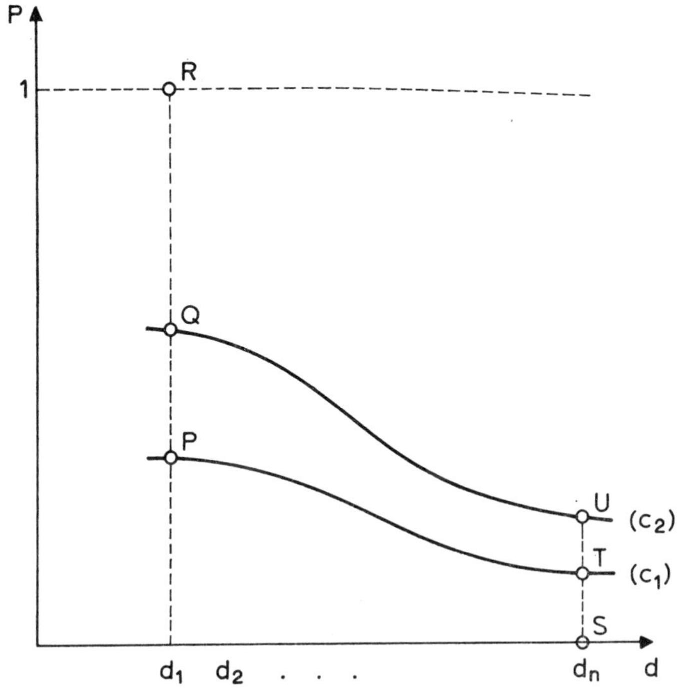

Fig. 3b.

To prove Theorem 2, consider first two scientists with abilities c_1 and c_2, and assume that $c_1 < c_2$. For fixed t, the functions $p(c_1, d, t)$ and $p(c_2, d, t)$ are given on Figures 3a and 3b. Here $d_1, \ldots, d_n$ denote the levels of difficulty of the problems of the set from which the choice is to be made (so that d_1 corresponds to the easiest, and d_n to the most difficult problem).

On Figure 3a we have the situation in which the less able scientist (with ability c_1) chooses the easiest problem: it is nearer from P to R than from T to S. Then for the more able scientist (c_2), it must also be nearer from Q to R than from U to S – which shows that he will also choose the easiest problem.

On Figure 3b we have the situation when the more able scientist (c_2) chooses the most difficult problem: it is nearer from U to S than from Q to R. But then it must also be nearer from T to S than from P to R – hence the less able scientist (c_1) will choose the most difficult problem too.

The proof of assertion (ii) is based on analogous reasoning: it suffices to consider figures similar to 3a and 3b where now the lower curve will correspond to a shorter period of time, and the upper – to a longer period of time. The rest of the proof is similar to the one given above, and will not be repeated[4].

5. An alternative choice model

The model of choice considered in the preceding section involved time limit t as a controllable variable, and consequently, the results were expressed in terms of the probabilities $p(c, d, t)$ of solving the problem (with difficulty d, by a scientist with ability level c) before time t. This model was therefore applicable to situations when the scientist in his choice is confronted with the time schedule for solutions, imposed by the general research plans, etc.

There are, however, situations when there are no time limitations imposed by external conditions. For such cases, one can design a different model, taking into account some other variables, such as persistence, and to some extent, also the changes in the ability level.

We shall then assume that time is measured in appropriate units (depending on the context, these may be years, months, days, etc.), and that – given that the scientist did not succeed in solving a chosen problem in a unit of time – his probability of solving it in the next unit of time may change.

Typically, the probabilities of solution should initially increase, due to gain in experience, elimination of approaches which have been tried and discarded, etc., and then they should decrease, as a result of increasing difficulty in finding new ideas, discouragement, frustration etc. One may also interpret this change as due to the change in the level of creative ability which initially increases, and then begins to decline (the latter interpretation is, of course, applicable only in the situation when one considers the life-histories of scientists, extended over periods of many years spent on attempts to solve a particular research problem).

The choice model will now be described as follows: we retain the assumption relating the probability of solution of the selected problem with the constant C characterizing the balance between need of achievement and fear of failure (i.e. we retain the maximization principle (1)). This time, however, the value p will depend on the parameters c, d and k, where c and d are as before (creative talents and difficulty of the problem), while k characterizes the persistence of the scientist.

To determine the probability $p=p(c, d, k)$, we assume that:

(a) The scientist works on a given problem for at most k units of time (i.e. if he does not succeed in solving it before the time k, he abandons the problem and starts working on a new one);

(b) In the first unit of time, the probability of solution is $p(c, d)$, depending on the creative talent c, and difficulty d as before (so that $p(c, d)=p(c, d, 1)$ in the previous notation). We assume that $p(c, d)$ increases with c and decreases with d.

(c) Given that in the nth unit of time the probability of solution was q, and the problem was not solved, if the scientist continues working on this problem in the next unit of time, his probability of solution is $f_{c,n}(q)$. We assume that $f_{c,n}(q)$ increases in arguments c and q for any n, while we make no assumption regarding the form of dependence of this function[5] on n.

Thus, in the first unit of time, the probability of solution is $p(c, d)$; in the second it equals $f_{c,1}(p(c, d))$, in the third it is $f_{c,2}(f_{c,1}(p(c, d)))$, and so on. It will be convenient to denote $p_1=p(c, d)$, $p_2=f_{c,1}(p_1)$, $p_3=f_{c,2}(p_2)$ etc., so that p_n is the probability of solving the problem in the nth unit of time, if it was not solved before.

Consequently, the probability that the problem will be solved equals

$$(2) \qquad p(c, d, k) = 1 - (1 - p_1)\dots(1 - p_k),$$

with $p_1, \dots, p_k$ defined above.

Now, from the hypotheses (a)–(c) it follows that the function $p(c, d, k)$ increases in c and k, and decreases in d.

Indeed, the fact that $p(c, d, k)$ increases with k is obvious from (2), since $(1-p_1)\dots(1-p_k)>(1-p_1)\dots(1-p_k)(1-p_{k+1})$.

Let now $c_1<c_2$ and denote by $'$ the probabilities relating to c_1, and by $''$ the probabilities relating to c_2. We have then $p_1'=p(c_1, d)<p(c_2, d)=p_1''$ by assumption (b), and $p_2'=f_{c_1,1}(p')<f_{c_2,1}(p_1'')=p_2''$ by assumption (c). Easy induction shows now that $p_n'<p_n''$ for all n. It follows that $1-p_n'>$ $>1-p_n''$, hence $(1-p_1')\dots(1-p_k')>(1-p_1'')\dots(1-p_k'')$, and we obtain $p(c_1, d, k)<p(c_2, d, k)$ as asserted. The argument showing that $p(c, d, k)$ decreases in d is analogous.

We may now formulate some theorems, similar to those in the preceding section. We mention here only the most important of them, and omit its proof, which rests on the same considerations as that of Theorem 2.

THEOREM 3. *Imagine two scientists with* $C<0$, *i.e. both with the property that fear of failure exceeds their need of achievement. Assume also that their ability levels are the same. Then, if the more persistent scientist chooses the most difficult among the problems, so does the less persistent scientist; and equivalently, if the less persistent scientist chooses the easiest problem, so does the more persistent one.*

6. The Scientific Authority Process $A(t)$

Consider now the sequence of events $E_1^{(1)}, E_2^{(1)}, \ldots$ occurring at times $t_1^{(1)}, t_1^{(2)}, \ldots$ for a given scientist. By definition, these are events consisting of solving research problems, and they are of varying levels of difficulty. Each of these events may therefore be assigned the value[6], say $q(E_1^{(1)})$, $q(E_2^{(1)}), \ldots$ reflecting the relative importance of a given event as a contribution to science, or equivalently, reflecting the scientific authority of the scientist, who solved the corresponding research problem.

The total scientific authority of a scientist at time t, denoted by $A(t)$, will be equal to the sum of all authorities due to the solutions of all problems which he accomplished prior to time t, with the provision that the authority is subject to a decay in time[7]. Thus, a contribution at time t of the authority due to the solution of a research problem, $E_k^{(1)}$, at an earlier moment $t_k^{(1)}$, equals

$$q\left(E_k^{(1)}\right) e^{-\alpha\left(t-t_k^{(1)}\right)},$$

where α is some positive constant reflecting the decay rate. Thus,

$$A(t)=\sum_{j: t_j^{(1)} \leqslant t} q\left(E_j^{(1)}\right) e^{-\alpha\left(t-t_j^{(1)}\right)}.$$

7. The Social Power Process $M(t)$

For defining the social power process $M(t)$, induced by the sequence of events $E_1^{(2)}, E_2^{(2)}, \ldots$, it will be assumed that to each position there corresponds a certain amount of social power. The latter concept will be understood as the possibility of exerting a direct and selective influence on the fate of others.

When a scientist advances to a given position, his social power increases by an appropriate amount, and when he stops occupying this position, his social power decreases by the same amount.

Denote by $m(E_j^{(2)})$ the change in social power associated with the occurrence of $E_j^{(2)}$; here the value $m(E_j^{(2)})$ may be positive or negative, depending on whether one advances to a given position, or looses it. We assume no decay in this case, so that

$$M(t) = \sum_{j:t_j^{(2)} \leqslant t} m(E_j^{(2)}).$$

Since obviously one cannot loose a position unless one first gets it, positive jumps in $M(t)$ must precede the corresponding negative jumps, and consequently, $M(t)$ is always nonnegative.

8. The recognition process $R(t)$

Consider finally the third sequence of scientific events, $E_1^{(3)}, E_2^{(3)}, \ldots$, occurring at times $t_1^{(3)}, t_2^{(3)}, \ldots$. Here with each event we may associate a value $r(E_k^{(3)})$ reflecting the degree of recognition associated with the occurrence of $E_k^{(3)}$. We assume no decay in this case either, and take again a simple additive model

$$R(t) = \sum_{j:t_j^{(3)} \leqslant t} r(E_j^{(3)}).$$

We shall mostly be interested in the increments of the process $R(t)$; clearly, $R(t+t') - R(t)$ reflects the total recognition received by the scientist between the times t and $t+t'$.

9. Problems of measurement

Clearly, the values $q(E_j^{(1)})$, $m(E_j^{(2)})$ and $r(E_j^{(3)})$ for events from classes Z_1, Z_2 and Z_3 need to be measured before one can, given the detailed biography of a scientist, plot his functions $A(t)$, $M(t)$ and $R(t)$. The appropriate scaling techniques are based on methods developed in psychophysics, and known in measurement theory under the name *cross-modality matching* (see Krantz *et al.*, 1971). These methods will be discussed in the last paragraph. Inexact as they may be in the present case, they provide at least rough estimates of the relevant values, and thus enable us to plot the functions $A(t)$, $M(t)$ and $R(t)$ measured on the same ratio scale. Thus, all three functions may be plotted on the same diagramme, with some conventionally chosen unit on the vertical scale. A typical example might look as follows (see Figure 4).

It ought to be remarked that $R(t)$ is always increasing in jumps, $M(t)$ changes in jumps either up or down, while $A(t)$ increases in jumps and decreases exponentially between them.

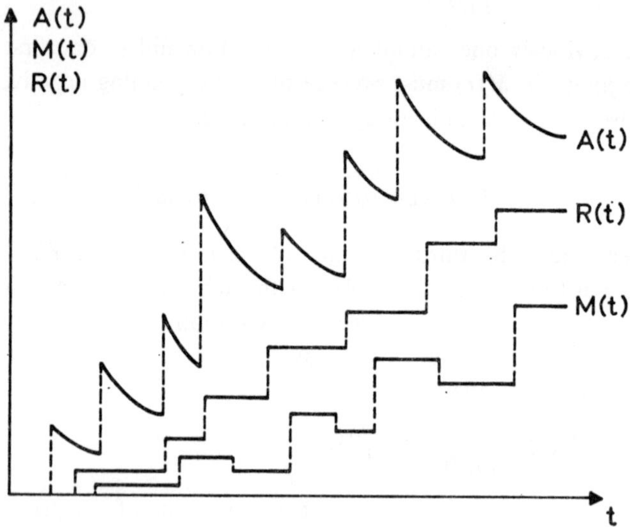

Fig. 4.

10. Classification of Careers

We may now posutlate that in 'normal' circumstances, the change in $A(t)$ should induce the corresponding changes in the other two processes: when a man's scientific authority grows up during a certain period of time, his recognition process $R(t)$ should also increase appropriately during this time, and the same ought to be true for the social power process $M(t)$. There may be a time lag involved here: the increase in $R(t)$ and $M(t)$ should start occurring some time after the increase in $A(t)$.

These qualitative considerations allow us to describe the individual career, and point out possible pathologies expressed in terms of mutual relations between functions $A(t)$, $M(t)$ and $R(t)$.

The 'pathological' cases would involve situations when either the growth of $A(t)$ is not matched by the growth of $M(t)$ and/or $R(t)$ (which might be termed 'under-career'), or vice versa, when the growth of $M(t)$ and/or $R(t)$ is not matched by the corresponding growth of $A(t)$ (which might be termed 'over-career').

11. Taxonomy of Scientists

We shall now introduce a taxonomy of scientists, and try to deduce some predictions connecting the type of the scientist with the type of his career.

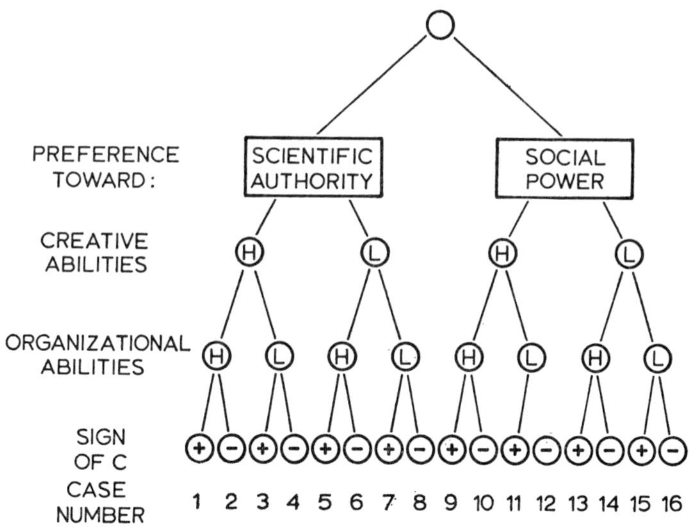

Fig. 5.

We shall introduce four dichotomous classifications of scientists, leading therefore to 16 taxonomical categories.

The first dichotomy (see Figure 5) will concern the direction of preference between scientific authority and social power. We shall then distinguish those scientists who attach more weight to collecting the individual authority than to attaining social power, and those for whom the situation is opposite.

In great simplification, from the psychological point of view, the main factors which cause that a scientist concentrates his attention more on scientific authority than on social power are curiosity, permanent need for self-realization in consecutive research problems, combined with relatively small social interest. Here the reward is simply the satisfaction from solving a research problem[8].

The scientists who concentrate rather on social power than scientific authority differ from the preceding ones as follows: besides curiosity and permanent need for self-realization in research problems, they have higher social interests, the need for influencing others, and the need of creating and executing wider research programmes.

The direction in which this need for influencing others will be executed depends in turn (again, in an oversimplification) on factors such as conformism vs. non-conformism, and self-orientation vs. orientation on the problem.

The second categorization concerns the creative talents – two categories will be distinguished here: those with high, and those with low creative talents. Similarly, we shall consider also organizational talents, and again, we shall simplify the situation by considering only two categories (high and low organizational talents).

Finally, the last categorization will concern the sign of the constant C, which appears in the assumption about the choice of the level of difficulty of the problem (scientists with positive C are those for whom the need of achievement exceeds fear of failure, while scientists with negative C have higher fear of failure than need of achievement).

The full list of all 16 categories is presented in Figure 5.

12. Prediction of Careers

In the present section we give some predictions of careers for scientists from different categories. These predictions will be expressed in terms of functions $A(t)$, $M(t)$ and $R(t)$.

Case 1. Because of general preference towards scientific authority, the scientist will concentrate mainly on research problems. Since $C>0$, he will choose problems which have a probability of success (for him) close to $\frac{1}{2}$, hence he may be assured of success in about half of the cases. As his creative talents are high, the problems which for him have a probability of success close to $\frac{1}{2}$, hence are of moderate difficulty, will be considered as rather difficult for others. Consequently, his authority function $A(t)$ will increase fast, and by fairly big jumps.

Since his authority will increase, there ought to be a corresponding increase in the social recognition function $R(t)$, and since he has high organizational talents, he will also quickly advance in occupying higher and higher positions, hence his social power function $M(t)$ will also increase.

This leads to a career of the type with high values of all three functions $A(t)$, $M(t)$ and $R(t)$.

Case 2. This scientist differs from the preceding one only by the sign of C (negative). Thus, as in Case 1, the scientist will concentrate on research problems, except that he will select those problems which are either very easy for him, or extremely difficult for him. Since his creative talents are high, the problems which are easy for him are considered of moderate difficulty for the majority of others, while those problems which are extremely difficult for him, will pose an ernormous difficulty for the majority of scientists.

The career depends now on whether the scientist in question will manage to solve at least one of the extremely difficult problems or not. If he does solve such a problem, his authority function $A(t)$ goes up by a very high jump [9], and, even considering the decay, will stay on high levels for a considerable period of time (a Nobel prize scientist, who does not produce much after getting his prize-winning result, yet still retains his authority, may serve as an example).

The function $R(t)$ goes up accordingly, and so does $M(t)$ – the latter due to the fact that the scientist in question has high organizational talents.

After such a big jump in authority, several possibilities are open. The scientist may continue his efforts working on extremely difficult problems, if he feels that the results on a lower level would not satisfy him, and may even be thought of as detrimental to his authority (as they are bound to be compared with the main achievement). Another possibility is that he may concentrate on easy problems (i.e. the problems of moderate difficulty for other scientists), and shift gradually away from pure research to organizational projects (where he also is assured of a success, having high organizational talents).

On the other hand, if the scientist does not manage to solve any of the extemely difficult problems, he will continue solving problems which are easy for him. These being of moderate difficulty for the majority of scientists, will cause that his authority function $A(t)$ will continue to increase, and there would be corresponding increases in $R(t)$ and $M(t)$. The scientist will consider the latter somewhat unjustified, as in his own eyes, the increase is not matched by a corresponding increase in $A(t)$.

Thus, Case 2 is not so easily predictable as in Case 1, since the course of the career depends here on the occurrence of events with small probability.

Case 3. As in case 1, the growth of the authority function $A(t)$ will be fast, since the scientist will choose problems with probability of success close to $\frac{1}{2}$, hence will solve a fair number of them. These problems are considered of high difficulty by the majority of scientists (as the scientist in question has high creative talents).

Now, the growth of $A(t)$ will be matched here by the corresponding growth of $R(t)$, but not necessarily of $M(t)$: the latter function is likely to grow up initially, and then either drop down, or – more likely – to stay on more or less the same level (the reason being that with low organizational talents, the scientist will either refrain from accepting higher positions, or will be recognized as organizationally inept, and not suitable for promotion).

Case 4. The situation here is similar to that in Case 2. If the scientist manages to solve one of the extremely difficult problems, he is assured of an enormous jump in $A(t)$, and consequently, in $R(t)$. The function $M(t)$, after an initial growth will cease to increase, since the man has no organizational talents.

If, on the other hand, the scientist does not solve any of the very difficult problems, he will build up his authority by solving a large number of problems easy for him. This will cause a steady increase of $A(t)$ and $R(t)$, but this increase will be not too fast. The function $M(t)$ should not grow up at all.

Cases 5–8 correspond to scientists with low creative talents, and are therefore of less interest. Only one example will be discussed, namely that of Case 5.

Here the scientist will attack problems which have the probability of solution close to $\frac{1}{2}$, hence he should solve about half of them. However, since his creative talents are low, these will be problems rather easy for most of the scientists, and there will occur a phenomenon opposite to that in Case 2: the scientist's authority, as perceived by others, will grow up slowly, if at all, while in his own opinion, it will increase rather fast. The function $R(t)$ will not grow up, as its growth depends mostly on the social recognition of the scientist's authority. Thus, after a period of disappointment, the scientist in question is likely to turn to the area where he is assured of a reasonable success, namely to advancing in positions (he has high organizational talents). We shall have here a situation when $A(t)$ and $R(t)$ grow up slowly, if at all, while $M(t)$ increases.

Case 6 is similar to Case 5, except that $A(t)$ will be based mostly on solutions of problems easy for the scientist in question, hence considered as trivial or almost trivial by the majority. Thus, it will assume even smaller values than in Case 5, leading to a corresponding small increase in $R(t)$, and again, the scientist might try to build up his function $M(t)$ where he is assured of a limited success.

Cases 7 and 8 correspond to scientists with low creative and organizational talents, and their careers are easily predictable.

Let us now turn to some of the cases of scientists whose general preference is towards social power over scientific authority. Here the interesting Cases are 9 and 10, those of scientists with high creative and organizational talents (Case 9 differs from Case 10 only by the sign of C, determining the level of difficulty of the attacked problems).

In both cases, after an initial build up of the authority function $A(t)$, which in any case is needed to start the career, both scientists will turn to organizational activities, where they are both assured of success, because of high organizational talents. The difference between Cases 9 and 10 lies in the fact that while in Case 9 the scientist will concentrate on goals which assure him of about 50% chance of success, in Case 10 the scientist will be, on the one hand, less ambitious in his goals, and on the other hand, he may pursue a perhaps somewhat unrealistic, or at any rate, extremely difficult goal.

Thus, in Case 9 we shall have the following picture: the function $A(t)$ grows up fast, and then slows down or even ceases to grow, when the man starts to become involved in organizational rather than research problems.

The function $R(t)$ behaves in the same way, with some time lag, while $M(t)$ begins to grow up quickly.

In Case 10 we have the same picture, except that the growth of $M(t)$ is less predictable: it is either slower than in Case 9, or else, at some moment it begins to climb up rapidly.

The remaining Cases 11–16 can be analysed in much the same way, and will not be discussed here.

13. The Empirical Perspectives

As already mentioned, the present model is an extension of a model of scientific career suggested earlier. In this earlier model, the situation considered was to a large extent similar, except for some simplifications. It was assumed that each scientific event may be assessed on three scales: scientific authority, social power, and general value. The latter served mainly to determine whether the scientist puts more weight to scientific authority or social power (the empirical procedure involved determination of the three values for each event from a certain class of events, and then determining the correlation between general value and social power on the one hand, and between general value and scientific authority on the other hand. These two correlations served then as an indication of the general direction of preference). For some results, see Table I (pp. 318–320).

The events considered were put in rather general terms, e.g. "To become a member of an editorial board of a leading foreign journal", "Publish a paper in a Polish journal," etc.

A reasonable degree of success in preliminary studies on determination of such values made it possible to suggest a more ambitious, and at the same time, more exact measurements, where scaling would concern specific events (e.g. nominations or elections which have really occurred, particular publications, etc.).

Such a procedure made it also possible to suggest considering three functions, $A(t)$, $R(t)$ and $M(t)$ describing the career, instead of only $A(t)$ and $M(t)$ as in the preceding model. Thus, it is possible to have more subtle predictions, hence also a greater chance of testing the model.

Of course, these possibilities are obtained at the price of much more difficult experiments: in order to obtain judgements of, say, values of particular scientific papers of a scientist given by someone else, one needs to find a scientist who is familiar with the whole scientific production of the subject. This is likely to restrict severely the scope of research, either to scientists whose production is not too rich, or to scientists in research teams who cooperate closely enough to be well familiar with all achievements of other members of the team.

14. Some additional hypotheses and tests based on psychological variables

In this section we shall assume that in addition to the values $A(t)$, $R(t)$ and $M(t)$ for a given scientist, it is also possible to measure two psychological variables: the level of anxiety X and the level of one's satisfaction from work Y. These may be measured by appropriately constructed questionnaires.

A partial validation of the model may be obtained by testing appropriate hypotheses relating the values A, M, R, X and Y. Some of these hypotheses are:

HYPOTHESIS 1. The level of anxiety X should be positively associated with each of the differences $|A(t)-M(t)|$, $|A(t)-R(t)|$ and $|M(t)-R(t)|$.

The intuitive justification lies in the fact that each of these differences measures an aspect of 'anomaly' of one's career. Thus, for instance, large positive $A(t)-M(t)$ indicates that one's authority is not reflected adequately in social power, while large negative values of this difference show that one is placed in a position one does not really deserve because of insufficient scientific authority. Both of these should be reflected in higher anxiety.

In a similar way, the difference between $A(t)$ and $R(t)$ shows to which extent one's authority is matched by social recognition, and large positive or negative values, being not normal, should be associated with anxiety.

In a similar manner one could argue about the difference between $R(t)$ and $M(t)$.

HYPOTHESIS 2. The level of anxiety X should be negatively associated with the level Y of one's satisfaction from work.

The intuitive justification of this hypothesis requires no comment.

Another intuitive hypothesis is:

HYPOTHESIS 3. The level of anxiety ought to be negatively associated with the ratio $A(t)/(t-t_0)$, where t_0 stands for the moment of the beginning of the career.

The justification of this hypothesis might be illustrated by Figure 6. It presents the authority functions $A_1(t)$ and $A_2(t)$ of two scientists, who reached about the same level at time t. One of them, however, used 20 years to attain the level which the other reached in 8 years. One could expect that the second scientist has less reasons to be anxious.

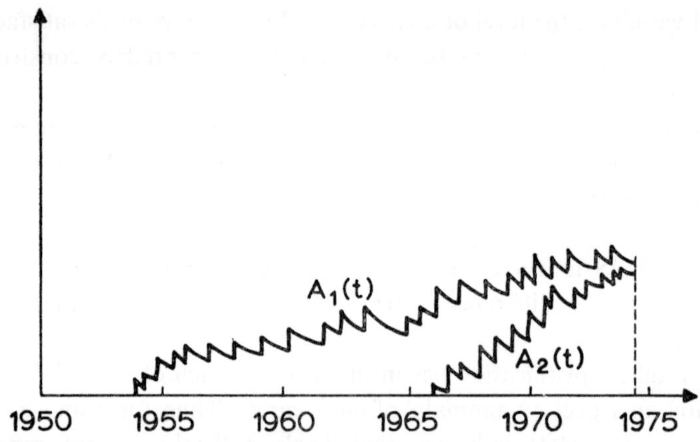

Fig. 6.

At present, extensive research on scientific careers is being conducted along the lines outlined above. Besides the measurements of parameters of the career, level of anxiety and level of satisfaction of one's work, the other variables measured refer to self-evaluation and evaluation by others, especially within the selected scientific teams, with respect to different criteria pertaining to the efficiency of individual and team research, etc. The results of this study will be published elsewhere. Here only some preliminary data are presented, concerning evaluations of scientific events on scales of authority, social power and general value, by three scientists: scientist A is after Ph.D. and before habilitation, scientist B is after habilitation, while scientist *C* is a professor and the director of a scientific institute.

The numerical values here are of independent interest; the reference point is habilitation, which is assigned the value 1000 on each of the three scales for all three persons.

One of the points worth noting in the data presented in Table I are the values assigned to the basic sequence: Ph.D. – habilitation – professorship – election to the Academy of Sciences. Other notable differences concern the relations between social power, scientific authority and general value for the three persons tested. It is also worth observing which events are assigned the maximal values on each of the scales, by different subjects.

The above data are, of course, only preliminary, and not sufficient to derive any justifiable conclusions. However, it seems that one might formulate the following tentative conjecture: the general values of scientific events are overestimated in the case of events which one should want to achieve, and underestimated in the case of events one has achieved. In particular, this may be expected for persons with low self-evaluation and high level of aspiration.

This conjecture could be tested by studying intra-individual changes of evaluations by the same person during several years.

One can also formulate the following conjecture: the interindividual variability of evaluations is higher for persons with different social positions than for those with the same social position.

Perhaps the relatively low evaluation of those events which have already been attained (see in particular person *C*) are connected with a certain value system, according to which one should rather hide one's own achievements and one's own satisfaction, since otherwise the behaviour is perceived as lack of modesty.

15. Discussion. Methodological Remarks

The discussion below will concern, firstly, the cognitive advantages and disadvantages of the suggested modelling approach to the concept of scientific career, and secondly, the problems of measurement of the relevant values of scientific events (authority, social power, and recognition).

Regarding the first problem, it is obvious that the described model oversimplifies considerably the mechanisms underlying the scientific career. It needs in fact little imagination to point out nearly endless lists of factors which in some circumstances might and do play an essential role, and which were omitted in the assumptions. However, concentrating on some mechanisms only and stating in a formal way the mode of their operation (which is the essence of every model) has some distinct advantages. The most important among them is that such a procedure allows going beyond purely descriptive analyses, and formulating testable hypotheses, the latter obtained deductively from some clearly stated premises. The eventual truth or falsity of these hypotheses is of much less importance than the very fact that they have been deduced from premises and that they may be, at least potentially, falsified.

If the hypotheses are rejected, one may examine particular underlying premises to detect that one, or ones, which are not satisfied in the practical situation at hand. If the model is not completely and unambiguously falsified, one can then assess how well it fits the real data. This enables us to assess, in turn, the contributions of other factors, not included in the model.

The above considerations should, hopefully, give enough justification for omitting in the considerations a number of important and well known factors influencing the scientific career.

Finally, the last problem which needs clarification, crucial for the whole consideration, concerns the measurement of pertinent aspects of scientific events (scientific authority, social power, and social recognition).

The methods employed here were based on psychophysical scaling theory. Since they may not be too widely known outside psychology, it is necessary to outline here at least the main principles.

The techniques are based on the theorem, which may be found in Krantz *et al.* (1971), concerning the so-called *cross-modality matching*.

In a simplified version, imagine that we have two sets of some objects, say A and B, and a weak ordering (i.e. a binary transitive and connected relation) $\succcurlyeq$ defined on the set $A \times A \cup B \times B$. Thus, one orders here pairs of pairs, the elements within each pair coming from the same set, either A or B. The symbol $(a_1, a_2) \succcurlyeq (b_1, b_2)$ denotes the fact that the pair (a_1, a_2) precedes the pair (b_1, b_2) (naturally, both $(a_1, a_2) \succcurlyeq (b_1, b_2)$ and $(b_1, b_2) \succcurlyeq (a_1, a_2)$ are possible, in which case we write $(a_1, a_2) \sim (b_1, b_2)$).

Now, the empirical nature of the relation $\succcurlyeq$ is irrelevant for the assertion of the theorem; in the case for which the techniques were applied, and which actually inspired the research on this theorem, $(a, b) \succcurlyeq (c, d)$ means that objects a, b are judged in some sense to be 'more apart' than the objects c, d. Thus, $(a, b) \sim (c, d)$ means that the 'difference' between a and b is the same as the 'difference' between c and d.

Whatever the empirical meaning of the relation $\succcurlyeq$, for the assertion of the theorem it is required that this relation satisfies certain formal conditions (axioms of the corresponding measurement system). We shall not list all of these axioms here; instead, only some of them will be presented.

Among others it is required that the transitivity in the following form holds: if $(a, b) \succcurlyeq (c, d)$ and $(b, e) \succcurlyeq (d, f)$, then $(a, e) \succcurlyeq (c, f)$. Another condition is 'solvability' in the form: for every pair (a, b) of elements of A, there is a pair (c, d) of elements of B such that $(a, b) \sim (c, d)$. This means that the set B must be 'rich': every pair in the set A must be matched by a pair in the set B.

Given that the axioms are satisfied, one can prove that there exist two functions, f_1 and f_2, real valued and positive, defined respectively on A and B, such that for all $a, b \in A$ and $c, d \in B$, we have

$$(a, b) \succcurlyeq (c, d) \text{ if, and only if } f_1(a)/f_1(b) \succcurlyeq f_2(c)/f_2(d)$$

(i.e. the ratios of values of f_1 and f_2 reflect the order $\succcurlyeq$). Taking logarithms, one can convert this expression into the one involving differences.

The functions f_1 and f_2 are unique up to similarity, that is, the 'scale values' f_1 and f_2 are unique up to the choice of unit (the measurement is on a ratio scale).

The theorem is relatively new, but the technique which it describes dates back to Fechner, and was later developed by Stevens. In standard psychological applications, elements of A and B are perceived stimuli, such as length, weight, brightness, loudness, etc.; in some other applications elements of one of the sets were taken as psychophysical, and the other – as psychological continuum (seriousness of crimes being one of the best known examples). In case of reported research, each of the psychological continua such as authority, social power, etc. was matched with the psychophysica continuum of perceived length. The choice of the latter continuum was dictated by the simplicity of the procedure: the subject was to produce two line segments (by marking the appropriate lengths on given lines) whose perceived ratio of lengths would correspond to his perception of the ratio of scientific authorities (resp. social power, etc.).

The procedure was not validated, and consequently, nothing is known as yet about the errors of measurement in the present case. The validation involves repetition of the experiment using another continuum, say, brightness, and comparison of the results (the relations between perceived brightness and perceived length are known, hence it is possible to compute the values which ought to appear in matching with brightness).

Such a validation with brightness requires, of course, somewhat more complex equipment than that used for experiments with matching the perceived length (a subject must have the possibility of setting the lights so as to produce the ratio of intensities equal to the perceived ratio of scientific authorities, etc.).

The results presented in Table I are only preliminary, and the report of the full experimental findings will appear elsewhere, after the results are collected and analysed.

TABLE I

Scientific authority $q(E)$, social power $m(E)$ and general value $w(E)$ for selected scientific events E, and three subjects, A, B and C

E	$q(E)$			$m(E)$			$w(E)$		
	A	B	C	A	B	C	A	B	C
Habilitation	1000	1000	1000	1000	1000	1000	1000	1000	1000
Co-authorship of a paper published in Poland	90	480	2000	0	0	0	40	130	210
Publishing a book in the West	700	1020	2100	0	0	0	630	740	740
Publishing a paper in a leading foreign journal	850	830	9800	0	0	0	550	400	2100
Publishing a book in Poland	510	860	1800	0	0	0	360	420	315
Co-authorship of a paper published in a leading foreign journal	150	690	1800	0	0	0	300	230	320
Publishing a paper in a leading Polish journal	190	730	9800	0	0	0	190	310	2100

Table I (Continued)

E	*q(E)*			*m(E)*			*w(E)*		
	A	*B*	*C*	*A*	*B*	*C*	*A*	*B*	*C*
Enthusiastic citation by a leading foreign scientist	940	1040	1000	0	0	0	1140	690	960
Membership in a committee of experts	720	950	2800	820	2270	9000	440	90	4500
Membership in a foreign scientific society	20	220	610	0	0	1000	40	30	630
Becoming an expert of UNESCO	1090	980	1320	1240	2980	3000	1190	400	1890
Becoming a member of a scientific committee in the Academy	450	870	450	170	500	6000	690	90	450
To give interviews in TV, etc.	30	310	280	270	280	1400	80	45	450
To become the head of a team	250	1120	5000	840	1150	9000	540	75	2000
To become the director of an institute	1260	1440	1100	2240	1760	12500	1170	175	500
To be elected President of the Academy	1850	1800	1150	2510	3120	1250	1290	2250	200
To become Associate Professor without habilitation	810	200	550	570	1000	1500	675	0	600
To become a professor	1160	1220	1150	1510	1300	1870	1200	1875	1700
To get a Ph.D.	650	310	2000	300	390	1620	480	200	2000
To be elected to the Academy	1620	1790	900	1700	1420	1870	1350	2500	950
To become a member of the editorial board of a leading foreign journal	1350	1610	5500	1800	90	720	1290	1650	3500
Refereeing for a publishing house	750	620	1100	760	790	1890	720	120	700
To be recognized by the whole scientific community in Poland	1850	1660	2750	1490	0	3600	1440	3090	3500

Table I (Continued)

E	*q(E)*			*m(E)*			*w(E)*		
	A	*B*	*C*	*A*	*B*	*C*	*A*	*B*	*C*
To become a member of an editorial board of a Polish journal	430	900	550	540	1100	590	370	90	370
To referee a Ph.D. or habilitation thesis	800	1040	720	940	1580	7200	950	90	520
To be the editor-in-chief of a scientific journal	920	1330	1100	1620	1750	9000	1300	90	1920
To be invited to a conference abroad	920	1200	5500	0	0	0	1380	800	5500
To receive a scientific prize of the institute	800	500	1260	0	0	0	560	160	825
To receive an honorary doctorate	1600	1540	3000	0	220	0	280	370	3630
To lecture abroad	1270	1570	5500	0	0	0	1720	1060	5500
To become an honorary member of a foreign scientific society	1140	1550	5500	0	0	0	1800	430	1000
To be invited to a conference in Poland	100	230	1100	0	0	0	840	50	2200
To receive the state prize	1440	1185	3350	0	350	0	1700	260	2800
To be a consultant in a scientific institute	250	915	450	810	780	1000	760	45	500

Institute of Praxiology,
Polish Academy of Sciences,
Warszawa

NOTES

[1] The level of creative talent may, of course, change in time, depending on the scientist's age, number of problems solved, etc. A model which takes this effect partially into account will be presented in the next section.

[2] In the theory of Atkinson, the number *C* is the difference $M_s - M_f$ between the

achievement motive M_s and fear of failure motive M_f. For the purpose of this paper, only the sign of this difference is essential.

Incidentally, the contemporary studies on motivation towards achievement (see in particular Heckhausen, 1967, 1968) indicate that in the choice of subjective probability of success, the essential role is played by such factors as need of affiliation and fear of rejection. Taking these factors into account in the model of scientific career might perhaps lead to more subtle differential predictions. The essential simplification in the assumption of the paper is that solving a research problem is treated in a binary way only.

[3] The time horizon may be imposed by the general plan of the institution employing a given scientist, length of the research grant, etc.

[4] Incidentally, the above theorems (especially the second) may serve as a basis for testing the adequacy of the model of Atkinson. One could namely design simple experiments for checking the predictions given in Theorem 2.

[5] As stated, typically this function should first increase and then decrease with n. This property, however, is not intervening in the proof, and there is no need to assume it explicitly.

[6] These values, as well as the values of social power and recognition connected with particular events, are measured on ratio scale. One should point out that these values reflect the subjective judgments of a particular scientist, and are not to be treated as representative for group judgments, etc. Thus, $q(E)$ gives the assessment of scientific authority connected with the event E (more precisely, the increase of authority of the author of E), as perceived subjectively by the evaluating person.

[7] Inclusion of the phenomenon of decay of scientific authority has been suggested by R. Taagepera (personal communication). It is assumed that the decay is constant in time, i.e. if $f(t)$ is the contribution to scientific authority at time t from an event which occurred some time earlier, then $df/f = -\alpha$, where α is some positive constant (depending on the discipline). This yields the exponential decay given below.

[8] Of course, other factors may enter here, such as the desire to achieve success in competition with other scientists, prizes, honoraria, and finally – simple vanity.

[9] More precisely, this jump occurs not at the moment of the solution of the problem, but at the moment when it is recognized.

BIBLIOGRAPHY

Atkinson, J. W.: 1964, *An Introduction to Motivation*, Van Nostrand, Princeton.

Frost, P. A. and Whitley, R. D.: 1971, 'Communication Patterns in a Research Laboratory', *R and D Management*, **1**, 71–78.

Heckhausen, H.: 1967, *The Anatomy of Achievement Motivation*, Academic Press, New York.

Heckhausen, H.: 1968, 'Achievement Motive Research: Current Problems and Some Contributions towards General Theory of Motivation', in W. J. Arnold (ed.): *Nebraska Symposium on Motivation*, Nebraska Univ. Press, Lincoln, Nebraska, pp. 103–174.

Klima, R.: 1973, 'Bemerkungen zur Theorie der "Scientific Authority"'. Manuscript of the lecture given in the Sozialwissenschaftlichen Ausschuss der Vereins für Sozialpolitik, Hofgeismar/Taunus.

Klima, R.: 1974, 'Scientific Knowledge and Social Control in Science: the Application of a Cognitive Theory of Behaviour to the Study of Scientific Behaviour', in R. D.

Whitley (ed.) *Social Processes of Scientific Development*, Routledge and Kegan Paul, London, pp. 96–121.

Krantz, D. L., Luce, R. D., Suppes, P., and Tversky, A.: 1971, *Foundations of Measurement*, Academic Press, New York.

Lemaine, G., Lecuyer, B. P., Gromis, A., and Barthelemy, C.: 1973, *Les Voies du Succès*, Editions de la Société des amis du Centre d'Études Sociologiques, Paris.

Mullins, N. C.: 1974, 'Aspects of a Sociological Theory of Revolutionary Science', paper presented at the Conference on Determinants and Controls of Scientific Development, Graz (see Section B of this volume).

Nowakowska, M.: 1973, 'A Model of a Scientific Career' (in Polsih), *Zagadnienia Naukoznawstwa*, **4** (36), 503–520.

Nowakowska, M.: 1975, 'A Logical Structure of the Development of a Scientific Domain', in *Kölner Zeitschrift für Soziologie und Sozialpsychologie*, special issue 'Wissenschaftssoziologie', Westdeutscher Verlag, Opladen.

Weingart, P.: 1974, 'On a Sociological Theory of Scientific Change', in R. D. Whitley (ed.), *Social Processes of Scientific Development*, Routledge and Kegan Paul, London, pp. 45–68.

Whitley, R. D.: 1974a, 'Competition, Division of Labour and Autonomy in Scientific Development', unpublished manuscript, Univ. of Manchester.

Whitley, R. D.: 1974b, 'Cognitive and Social Institutionalization of Scientific Specialities and Research Areas', in R. D. Whitley (ed.), *Social Processes of Scientific Development*, Routledge and Kegan Paul, London, pp. 69–95.

JERRY GASTON*

SCIENTISTS FROM RICH AND POOR COUNTRIES

1. INTRODUCTION

The state of knowledge regarding the factors promoting or retarding the growth and development of science is still in a condition of ignorance. In the first place, when we talk about growth and development we are almost always talking about the economic, political, and sociological factors that promote development. These are presumed to be the most crucial determinants, because we certainly believe that science is *possible* in all groups of peoples, regardless of their geographical location. The relative neglect by experts in these scholarly disciplines has resulted in our low level of knowledge.[1] Fortunately that situation is beginning to change as the sociology of science is beginning to grow and mature. At this time, however, our general impressions about the state of science and of scientific development in poor countries come from unsystematic observations.[2] These observations are not incorrect, but they are necessarily limited in their generalizability and applicability.

If science is poorly developed in poor countries, what does this mean? First, it means that science is not well organized; second, that financial support is inadequate; and, third, that the prospects for a scientific career are not great. Consequently, not many discoveries are made in such countries, and on average, the science that is done is thought to be pretty poor.

Professor Abdus Salam (1966:461) tells the story about Saif-ud-din Salman who gave a good account of what it's like to be an intellectual in a country that does not understand or appreciate his interests and talents:

Five hundred years ago – around 1470 A.D. – Saif-ud-din Salman, a young astronomer from Kandhar, working then at the celebrated observatory of Ulugh Beg at Samarkand, wrote an anguished letter to his father. In eloquent words Salman recounted the dilemmas, the heart-breaks, of an advanced research career in a poor, developing country:
"Admonish me not, my beloved father, for forsaking you thus in your old age and

Knorr, Strasser, and Zilian (eds.), Determinants and Controls of Scientific Development, 323–342.

sojourning here at Samarkand. It's not that I covet the musk-melons and the grapes and the pomegranates of Samarkand; it's not the shade of the orchards on the banks of Zar-Afshan that keeps me here. I love my native Kandhar and its tree-lined avenues even more and I pine to return. But forgive me, my exalted father, for my passion for knowledge. In Kandhar there are no scholars, no libraries, no quadrants, no astrolabes. My star-gazing excites nothing but ridicule and scorn. My countrymen care more for the glitter of the sword than for the quill of the scholar. In my own town I am a sad, a pathetic misfit. It is true, my respected father, so far from home men do not rise from their seats to pay me homage when I ride into the bazaar. But some day soon all Samarkand will rise in respect when your son will emulate Biruni and Tusi in learning and you too will feel proud."

Salam later gives his own experience as a young scientist in his native Pakistan. He writes that:

The greatest obstacle of all lies in the very low probability of having the opportunity to work with the few men – in the case of India and Pakistan, the Siddiquis, the Usmanis, the Menons, the Sarabhais, the Seshachars – at the few centres of excellence, who appreciate at all the demands of a research career and who run laboratories which are reasonably well equipped. There are just too few scientists who retain the creativity of which they gave promise when young and there are therefore too few to train younger scientists through a fruitful master-apprentice relationship. ... The colleges which provide a very large proportion of undergraduate education in India and Pakistan have grown up in a tradition of concentrating such resources as they have on the instruction and moral formation of undergraduates. I shall always remember my first interview with the head of the premier college in Pakistan, which I joined after a spell of theoretical work in high-energy physics at Cambridge and Princeton. My chief said: We all want research men here, but never forget we are looking more for good, honest teachers and good honest college men. This college has proud traditions to uphold. We must all help. Now for any spare time you may have after your teaching duties, I can offer you a choice of three college jobs: you can take on wardenship of the college hostel; or be chief treasurer of its accounts; or if you like, become president of its football club. As it was, I was fortunate to get the football club (Salam, 1966:462–3).

There are numerous articles about science in poor countries.[3] This paper will not deal with general observations of science in poor countries, but rather it will address two specific questions:

(1) What are the actual career patterns of scientists in poor countries?

(2) How are these patterns different from scientists in rich countries?

I am not going to argue that scientists in poor countries do not have great obstacles to overcome. Neither am I going to argue that scientists in poor countries are as well or better off than scientists in rich countries. What I intend to do is to raise some questions about the notion that science is virtually impossible among our colleagues in poor countries. I shall do this first by presenting some data on scientific careers and then I shall make some comparisons between scientists in rich countries with these scientists from poor countries.

The information on scientists in poor countries was obtained from data collected by The International Center for Theoretical Physics (Trieste, Italy).[4] In an attempt to locate scientists and publish a directory about their educational backgrounds, research interests, and research accomplishments, The International Center for Theoretical Physics collected information on as many scientists as they could locate. These scientists obviously do not represent all science in poor countries. The group consists only of theoretical physicists, but a strong case can be made for analyzing data on one particular group such as this. It eliminates or greatly decreases the variability due to different disciplines.[5] Furthermore, theoretical scientists by and large do not require high technological intensity to do good work. Most require only pencil, paper, and an adequate library. It's questionable whether an adequate library exists in most institutions in poor countries, but the facilities are probably reasonably comparable among nations because libraries are a first requirement of any science.[6] The analysis of these scientists may be especially significant, for as Abdus Salam (1966:463) tells us:

> ... Theoretical physics happens to be one of the few scientific disciplines which, together with mathematics, is ideally suited to development in a developing country. The reason is that no costly equipment is involved. It is inevitably one of the first sciences to be developed at the highest possible level; this was the case in Japan, in India, in Pakistan, in Brazil, in Lebanon, in Turkey, in Korea, in Argentina.

I do not know who or how many scientists were in the original universe; therefore, it is not possible to judge the representativeness of this sample. The one thing about which I am reasonably sure, however, is that The International Center for Theoretical Physics is likely to have collected data as good as any individual could have obtained, because scientists are aware of and know the motives of the Center whereas they may be less certain about individual social scientists, regardless of their institutional or financial sponsors. These data must be taken as tentative, however, and only suggestive for future research.[7]

2. The Scientific Career in Poor Countries

The most important concern of scientists in poor countries is their ability to perform the research role. They need to conduct research, in the broadest sense, publish the results, and have their work accepted by the world scientific community. The extent to which scientists in any country are able to do this more successfully as time passes is an indication of scientific development in that society.

Let us look then at this group of theoretical scientists from the perspective of their ability to publish research, be accepted into the scientific community, and move ahead in their careers. To do this, scientists must have sufficient educational qualifications, opportunities for research, contact with other scientists, and some kind of institutional position that recognizes their qualifications and enables them to pursue their research interests.

These scientists from poor countries have high qualifications with 18% (75) having a B.A. or M.A. degree or equivalent, and 82% (333) having a Ph.D. or equivalent. The last degrees that the scientists took were obtained to a large extent in three countries, India, the United Kingdom, and the United States. India awarded 28% (113), the United Kingdom awarded 20% (82), the United States awarded 21% (85), and all other countries combined awarded 31% (126).[8] While 28% did not have information on their current institutional position, of those who did, 4% (14) are lecturers, 30% (94) are senior lecturers, 31% (99) are professors, 27% (85) are 'research' and 7% (23) indicated they have positions involving administration and research.

A scientist must have connections, either formal or informal, with other scientists in order for him to develop intellectually and professionally. Isolation from other scientists tends to cause intellectual stagnation. It may be that failure to experience a continuing sense of community and belonging to an intellectual tradition is as important as the stimulation of ideas. That is, the social aspect may be as important as the intellectual. In any case, there is widespread support for the notion that contact is vital. Some indicators of connections with other scientists include whether or not scientists have held positions in other countries that would have allowed them considerable opportunity for rewarding social and intellectual experiences. Another indicator is whether scientists have attended major conferences, a briefer exposure than a position to be sure, but nevertheless important. Traveling to a foreign country to attend a major conference is still more important perhaps than attending a conference in one's home country.

A significant number of scientists have held positions abroad. While 35% (151) have held no such position, 28% (123) held one position outside of the country; 19% (83) held two; 10% (45) held three, and 8% (34) held four or more positions abroad. Although information on conferences attended abroad was not available for most (309) of the scientists, among the 129 for whom information was available, 55% (27) had attended two, 16% (20) had attended three or four, and 9% (11) had attended five or more major conferences. Of the 129 who had attended a conference, 7% (9) had not travelled out of the country, but 93% (120) had traveled to a foreign conference.

The major consideration about research productivity is the extent to which a scientist has published his research. Researching without publication is hardly fulfilling the scientific role. No scientist indicated zero publications although 45 omitted data on the question and many of them probably had not published anything. In any case, that only would be 10% of the whole group of 438 without publications. Excluding those, 31% (123) had published 1–4 papers, 30% (119) had published 5–9 papers; 15% (59) had published 10–14 papers, 23% (92) had published more than 14 papers. Some had published dozens of papers.

Absolute numbers of publications do not reflect career activity because an older scientist may have published four papers over his entire life while a younger scientist might have published one or two each year for three or four years. It is appropriate, therefore to consider the rates of productivity that take professional age into account. Professional age is the time since a scientist received his last degree. (In this analysis 53 are excluded, the 45 without publication data and 8 others without data on professional age.) Rates of productivity show considerable variation in productivity with 34% (127) having a rate of from 0.00 to 0.99 papers per year; 26% (98) having a rate of from 1.00–1.99 per year; 17% (63) having a rate of from 2.00–2.99 per year; 9% (34) having a rate of from 3.00–3.99, and 14% (53) having a rate of 4.00+ papers per year. The gamma correlation in the table cross-tabulating professional age and publications is 0.38, showing a mild positive relationship. The gamma correlation in the table cross-tabulating professional age and rates of productivity is −0.31 indicating a mild negative relationship. Thus, one interpretation is that while a scientist adds to the total number of publications as he gets older, his average yearly production is reduced because he 'slows' down.[9]

After a scientist has produced research results, the next concern is whether he will receive acknowledgement from one's scientific colleagues that the research is worthwhile.[10] There are various ways that this happens, ranging from the most elusive of all having the Nobel Prize awarded for research, down to a simple request from a colleague for a reprint of a recent paper. Somewhere in between this is having other scientists refer to published research. Because of the publication of the *Science Citation Index*, it is relatively easy to get an objective indicator of the quality or usefulness of a scientific paper.[11]

Taking the number of citations in 1970 to each scientist's cumulative publications as an indicator of the usefulness or quality of his published research, we find that 51% (225) of the scientists had no citations, 19% (85) had one to three, 9% (41) had four to seven, 10% (42) had eight to fifteen, and 10% (43) had sixteen or more citations. It is obvious that many scientists' work had not be appreciated to the extent that some had been. It will be remembered that no information was available on the publication record of 45 scientists, 31% of them had citations, so 69% (31) probably had no publications yet and thus could not be cited.

TABLE I

Number of publications and number of citations to published research

Number of citations		Number of publications				Total
		1–4	5–9	10–14	15+	
0		73	51	42	22	50
1–3		15	25	25	17	20
4–7		7	9	10	14	10
8–15		3	11	14	16	10
16+		2	3	9	30	10
	Totals	100	99[a]	100	99[a]	100
	(*N*)	(123)	(119)	(59)	(92)	(393)
gamma = 0.36						

[a] Totals in this and other tables not equal to 100% due to rounding.

Before attempting to explain some of the variation in career achievements, it is important to see whether publications and citations are related, that is, whether scientists who have published a significant amount of research have also received a significant number of citations by other scientists. Publications and citations are highly correlated (see Table I). Whereas 73% with only 1–4 publications have received no citations, only 22% with 15 or more publications have received no citations. Similarly, whereas only 2% of the scientists with only 1–4 publications have 16 or more citations, 30% of the scientists with 15 or more publications have 16 or more citations.

The rate of scientific productivity is also related to number of citations (see Table II). This means that the number of publications, spread over

TABLE II

Average reates of productivity and number of citations to published research

Number of citations		Average rates of productivity					Total
		0.00–0.99	1.00–1.99	2.00–2.99	3.00–3.99	4.00+	
0		66	48	41	35	32	50
1–3		20	25	24	15	15	21
4–7		7	12	10	9	11	10
8–15		5	10	13	21	11	10
16+		2	5	13	21	30	10
	Totals	100	100	101	101	99	101
	(*N*)	(127)	(98)	(63)	(34)	(53)	(375)
gamma=0.37							

several years of professional life, is as important as the average rate of productivity over one's professional career. Whereas 66% of the scientists with less than one paper per year received no citations in 1970, 32% with more than four papers received no citations; similarly, whereas only 2% of the scientists with less than a paper a year received 16 or more citations, 30% of the scientists with four or more papers a year received 16 or more citations. Because of this relationship between number of publications, number of citations, and rates of productivity, subsequent data will focus on only productivity rates because professional age is taken into account when rates are calculated.

TABLE III

Geographical location and average rates of productivity

Average rates of productivity		Geographical location						Total
		South America and Mexico	Africa	Indian Subcontinent	Far East	Philippines and Islands	Middle East and Greece	
0.00–0.99		38	39	29	48	11	42	34
1.00–1.99		23	26	25	30	67	27	26
2.00–2.99		15	17	19	0	11	21	17
3.00–3.99		8	9	11	9	0	6	9
4.00+		17	9	17	13	11	3	14
	Totals	101	100	101	100	100	99	100
	(*N*)	(79)	(46)	(185)	(23)	(9)	(33)	(375)

TABLE IV

Geographical location and number of citations to published research

Number of citations		Geographical location						Total
		South America and Mexico	Africa	Indian Subcontinent	Far East	Philippines and Islands	Middle East and Greece	
0		53	67	44	70	64	57	51
1–3		25	16	20	17	9	14	19
4–7		7	10	11	0	27	6	10
8–15		8	3	13	0	0	9	10
16+		7	3	12	13	0	14	10
	Totals	100	99	100	100	100	100	100
	(*N*)	(85)	(58)	(224)	(23)	(11)	(35)	(436)

TABLE V

Research specialty in physics and average rates of productivity

Rates of productivity		Research specialty in physics								Total
		High energy	Low energy	Solid state	Fundamental theory	Combination of LE & HE	Other	Plasma	SS & Other	
0.00–0.99		30	38	38	64	42	29	14	32	34
1.00–1.99		25	27	32	21	13	43	36	19	26
2.00–2.99		22	13	17	7	16	14	14	15	17
3.00–3.99		12	5	6	0	13	10	14	8	9
4.00+		12	17	8	7	16	5	21	26	14
	Totals	101	100	101	99	100	101	99	100	100
	(*N*)	(122)	(60)	(53)	(14)	(31)	(21)	(14)	(53)	(368)

TABLE VI

Research specialty in physics and number of citations to published research

Number of citations		Research specialty in physics								Total
		High energy	Low energy	Solid state	Fundamental theory	Combination of LE & HE	Other	Plasma	SS & Other	
0		49	50	64	63	38	52	53	51	51
1–3		25	8	13	19	22	26	20	21	19
4–7		8	14	10	6	9	4	20	8	9
8–15		12	14	8	13	9	9	0	5	10
16+		7	15	5	0	22	9	7	15	10
	Totals	101	101	100	101	100	100	100	100	99
	(*N*)	(146)	(74)	(61)	(16)	(32)	(23)	(15)	(61)	(428)

The rates of scientific productivity have been shown to vary, and one interesting difference is related to geographical area in which the scientist is located (see Table III). Scientists in Africa and the Middle East generally produce research at a slower rate than scientists located elsewhere. It is not the case, however, that citations accrue to scientists to the same extent in spite of the fact that rates of productivity and citations are related (see Table IV). Whereas scientists in (1) South America and Mexico, and (2) the Indian subcontinent produce at a higher rate than average, scientists in (1) the Middle East and Greece, (2) Far East, and (3) the Indian subcontinent appear to receive an above average number of citations to published research.[12]

All of these scientists are theoretical physicists, but even with this homogeneity, there are considerable differences in research specialties and rates of productivity (see Table V). Scientists in plasma physics and those with two specialties including solid state theory are by far the most productive. The group of plasma physicists is small and caution should be used before making any sweeping generalizations.

The pattern changes when citations to published research are considered. Scientists with research specialties in some combination of high energy physics and low energy physics are twice more likely than the average to have received 16 or more citations to their published research (see Table VI).

3. Scientific careers and economic development

It is commonplace that there is more science, and presumably better science, in the economically advanced countries.[13] This suggests that scientists in poor countries should be more productive than scientists in still poorer countries. A useful index of development has been suggested by Harbison and Myers (1964).[14] They divide countries into four categories of development from 'underdeveloped' to 'advanced'. Rates of scientific productivity are only slightly correlated with levels of development (the gamma equals 0.16). This is considerably less than I would have expected, but these results must be considered suggestive rather than definitive.[15] If scientists are slightly more productive on average in more developed countries, they are even more likely to have their research cited by other scientists (the gamma for level of development and number of citations is 0.24).[16] This effect is not large enough to suggest 'discrimination' against the work of scientists in poor countries, but it is suggestive for future research.

In the Harbison and Myers (1964:39) study, the GNP per capita (U.S. dollars) correlated −0.82 with the percentage of the population in agriculture and 0.83 with number of engineers and scientists per 10000 population. These measures are pertinent to the index of development because GNP correlated 0.89 with the index. The index correlated −0.81 with percentage of the population in agriculture and 0.58 with number of engineers and scientists per 10000 population.[17]

One indicator of economic development is the percentage of population *not* in agriculture. The higher that percentage, the more the population is engaged in industrial-type activities. Examining this sample of theoretical scientists using percentage of the population not in agriculture as an independent variable shows that rates of productivity for scientists in countries with smaller percentages not in agriculture are higher than for scientists where larger percentages are not engaged in agriculture (see Table VII). In contrast to some other comparisons between the rates of

TABLE VII

Percentage of population not in agriculture and average rates of productivity

Average rates of Productivity		% Not in agriculture			Total
		10–29%	30–39	40+	
0.00–0.99		31	33	41	34
1.00–1.99		23	33	25	25
2.00–2.99		17	16	14	16
3.00–3.99		14	3	6	10
4.00+		16	15	14	15
	Totals	101	100	100	100
	(*N*)	(181)	(67)	(98)	(346)
gamma = −0.14					

productivity and the number of citations to published research, in which the patterns between the two measures were not similar, the pattern is similar in the case of percentage not in agriculture (see Table VIII).

For the final presentation of data comparisons, attention is focused on what many would agree to be very important to the atmosphere or climate

TABLE VIII

Percentage of population not in agriculture and number of citations to published research

Number of citations		% Not in agriculture			Total
		10–24	30–39	40+	
0		45	64	53	51
1–3		20	15	23	20
4–7		10	9	7	9
8–15		13	4	8	10
16+		12	8	8	10
	Totals	100	100	99	100
	(*N*)	(220)	(78)	(106)	(404)
gamma = −0.18					

for research. We hear much about the critical mass necessary for the functioning of a scientific community. Critical mass is a difficult concept to deal with empirically because whereas we use it frequently it may be more a rhetorical phrase than a social scientific concept. Regardless of its ambiguity, there is a basic notion that the more scientists there are in the population, certainly to a point, the more the society approves of and supports science. Consequently, there are more opportunities for scientific exchange, mutual criticism, and intellectual affairs. Based on scientists (and engineers) per 10000 population, these data show that the more

TABLE IX

Scientists per 10000 population and average rates of productivity

Average rates of productivity		Scientists per 10000 population			Total
		1–4	5–9	10+	
0.00–0.99		26	35	37	29
1.00–1.99		24	31	33	26
2.00–2.99		18	15	10	17
3.00–3.99		13	4	3	11
4.00+		18	15	17	18
	Totals	99	100	100	101
	(*N*)	(160)	(26)	(30)	(216)
gamma = −0.20					

scientists there are, the *less* the rate of scientific productivity (see Table IX).[18] As was the case for percentage of population not in agriculture, the pattern between rates of productivity and citations to published work is consistent (see Table X). Scientists in countries with lowest levels of

TABLE X

Scientists per 10000 population and number of citations to published research

Number of citations		Scientists per 10000 population			Total
		1–4	5–9	10+	
0		42	57	61	46
1–3		21	14	13	19
4–7		10	14	0	10
8–15		13	4	13	12
16+		13	11	13	13
	Totals	99	100	100	100
	(*N*)	(201)	(28)	(31)	(260)
gamma = −0.19					

scientists and engineers per 10000 population are as likely to receive 16+ citations and less likely to receive none than scientists with highest levels. This results in a negative sign for the gamma coefficient (−0.19).

The explanation for the negative relationship between rates of productivity and both percentage of population not in agriculture and number of scientists per 10000 population is not obvious. One possibility is that when a country has few scientists, they are left to do science, and they do it well, but when a country increases its stock of qualified manpower the best scientists are drafted into administrative roles. There are no adequate data to test this explanation. One hint that administration may be the province of more successful scientists is the age factor. Scientists in administration (and research) are much older than the others. Whereas 7% of the sample are involved in administration and research, 33% of these scientists with a professional age of 14+ years are involved, 4% with 9–13 years, 5% with 5–8 years, and 4% with 1–4 years are involved in administration. This does not prove that *competent* scientists are given the burdens of administration but certainly in the more advanced countries, the more successful scientists often find themselves coopted into administrative roles. A good example is the general historical case of Britain where the administrative head of the department was expected to be the leading scientist in the department. Whatever the explanation really is for this apparently peculiar finding, it should provide some questions for further research.

4. Some comparisons from rich countries

The picture that emerges from the data presented on scientists from poor countries is that one must question the notion that scientists in these countries suffer as much as we might think. Taking an admittedly limited indicator – research productivity – the fact that only 10% of these scientists at the most had never published, and that over half produced at a rate of more than a paper a year, is hardly cause for alarm. I do not have exactly comparable data on other scientists, but some data that are available are instructive. Price (1963:45–46) states that the average life-time productivity of scientists is 3.54 papers. Clearly these theoretical scientists are well above average. Halsey and Trow (1971:297) report data on British natural scientists which show that 2% of them have not published research papers during their entire careers. Only 1% of the scientists in my study of British scientists (Gaston, 1973) had never published. While these figures may seem to support the claim that scientists in poor countries have a more difficult time, one must remember that presumably all these British scientists had a somewhat easy time of it. What this probably shows for theoretical scientists in poor countries is that they are able to perform research that is competitive with that produced by other scientists in the international community.

Further data suggesting that the scientists in poor countries are not doing too badly refer to data on Ph.D. holders in the United States. Fulton and Trow (1974:46) report that 50% of American academics in the physical sciences had not published an article in the two years previous to the inquiry (the low was 90% for junior colleges to 14% at the highest prestige universities). The Carnegie Commission and the American Council on Education study of over 60000 respondents in 1968–69 shows that 44% of the faculty have never published a professional article (Bayer, 1970:15). That finding held for a follow-up survey in 1972–73 for which Bayer (1973:28) reports that 43% had never published a professional article. To be sure the figure is probably much lower for scientists but if it is zero or 5%, the scientists from poor countries are still doing comparatively well.

5. Some concluding remarks

Representatives of economically advanced countries sometimes advise poor countries to reexamine their goals for economic development because, they argue, economic development will inevitably produce urban social problems and environmental pollution. In contrast to that advice, scientific spokesmen advise their potential colleagues in scientifically poor countries to develop as fast as circumstances allow. The key concept here is circumstances, because therein lies the basis for most social science interest in the development of science. Science is going forward in other parts of the world, and poor countries that remain underdeveloped in their scientific work do so not because of the lack of intellectual problems to be solved or the potential intellectual capabilities of solving them.

The dimensions of scientific development that are the most appropriate for study are (1) intellectual or cognitive development, and (2) institutional development. These are important independent variables, for example, in any study on the effects of scientific development. They are dependent variables when one is attempting to determine the cause of development itself. The relationship between cognitive and institutional development also is of great interest to the sociology of science.

What is intellectual development? In its simplest form, intellectual development is the difference in the *state* of scientific knowledge between Time 1 and Time 2. Intellectual development is the increase in 'certified knowledge' or 'public knowledge'[19] that obtains after some passage of time. No one argues that science is less developed today than it was 50 years ago. We know and are (tentatively) certain about much more than we were then. The notions of 'development', 'breakthroughs', and 'advances' are inherently a part of scientific research. If one uses terms such as more, better, truer, to describe development or progress, then science can be quantitatively measured between two points in time. What indicators are used, and how they are used, is yet another question.

What is institutional development? In its simplest form, institutional development is also the difference between two points in time. At one point, there may be three research institutes; later there may be five. At one point, there may be no government financial support; at a later point there may be some. What is not clear, of course, is how intellectual development and institutional development are related. The consensus is that some institutional development – at the minimum a community of scientists with some visibility and identification, or some *opportunity* for such a community – must be present to foster intellectual development.

The question of standardization or the problem of comparability must necessarily enter any discussion of scientific development in more than one country or at different historical periods. For example, in one of the few studies available on the effects of social structure on development and progress, Ben-David (1965) argued that the decentralized German university system of the nineteenth century produced the great achievements in medical research that were not achieved in France and England at that time.[20] He did not ignore the problem of standardization when comparing the accomplishments of several groups, but he argued that it was not necessary to do so. Perhaps it was not necessary, but I think it was. In any case, comparing Germany, France, and England during a time when each may have been approximately equal in other areas of development may have been legitimate, but we can not assume without considerable further study, that countries with twice the scientific population are comparable without some kind of standardization. My point is that when we look at some aspect of the development of science and we see that one country has twice the development as another country while neglecting the fact that the country has twice the active scientists as another, we should not hastily conclude that the first country is more developed scientifically. I believe we should consider structural and demographic factors to a greater extent than we do.

The development of science today in poor countries occurs in a world in which other countries are highly developed economically and scientifically. This is very different from earlier scientific development where all countries were underdeveloped. Thus the study of science in developing countries can offer a unique opportunity to examine our notions about historical developmental processes. On the one hand, it could be argued that the processes would be different because poor countries today can import scientific knowledge, an obvious impossibility historically. On the other hand, one could argue that the structure of the processes of scientific development, if not the content of the processes, probably remains constant throughout history. Thus, if the structure of social conditions apparently necessary for scientific development in the seventeenth century are not present in contemporary societies of poor countries, but science advances nevertheless, then one must question the explanation for the development of science in the seventeenth century.

Scientists in the 1700's and 1800's worked in a context of uncertainty. There was uncertainty about the social support of their activities. What they did was not widely understood or appreciated. There was uncertainty about the fruits of their labors. As pioneers at the frontier of learning, they had to trust strongly that their efforts would result in some kind of success.

While the level of the development of science in many poor countries may be at levels similar to those in scientifically advanced countries in the 1800's, scientists in poor countries today have a distinct advantage over their historical colleagues: they can be rather certain that science can be fruitful. They are not embarking upon some unchartered institutional activity. They know that scientific research can produce results, and they realize if others can do it, why can't they?

Price (1973:153) writes that "In no ways, especially historically, can we find technology arising as applied science." If this is true, then we should not expect the development of science to result in applied science, and that, in turn, in technology. If we want to study the development of science as an intellectual and academic question, that is one thing. It is another thing if we want to study the development of science in order to change the direction or rate of development so that scientific development can enhance technological and economic development.

We should first determine what relationship, if any, there is between science and potential economic development. While some interests may lean toward the academic questions, we should be careful not to suggest that solutions to economic problems are possible, either directly or indirectly, through development of pure science. Perhaps we should determine first what we really want to accomplish in this area, and then address the questions that are really pertinent to accomplishing those goals.

Southern Illinois University,
Carbondale, Ill.

NOTES

* The author acknowledges and greatly appreciates generous support of this research by Carnegie Endowment for International Peace and the National Science Foundation (Grant ≠ GS-27291).

[1] Two notable exceptions to this are Merton (1970) and Barber (1952).

[2] Among the exceptions to this is Price (1969).

[3] For example, *Science*, *Minerva*, *Science Studies*, and *Impact of Science on Society* are among the periodicals that occasionally publish articles on science in poor countries.

[4] The foreword of the directory (International Center for Theoretical Physics, 1970) states that "Since its inception, the International Centre for Theoretical Physics has compiled a card index of theoretical physicists from developing countries for internal use. As this card index proved to be so valuable to institutes and individuals who requested lists of these physicists, it was decided to publish this Directory which includes summarized curricula vitae of 446 physicists from 39 developing countries."

[5] For one of the few studies of discipline differentiation see Hargens (1971).

[6] This is to say that if anything is going to be developed, the library is an obvious place to begin. To most scientists anyplace in the world, it is probably the case that every library is not quite adequate – there is always something else that would be helpful.

[7] The Center was planning later editions of the Directory, so other researchers will have opportunities to replicate research like this if they desire to do so.

[8] I beg the reader's indulgence in presenting statistical data in this form, but tabular material is expensive to print and this is an economical format for presenting figures on marginal distributions.

[9] There are other interpretations and this does not prove that as age increases the quantity or quality of research is reduced. Scientists could be slow producers all their lives which would give a positive correlation between total productivity and professional age, and if there were some high producers and low producers among the older age categories, the correlation between professional age and rates of productivity would tend toward zero. It would depend on the proportions in both younger and older groups who were high or low producers.

[10] One might be criticized for looking at science at this kind of empirical level. Criticisms may include the notion that there is more to science than just papers. I agree. However, as a sociologist I feel somewhat like the preacher in the tale in which he had a critical church member who complained that the preacher spoke too much about sin, to which the preacher responded that when the members stopped sinning, he would stop preaching about it. When scientists stop publishing or stop placing importance on publishing behavior, then it will be time for social scientists to neglect this aspect of science.

[11] Price (1973:155) writes that "... it is now extremely well known that the frequency of citation is a very good guide to quality. In fact, any expert witness' statement of quality correlates better with the citation count than any expert witness correlates with any other expert witness." I am not going to enter here the debate on the use of citations for sociological indicators. I have heard of no scientists or other scholar denying that he appreciated someone citing his own work. Until at least that aspect of citations is eliminated I shall continue to use citation counts where they seem appropriate.

[12] The data are grouped and it cannot be determined by this method of analysis whether there are outstanding differences between individual scientists. That is, this does not tell

us whether an individual scientist who produces many papers is ignored or whether a scientist producing few papers received many citations. Further study using more sophisticated procedures will be completed that will deal with this question.

[13] Vanderpool (1971) indicates that the United States is generally acknowledged as the center of research by foreign scientists in his study. There are endless indications that most science is done in advanced nations.

[14] In anticipation of questions regarding the use of this particular index, I do not wish to justify its preference over any other, but it was used for several reasons: (1) Vanderpool (1971) used this to categorize scientists in his study; (2) it is largely based on GNP per capita, a significant factor in scientific productivity according to Price (1969); and, in spite of its limitations, it is a published source and not one that has the limitations of an individual researcher's peculiar biases.

[15] Some examples of the data, not shown in tabular form, are as follows: 47% in level one (underdeveloped) and 35% in level four (advanced) had a rate of from 0.00 to 0.99 papers per year; 0% in level one but 20% in level four had 4.00+ papers per year. In this sample, only Argentina was 'advanced'. Eliminating level four, then, gives figures for level three (semiadvanced) of 31% and 18% for the respective rates of productivity.

[16] In level one, 70% received no citations, as did 63% in level two (partially developed), 43% in level three, and 64% in level four. In level one, 0% received 16+ citations, 8% in level two, 12% in level three, and 10% in level four.

[17] Not all countries had reliable statistics on these social indicators and many had no statistics at all. Thus, any analysis using them must be conservative. The dependent variables are probably much more valid and reliable than the independent variables in this case.

[18] The data on scientists and engineers per 10000 population are available for only about half the sample. The index of development was available for nearly all the sample so these particular data were simply not available to Harbison and Myers (1964). One might argue that the unknown data would fall in the lowest category. If so, one might further argue that adding them would greatly dilute the rates of productivity for that category because, the argument might go, countries with data are special cases of well-organized governmental bureaucracies that use very well even the few scientists they have.

[19] See Merton (1957) and Ziman (1968) for discussion of these concepts.

[20] Ben-David's (1960; 1968; 1971) work on growth and change also deal with the same kind of structural effects on scientific development and progress.

BIBLIOGRAPHY

Barber, B: 1952, *Science and the Social Order*, Free Press, Glencoe, Ill.

Bayer, A.: 1970, *College and University Faculty: A Statistical Description*, American Council on Education, Washington, D.C.

Bayer, A.: 1973, *Teaching Faculty in Academe: 1972–73*, American Council on Education, Washington, D.C.

Ben-David, J.: 1960, 'Roles and Innovations in Medicine', *American Journal of Sociology* **65**, 557–68.

Ben-David, J.: 1965, 'Scientific Productivity and Academic Organization in Nineteenth Century Medicine', in N. Kaplan (ed.), *Science and Society*, Rand McNally, Chicago, pp. 39–61.

Ben-David, J.: 1968, *Fundamental Research and the Universities*, Organization for Economic Co-operation and Development, Paris.

Ben-David, J.: 1971, *The Scientist's Role in Society: A Comparative Study*, Prentice-Hall, Englewood Cliffs, New Jersey.

Fulton, O. and Trow, M.: 1974, 'Research Activity in American Higher Education', *Sociology of Education* 47, 29–73.

Gaston, J.: 1973, *Originality and Competition in Science*, University of Chicago Press, Chicago, Ill.

Halsey, A. and Trow, M.: 1971, *The British Academics*, Harvard University Press, Cambridge, Mass.

Harbinson, F. and Myers, C.: 1964, *Education, Manpower, and Economic Growth*, McGraw-Hill Book Company, New York.

Hargens, L.: 1971, 'The Social Contexts of Scientific Research', Unpublished Ph.D. dissertation, University of Wisconsin.

International Center for Theoretical Physics: 1970, *Directory of Theoretical Physicists from Developing Countries*, ICTP, Trieste, Italy.

Merton, R.: 1957, *Social Theory and Social Structure*, Revised Edition, The Free Press, New York.

Merton, R.: 1970, *Science, Technology and Society in Seventeenth-Century England*, Harper and Row, New York.

Price, D.: 1963, *Little Science, Big Science*, Columbia University Press, New York.

Price, D.: 1969, 'Measuring the Size of Science', *Proceedings of the Israel Academy of Sciences and Humanities* 4, 98–111.

Price, D.: 1973, 'The Relations Between Science and Technology and Their Implications for Policy Formation', in G. Strasser and E. Simons (eds.), *Science and Technology Policies*, Ballinger Publishing Company, Cambridge, Mass., pp. 149–172.

Salam, A.: 1966, 'The Isolation of the Scientist in Developing Countries', *Minerva* 4, 461–465.

Science Citation Index: 1970, Institute for Scientific Information, Inc., Philadelphia, Pa.

Vanderpool, C.: 1971, 'Center and Periphery in Science', paper presented at AAAS Meetings, Southwestern and Rocky Mountain Division, April 21–24, Tempe, Arizona.

Ziman, J.: 1968, *Public Knowledge: The Social Dimension of Science*, University Press, Cambridge, England.

WALTER HIRSCH

THE AUTONOMY OF SCIENCE IN TOTALITARIAN SOCIETIES: THE CASE OF NAZI GERMANY*

1. INTRODUCTION

The thesis that there is a special affinity between democratic social structure and the ethos and practice of science has been maintained by numerous philosophers and scientists. Yet it is also evident that many scientists have made important contributions in an environment that can hardly be called democratic, e.g., Kepler or Galileo. Hence the question has been raised whether the norms considered essential or conducive for scientific productivity may not be found in other types of societies, and whether the values and practices associated with non-democratic social structures may be equally or more instrumental in the flowering of science.[1]

This counterthesis, already promulgated by Marxists, became more generally accepted when the Soviet Union shot off its sputnik, even though it could be argued that this was a technological rather than a scientific accomplishment, and that the know-how involved was borrowed from Western scientists. The question of whether modern science could have developed as successfully under non-democratic auspices is, like most other historical problems, ultimately irresolvable. But we can come closer to a tentative answer by comparing scientific developments in contemporary 'democratic' and 'totalitarian' societies, thus minimizing the 'historical' component: Science existed in Russia before the revolution and in Germany before 1933. We can investigate to what degree the existing institutions were modified by the impact of the political and social changes effected by the revolutionary regimes. The general problem can be stated as follows: Does a totalitarian system necessarily restrict scientific development?

There are several methodological and theoretical problems which need discussion before we try to answer the question. First, the definition of totalitarianism (and, for that matter, democracy). For the present purposes, I characterize totalitarianism as a system *sui generis* which maximizes political power on the top and minimizes the power of all social

Knorr, Strasser, and Zilian (eds.), Determinants and Controls of Scientific Development, 343–366.

institutions. It is found in modern industrialized societies and those with a strong drive towards industrialization, e.g., China, and has no real counterpart in older forms of tyranny or dictatorship which have appeared through history and are endemic in South America (Spiro, 1968).

The next problem concerns the degree to which historical factors can be divorced from the new totalitarianism. The Soviet Union, Germany and the U.S.A. were all involved in a great war. To what degree can the developments in science be ascribed to military exigencies or, on the other hand, to the impact of totalitarian or democratic forms of control? The best we can do is to use appropriate comparisons and hope that we come close to controlling the 'experimental' factor.

Finally, there is the problem of values. In my own case I have a strong antipathy against totalitarianism. For the sake of objectivity I make a deliberate attempt to be the devil's advocate and it may be that in leaning over backwards in my search for evidence that my account is biased. I also refrain from analyzing the motives of the actors and from judging the morality of their acts. While recognizing the tremendous importance of individual and collective guilt in the conduct of scientists I don't treat these matters in the present investigation.

While Soviet and recent Chinese science have been the subject of fairly systematic and continued investigation, science under the Nazis has been relatively neglected, except for nuclear physics and rocket research (Hirsch, 1961). The social sciences and humanities have been dealt with more intensively; the analysis to follow deals only with the natural sciences. One reason for the neglect is probably the early demise of the "thousand year Reich." Very likely there are others. At any rate, here is the beginning of filling the existing gap.

2. The universities

German Universities and other institutions of higher learning (institutes of technology, teachers colleges, and other professional and vocational schools) were particularly vulnerable to political controls for the simple reason that they always were, and still are, organs of the state. Professors were civil servants, and the bulk of the financial support came from the *Laender* and the federal government. To be sure, the faculty, consisting of the full professors who were holders of chairs and directors of institutes had a high degree of autonomy as to what could be taught, by whom, and under what conditions. But the traditional *Lehrfreiheit* (freedom to teach) was never total. The state had the power of the purse and could influence the establishment of new departments; the ministry of education could exercise veto power over the appointment of professors. Typically the faculties would submit a list of three possible appointees, of which one or more – or none – could be approved.

After 1933 appointments were made from above by the ministries of the *Laender* (states) and increasingly by the Reich ministry in consultation with the party. The *Rektoren* (university presidents who were traditionally members of the faculty) were no longer elected by their peers but appointed. 'Non-aryans' and political opponents of the regime, many of whom had already resigned at the time of the appointment of Hitler as chancellor or shortly thereafter, often under pressure from demagogic attacks and demonstrations by Nazi students, were 'legally' removed from office under one of the first laws passed by the new regime, ironically entitled the "law for the reconstitution of the professional civil service."

One of the requirements of being a part of the university was membership in the official organization of teachers (*N. S. Dozentenbund*) or students (*N. S. Studentenbund*). Moreover it was understood that special privileges and support in the academic career of students and faculty such as fellowships, appointments and promotions were contingent on membership in one or more of the para-military party organizations and generally a 'positive' attitude to national socialism, including of course party membership. (The latter was not restricted to a highly selected elite, as in the Soviet Union.) And before a student could enroll in the university he had to absolve his labor service for six months, unless he could prove that he was medically unfit.[2]

These were some of the major structural reasons for the personal vulnerability of academic scientists. In addition, among the students particularly, Nazi ideology had been accepted enthusiastically for several years before the 'bloodless revolution'. This was less the case for the professors, who like most other professional and civil servants tended to be conservatively oriented and could not feel sympathetic to the goals and values of the Weimar republic. Yet many professors joyfully embraced the controls placed upon them, in the belief that a charismatic leader would produce the 'new order' which would eliminate unemployment, stop moral decay, and restore Germany to the glorious position it had occupied before the great war (Roegele, 1966; Ringer, 1969).

What was the role of science to be in the new Reich? No unequivocal answer can be given, because *Kulturpolitik*, or science policy in the narrower sense was not of one piece. On the other hand, there was the 'blood and soil', anti-rationalistic, anti-intellectual stance epitomized in Rosenberg's *Myth of the Twentieth Century* and in the famous phrase (often attributed to Goering) of the hero of a play written by the first leader of the Hitler youth: "When I hear the word 'culture' I release the safety catch on my gun." At the same time there was a sober appreciation of the services which science and technology had to provide for the industrial and military needs of the regime. Hitler had great respect for any expert who could get things done, as shown by the great powers he delegated to Speer, for example, but he hadn't the faintest notion about the meaning of pure research. (Among his plans for the future was the erection of a museum in Linz, to be dedicated to the cosmology of Ptolemy, Copernicus – and Hoerbiger, the originator of the *Welteislehre* according to which the center of the earth consists of ice, and which was the basis of one of the esoteric cults to which Himmler and other Nazi leaders were addicted (Fest, 1963: 88).)

The middle-class Nazi leaders also partook of a general respect (mingled with fear and distrust) for intellectuals, artists and other *Kulturtraeger*, bearers of the culture which made Germany the nation of poets and thinkers. The attitude is well expressed in the following citation:

> Germany's international prestige depends more than on its statesmen, soldiers and industrialists, on its artists, scholars, scientists and inventors (*Physikalische Blätter*, 1944:22).

The quote is from a speech given by none other than Goebbels. Discounting the fact that he made this speech at Heidelberg, his *alma mater*, it still represents the sentiments held by a significant element of the leadership. But with the possible exception of Goebbels, Speer, Rust and Funk, the top leaders were as ignorant as their *Fuehrer* about the scientific process. The task of the scientists then, along with other members of the *Volksgemeinschaft* was to give their all for the glory and power of the greater Reich. In the subsequent analysis I will try to demonstrate that the ignorance about science among the top leaders made it possible for scientists to work along traditional scientific lines and to preserve the autonomy of the scientific institution to a greater degree than earlier analyses have suggested.

The immediate impact of the new policy was stunning. Many of the leaders in certain disciplines disappeared from the universities; others could hang on for a few more years. The most 'visible' were usually among the first to go, e.g. Einstein and Fritz Haber. The latter, ironically, was considered a war criminal by the allies along with the Kaiser and General Ludendorff for his contribution to the German munitions production (including poison gas). Haber, who was a professor at Berlin University as well as director of one of the Kaiser-Wilhelm Institutes concluded his letter of resignation as follows:

> The decision to resign derives from the contradiction between the tradition concerning research which I have lived with up to now and the different point of view, which you... represent as bearer of the great actual national movement. My tradition requires that in my role as scientist I consider only the scientific capacities and personal character in the choice of my co-workers, without regard for their racial ancestry. You cannot expect from one in his sixty-fifth year to change the thought patterns which have guided him for the past thirty-nine years of life in the university, and you will understand that the pride with which he served his German homeland all his life demands that now I ask for my retirement (Max-Planck-Gesellschaft, 1961:190).

Some universities lost a substantial part of entire departments; in Berlin and Frankfurt 32% of the faculty was removed, though in others the loss was much smaller. Along with this came the elimination of entire institutes and disciplines, such as sociology, though this was rarely the case in the natural science. The student population was also decimated: They went into required labor service or military service, or voluntarily into industry and other non-academic areas.[3]

These measures, along with the re-introduction of universal military service, can be considered aspects of totalitarianism. But the outbreak of the war found its counterpart in democratic societies as well. As was mentioned already, it is necessary to disentangle these two variables. For the moment let me only make the point that certain totalitarian measures such as the genocide of the Jews or the euthanasia program could be implemented more efficiently under conditions of 'military' secrecy. On the other hand, the need to maintain the morale and cooperation of the civilian population made a loosening of the early restraints desirable to the regime during the latter years of the war – a fact which also affected university scientists.

The means available to academic scientists for maintaining their autonomy were largely indirect, involving the use of personal contacts and a flair for political machination. The Nazi 'system' was anything but the monolithic, omnipotent machine which the term 'totalitarian' evokes as has been documented by numerous accounts, such as Albert Speer's memoirs (Speer, 1970). Thus, if the person to be appointed university *Rektor* was considered unacceptable by the faculty for academic reasons, the fact could be made known to the ministry involved, which in many cases would substitute someone with whom the faculty could live. Professors who managed to climb into a chair because of their political stance would be ostracized by their colleagues, and students would often show their lack of confidence by demonstrative behavior in the lecture halls or by staying away from his seminars, if there was another choice available, or by choosing another one to direct their Ph.D. research. To cite a further example: Philipp Lenard, a Nobel laureate in physics, was about to retire at Heidelberg. He was considered one of the most powerful scientist-politicians since he had been a Nazi from way back, and along with Stark was the chief proponent of 'German' physics as superior over the 'Jewish' physics of Einstein and 'white Jews' like Planck, Heisenberg, and indeed the great majority of the leading physicists. The departmental search committee picked a slate of three to succeed him, which included the non-Aryan James Franck and two other eminent 'white' Jews. Lenard was furious and threatened to revoke his retirement. His complaint to his friend Rust, the minister of education, was of no avail. Eventually a choice was made satisfactory to both the faculty and the ministry, and Lenard withdrew into his tent, from where he occasionally sallied forth in the cause of 'German' physics.[4] (Another episode involving him will follow below).

3. Forschungsgemeinschaft and Forschungsrat

At the end of World War I, German scientists founded an 'emergency council' (*Notgemeinschaft der deutschen Wissenschaft*) to put German science back on its feet. The immediate needs included purchase of foreign books and journals, subsidies for kinds of research areas which had been neglected during the war, and financial support for students and young Ph.D.'s. The emergency organization gradually became permanent under the name of *Deutsche Forschungsgemeinschaft*. Its membership included universities and similar institutions, scientific academies, scientific societies, and the Kaiser-Wilhelm-Gesellschaft. Funds were raised from both private donors, and, for the most part, governmental sources.

With the coming of the Third Reich a power struggle ensued which was to last until the end of the war. In fact, it can be said to be a continuation of a struggle which had already gone on during the Weimar Republic (Schroeder-Gudehus, 1972), now on a level both less public and more acerbic. The major concern of the scientists was to maximize financial support from public funds and minimize state control, using industry as an ally as much as possible. In 1933 Schmidt-Ott, a civil servant who was very active in the field of cultural politics and had been the person most responsible to get the organization started along with Haber, was fired as president and replaced by Stark, the proponent of 'German physics' who was reputed to have ambitions to become the *Fuehrer* of German science. Rust, the minister of education, asked for a vote of confidence in him; he was accepted by the *Rektoren* of all the universities except Munich, rejected by all the academies of science except Heidelberg, while the KWG abstained (Zierold, 1968: 175–177). But Stark's triumph was short lived; his lack of tact and organizational ability were too much even for his supporters in the party, and he was replaced by Menzel, an official in Rust's ministry. Under his leadership the flow of governmental funds was kept on a fairly even level, ranging from six to nine million marks between 1937 and 1945 (Zierold, 1968: 234).

As part of the application of the 'leadership principle' the organization was restructured as a research council (*Reichsforschungsrat*) with considerably larger functions. Initially placed in the ministry of education, when in 1942 Rust lost out in a power struggle against Goering, it was transferred under the latter's aegis as part of the four-year plan for 'total war'. Eventually, Speer replaced Goering in that position (Zierold, 1968: 239). The heads of special sections were now appointed rather than elected; the new specialists ranged from incompetent opportunists to topranking scientists who were barely acceptable to the ministry because of their lack of positive attitude to the regime, such as the physicist Gerlach, who also turned out to be a very astute political manipulator.

TABLE I

Expenditures of the Forschungsgemeinschaft (thousands of Reichsmark)

Year	Humanities & social sciences	%	Natural sciences technology	%	Agriculture	%	Medicine	%	Other	%	Total	%	Administration
1935	1073	22	1838	37	983	20	174	3	904	18	4972	100	303
1936	1317	29	1139	25	1440	32	214	5	408	9	4578	100	365
1937	1146	21	1289	24	1907	35	793	15	258	5	5393	100	297
1938	1721	22	1970	26	2493	33	1088	14	414	5	7686	100	296
1939	1406	18	2482	31	2657	34	930	12	408	5	7883	100	?
1940	604	12	1786	36	1382	27	780	14	560	11	5040	100	300
1941	1068	17	2068	32	2141	33	796	12	391	6	6464	100	356
1942	1785	22	2489	31	1823	23	938	12	976	12	8011	100	313
1943	1744	23	1915	26	2749	37	853	11	229	3	7490	100	394

Source: Kurt Zierold, *Forschungsfoerderung in Drei Epochen*, Steiner, Wiesbaden, 1968.

As an organization which was supposed to establish and implement a national science policy the *Forschungsrat* was a flop. In fact, the council had only one full meeting during its existence; after the morning session it became apparent to the busy scientists that an afternoon meeting would be a waste of time. Koester called it "one of the sorriest outfits that ever existed," and Gerlach said openly that the research council doesn't have anything to do with research.[5] On the other hand, the traditional activities of the old DFG were continued. During the war, grants tended to become more mission-oriented, but the proportion of funds for various categories remained remarkably steady.[6]

Most important, men like Gerlach could use their position to see to it that men and money were channelled into scientifically valuable basic research (Zierold, 1968:272). This required considerable political savoir faire and the courage to ignore decisions emanating from the highest leaders who were kept busy with more pressing problems. An official charged with reviewing the work of the organization complained that of the 25 projects in physics which were supported only one was concerned with a problem of 'decisive military import'.[7] In 1940 Hitler had ordered that all research which failed to show results within a year was to be terminated (Zierold, 1968:271). To my knowledge this order was never revoked, but evidently it was not generally implemented.

While Goering was nominal head of the *Forschungsrat* he could be prevailed upon to issue ukases favorable to science. Thus in 1943 he got special food rations for scientists in the 'hardest working' (*schwerstarbeitenden*) category, and stated as his official opinion that "nothing is so sensitive to intrusion by the organization as art and science." Further, "freedom of research should be reactivated in as many fields as possible," and all disciplines in institutes of higher learning should be provided for because of the "frightening decline of investment in research." [8] (Actually, the decline was hardly frightening – but the scientists were able to scare Goering.)

Among the most important, and considering the political and military situation, astounding accomplishments of the scientist–managerial coalition was the so-called Osenberg enterprise. Osenberg, a professor of engineering, was first in the navy's research department and then became chief of planning in one of the ever-changing organizations for the total prosecution of the war effort. In 1943, with the support of Speer he was able to bring back 5000 scientists and technicians from the armed services, and as late as July 1944 he prevailed on Himmler to demand that 14600 potential draftees be kept at their scientific work – this time over the opposition of Speer, who wanted to mobilize more personnel for the armament industry, but eventually gave in (Zierold, 1968:254). Similar successful attempts to exempt scientists from the draft had already been undertaken by Gerlach, who had a powerful ally in Baeumler, who was responsible for research in the air transportation ministry.

In addition to these large-scale measures to insure the maintenance of normal scientific activity, individual department heads and institute directors tried their best to keep their students and collaborators, with varying success. Ultimately the army allowed drafted students to pursue their studies on special leave, and many who were considered unreliable benefited from the protection of the uniform. In one case, a chemistry student who had joined the SS some years earlier became disenchanted and asked to leave the organization. At SS headquarters he underwent a painful interrogation and his superiors decided that he must be insane. Since he also had a Ph.D. in psychology he was allowed to diagnose himself and 'officially' certified himself a psychopath, whereupon he was dismissed from the SS and joined the air force, which promptly gave him leave to continue his studies. This is hardly a typical case, but it illustrates the possibilities available (Pascher, 1966:68).

4. The Kaiser-Wilhelm-Gesellschaft

Among the scientific institutions outside the universities the Kaiser-Wilhelm-Gesellschaft (KWG) is the most important. Founded in 1911, it was the brain-child of Schmidt-Ott, whom we already mentioned in connection with the *Notgemeinschaft* and other 'statesmen of scholarship'. Its major aims were the establishment of institutes for the pursuit of pure research, emphasis on newly developing areas of scholarship, particularly interdisciplinary fields such as biochemistry, which had little chance to develop within the traditional university departmental structure, and the provision of research facilities of an applied nature for specific industries (Max-Planck-Gesellschaft, 1961). Funds for these activities were provided for the most part by industry and by state governments where specific institutes were located. With very few exceptions decisions concerning research operations were left in the hands of the scientists who were the institute directors. The establishment of new institutes was generally up to the interested parties: scientists, industrialists, and occasionally local or national governmental agencies. The society was governed by a board of financiers, industrialists, public servants and eminent scientists. An annual budget was drawn up for the entire organization, but individual institutes could get special grants independently.

When the Nazis came to power a lengthy tug of war ensued between the KWG and the new masters. The board of directors co-opted a number of new members, either because of their formal position, e.g. members of the ministry of education, or because it was hoped that because of their good standing with the party they could be politically useful. In exchange the Jewish directors resigned. Planck, the president, tried to intervene with Hitler to keep Haber as head of his Institute, to no avail. Other prominent scientists of Jewish ancestry, such as Lise Meitner and Meyerhof could stay for the time being, until they emigrated. A number of other non-aryans and others who had lost their university positions remained or found a new home in various KWG institutes, however.

The tactical need for compromise was subordinated to the maintenance of principle when the occasion seemed to warrant in it. Thus in January 1935 the senate of the KWG, along with the German Physical Society and Chemical Society decided to have a public memorial service for Haber, who had died the year before. Invitations were sent out to all who would normally be participants in such an occasion, including Rust, the minister of education and culture! Rust took umbrage, not surprisingly considering the invitation as a provocation; he was particularly incensed about the fact that the wearing of uniforms was suggested. Nevertheless he allowed the service to be held on a 'private' basis, not to be reported by the press. The attendance included many scientists and industrialists. Nobody appeared from Rust's ministry (he had specifically prohibited attendance to his underlings), but the physicist Schuman was there as official representative of the German army (Max-Planck-Gesellschaft, 1961: 192–194; Zierold, 1968: 194).

The 'political' stance of the KWG was also shown in the decision to give strong financial support to the Institute for Anthropology, Human Genetics and Eugenics, in order to "confront the nonsense about racial questions and eugenics which was being perpetrated with something scientifically tenable" (Glum, 1964:371). The director, Eugen Fischer, had previously demonstrated that mixed bloods in South Africa were not inferior to 'pure blooded' black Africans. (This finding was considered undesirable by 'race scientists' in the party and consequently Fischer was distrusted, even though he would be considered a 'racist' by American standards.)

In 1937, when Planck was to retire as president, the board picked Carl Bosch, an eminent chemist and head of the powerful I.G. Farben industry as his successor. Bosch was already a very sick man at the time and was replaced in 1940 by Albert Voegler, one of the chief leaders of the iron and coal industry. (He committed suicide in 1945 after his capture by U.S. troops.) These men were picked because of the prestige and power emanating from their economic position, but both had been associated with the KWG for many years, along with others of similar stature. The executive secretary, Friedrich Glum, was *persona non grata* with the party and was forced to resign in 1937; he was replaced by Ernst Telschow, his former assistant, who was a nominal party member but had the full confidence of the scientists; in fact, he was kept in this position after the KWG was reconstituted as the Max-Planck-Gesellschaft after the war.

The strategy of the leaders of the society was to counteract the power of the party by mobilizing members of industry who were involved in the military-industrial complex and could be utilized at critical times. Glum mentions that even during the Weimar republic he insisted that no public funds should be used for the general administrative purposes in order to maintain its independence from the state (Glum, 1964:287). The tactic he espoused later was 'voluntary *Gleichschaltung*' on matters of lesser consequences in order to maintain maximum autonomy where it really mattered. It was a dangerous game, and he became one of the losers.

How did the KWG maintain its original mission during the Nazi regime? The documentation for a conclusive answer is rather spotty, because of the loss of files during the war and the fact that the archives are not yet systematically ordered. Nevertheless, the data supplemented by information by members of the society, are sufficient to come to a tenable answer.

Between 1937 and 1938 the 'scientific personnel' rose from 264 to 331. In 1939 Telschow reported to Bosch that "naturally basic research will have to be somewhat restricted and greater emphasis put on applied research, due to the outbreak of the war." "Some institutes, e.g. anthropology and psychiatry, will be maintained on a limited scale only."[9] As it turned out, psychiatry got more support rather than less as late as 1943, including a special grant from an outside source for research on schizophrenia.[10]

In 1939–40 the income of the KWG was derived from the following sources:

Ministry of Science and Culture	RM 3552000
Ministry of Food	1274000
Ministry of Aviation	55000
Other ministries	684000
Other public sources	231000
Private donations to institutes	5268000
(Including RM 70000 from I. G. Farben to 6 different institutes)	
Total	11064000[11]

Some of the funds were given for general purposes, some earmarked for specific use but as a matter of principle industrial donors were discouraged from attaching, and generally did not attach, conditions to their donations.[12] In 1941, 25 of the 36 institutes were designated as organizations of importance to the army or the SS. The question arises what this meant in fact. A certain amount of research was clearly mission-oriented, but according to my informants the institutes which had been devoted to pure research maintained or even increased this kind of activity. In many cases research directors would 'piggy-back' their ongoing work on some war-related task (a procedure not entirely unknown among their colleagues in democratic societies). In 1941, 31% of the male *Gefolgschaftsmitglieder*, which included everybody from semi-skilled technicians to top researchers, was in the armed forces. Unfortunately figures for later years could not be found in the archives, but interviews with institute directors indicate that they fared fairly well in comparison with universities in retaining or recalling their most important co-workers. (Incidentally, in the universities professors were occasionally successful in getting rid of political trouble-makers by failing to attest to the urgent need for their services, and on the other hand others could escape the clutches of the *Gestapo* by volunteering for service.)

In general, the structure and operation of the KWG institutes remained unchanged after 1933. Three new institutes were founded: two for agricultural research (one of them in Bulgaria), and one in Rome for the strengthening of German-Italian cultural relations. The implementation of the 'leadership principle' depended on the confidence which the president of the society could engender among the directors of the institutes, who essentially maintained their established autonomy regarding scientific matters. The main exception concerned the right to choose their own personnel, where the party could exercise veto power in a number of cases. Support from public funds remained at roughly the same level, with the exception of a special grant from the air ministry for the Aerodynamic Institute in Goettingen. It is of interest that government money paid for an addition to the Physics Institute of Otto Hahn, who was known as being unsympathetic to Nazi ideology.[13]

According to one director, "we lived like kings."[14] Not all of them were equally fortunate, but the evidence indicates that the KWG was able to maintain itself with remarkable success in the face of political and military vicissitudes. Very few publications were restricted for reasons of military secrecy. Before the war members of the society could freely interchange reprints with their Soviet colleagues; this was usually not possible for university-based researchers.[15]

5. Scientific societies

Along with most other organizations scientific societies were required to belong to some umbrella organization which in turn was subordinated to a ministry for control. Somehow the German Physical Society was able to remain outside this structure. The physicists were outstanding in the degree to which they maintained their opposition to the regime.[16] Not many of them could emulate the courage of a Max Planck, lacking his renown, but there were others. In 1936 Max von Laue proposed to Planck that the Jewish Lise Meitner should be nominated for the Nobel prize.[17] In 1942 Planck wrote to Laue not to worry about mentioning relativity theory in a publication (though he thought it wasn't 'necessary' to mention Einstein by name; on the other hand, if no suitable circumlocution for the 'Londonian stream' was available, he advised him to use the eponym.)[18] The official requirement not to mention Jewish scientists in publications, or to use separate bibliographies for Aryan and non-Aryan authors, depending on whether circulation was to be on the domestic or international level, was often not observed.

Such individual acts of independence may seem insignificant, but given the general atmosphere of terror and fear, they symbolize the attempt to maintain the traditional status of science against *force majeure*. Of greater practical consequence was the waging of the *Kulturkampf* with the party ideologists on the matter of 'Jewish physics'. Stark and Lenard simply could not make headway with their ideas; the opposition to them among their peers was so strong that the party finally gave up its support. The showdown came in 1940 at the so-called 'religious conference' (*Religionsgespraech*) held at the party headquarters in Munich – right in the lion's mouth. The discussion resulted in a clear-cut acceptance of the validity of quantum theory and the special theory of relativity and the official establishment of a truce.[19]

Another important action was the establishment of a new journal, the *Physikalische Blaetter*, in 1944. It was the brainchild of Ernst Brueche, who became its editor and of its successor, the *Neue Physikalische Blaetter*, until his retirement in 1973. For this venture he enlisted the support of Speer, who made available the necessary high-quality paper at a time when it was in extremely short supply and other journals were forced to terminate or severely restrict their circulation. The journal was a forerunner of both the *Bulletin of the Atomic Scientists* and *Physics Today*, focusing on professional problems of the discipline and the social responsibility of scientists. In fact, it was considered an underground publication by its founders. Among its articles it featured discussion of the need for basic research, for increasing the quantity and quality of a new generation of physicists and other scientists (an entire issue is devoted to this).[20] Among the news items one is particularly interesting: according to a Swedish newspaper, in the U.S. experiments are being conducted to built a powerful bomb with uranium as the raw material. The item is headed 'Yet another Utopia'.[21] The new journal was very successful; within four months after the first issue it had 1500 subscribers, many of them high school teachers.

The other area in the natural sciences where ideology played a major role was human anthropology and heredity or 'race science'. Here the opposition to the official doctrine was sporadic and widely scattered and never reached the level of an open confrontation. I could not discover any communication among anthropologists or biologists regarding possible concerted action analogous to that of the physicists, though there were some individuals who had the courage to openly dispute the concepts of aryanism and of 'nordic' racial superiority. In 1935 the anatomist Fick gave a paper before the Prussian Academy of Science in which he rejected the validity of the concept of race altogether (Seller, 1961: 47–48). The Munich anthropologist Karl Seller was probably the most outspoken opponent of the *Rassenlehre* and was dismissed; he regained his chair after the war. High school teachers who had to use the official syllabus could at best offer passive resistance by absolving the material as quickly as possible; any open resistance could have meant the loss of job or the concentration camp.[22] The difference in the stance of the biologists and the physicists is striking. One likely reason is the relative "hardness" of the theoretical systems in the two sciences; moreover, the thesis of "Nordic" racial superiority had found wide acceptance among German and other intellectuals long before its official adoption by the Nazis.[23]

6. Conclusion

The evidence I have presented leads me to the following:

Some time after the traumatic impact experienced during the early stages of the Nazi takeover, the German scientific institutions were able to recover their traditional strength to a remarkable degree. The recovery was by no means general; it depended on the amount of loss of leading scientists in a specific field, the presence of highly motivated, energetic and politically astute leaders, and other factors. The short duration of the Third Reich leaves much in the realm of speculation, but it is my impression that there was sufficient momentum to further maximize the strength of the scientific institutions, assuming that Germany would have won the war.[24]

The closest approximation to the Soviet Union's Lysenko episode in genetics was the attempt to drive out 'Jewish physics', which was a dismal failure. This comparison leads us back to the general problem of the relation between totalitarianism and scientific growth. It would be foolhardy to defend the thesis that totalitarianism is a more fruitful environment for the growth of science than democracy, yet even this 'outrageous hypothesis' as Robert Lynd would have called it should not be dismissed out of

TABLE II

Students in the natural sciences in German scientific institutions [a]

	Math	Phys.	Math & Physics	Chem.	Life sciences	Geography	Geology	Other	Tot	% change
1932	4163	2231		4007	1821	913	217	379	13731	
1933	2942	1849		3504	1418	645	153	375	10886	—20.7
1934	1655	1437		3006	1009	480	139	264	7990	—26.6
1935			2330	2707		378		1089	6504	—18.6
1936			1741	2444		198	74	847	5304	—18.4
1937			1371	2669		134	80	650	4904	—7.5
1938	440	897		2912		113	73	599	5034	2.6
1939	186	519		2486	191	50	33	112	3577	—28.9
1940	300	590		3106	314	100	50	422	4882	36.5
1940	483	661		3617	502	177	46	408	5894	20.7
1941	400	564		2873	509	169	46	330	4891	—17.0

Source: Statistisches Bundesamt, VII B

[a] Winter semesters until 1939; then Fall and Winter trimesters.

TABLE III

Final examinations for Ph.D. and diplomas in German universities

Year	Earth sciences	Astron.	Anthrop. & genetics	Maths	Phys.	Chem.	Botany	Zool.	Pharm.	Tot.	% change
1932	132	22	–	48	150	334	103	106	8	903	
1933	155	25	4	54	153	336	94	97	4	922	2.1
1934	148	42	2	49	182	331	96	111	7	968	5.2
1935	143	51	2	38	219	280	96	113	6	958	−1.0
1936	147	30	8	47	177	329	99	101	15	953	−.5
1937	133	23	2	29	165	326	54	80	17	829	−13.0
1938	122	21	9	31	168	324	73	76	13	837	.9
1939	201	22	11	40	174	352	10	80	25	915	9.3
1940	66	17	6	18	83	225	48	42	18	623	−31.9

Source: Statistisches Bundesamt, VII B Wiesbaden

TABLE IV

Students in pure science in British universities

Year	Full time students		Advanced students		Degrees awarded	
	No.	% change	No.	% change	No.	% change
1933	8757		1170		2561	
1934	8784	.30	1141	−2.47	2724	−6.36
1935	8602	−2.07	1218	6.74	2750	0.95
1936	8102	−5.81	1209	−0.70	2662	−3.20
1937	7749	−4.38	1232	1.90	2659	−0.11
1938	7661	−1.14	1268	2.92	2460	−7.49
1939	6685	−12.74				
1940	5632	−15.75				
1941	6022	6.92				
1942	6081	0.98		No information		
1943	6057	−0.39		available		
1944	6783	11.99				
1945	9309	33.26				
1946						

Sources: Department of Education and Science, London (University Grants Committee)

TABLE V

U.S. Ph.D.'s in natural sciences

Date	Number	% change
1933	1229	
1934	1387	12.8
1935	1301	−6.2
1936	1359	4.4
1937	1386	2.0
1938	1342	−3.2
1939	1421	6.0
1940	1653	16.3
1941	1833	10.9
1942	1705	−7.0
1943	1332	−21.9
1944	995	−25.3
1945	772	−22.4

Source: Lindsey R. Harmon and H. Soldz, *Profiles of Ph.D.*'s *in the Sciences*, Washington, D.C. National Academy of Sciences, 1965, Table 3.

hand (Lynd, 1939). The 'heretical' point of view was held by Lundberg (1947a: 48), who asserted that "if we plot the course of scientific advances during the past two hundred years, the impressive fact is how little its main course has been deflected by all the petty movements of so-called 'social action' including the major political revolutions." And Niebuhr (1957: 7–8), who did not share Lundberg's positivist frame of reference, wrote that "on the whole, political despotism has no quarrel with pure science or with scientists who do not feel called upon to challenge the basic dogmas of the regime." These would seem rather hasty generalizations; in fact, Lundberg (1947b) himself attached the very essence of totalitarian dictatorship – the restriction on the free competition of ideas, which is considered essential also for the flourishing of science. But on the other hand, the following assertion also appears to be an overstatement:

> In fact, modern totalitarian dictatorships tend to separate the rational and empirical which science has united, forcing knowledge primarily to be responsive to ideological dogma rather than to empirical observation, and forcing empirical inquiry to develop in ways which are immediately applicable to ideological demands rather than to the evaluation of abstract systems (Richter, 1972: 76).

One of the major problems in this respect involves the question of whether a given situation exists *because of* or *in spite of* the totalitarian environment. This in turn takes us back to a valid definition of totalitarianism as a generalized concept applicable to all social systems regardless of their socio-cultural and historical idiosyncracies.

In this connection we should be sensitive to the fact that Germany was beginning to lose its scientific dominance in certain fields at a time when its social structure had already become more 'democratic', in the political realm at any rate. Thus Price (1963: 96) has shown that Germany's percentage of the total output of papers listed in *Chemical Abstracts* had been declining since 1910 (except for a brief spurt between 1940 and 1945!) The role of institutional factors having little direct relevance to the totalitarian-democratic dichotomy is documented by Ben-David's investigation of medical discoveries; in this case Germany lost its leadership as early as the latter part of the nineteenth century (Ben-David, 1962; 1968). Or, to cite a contemporary example showing differences between two democratic societies: "Some work in [British molecular biology] is of the highest class... but there is much which is... at the best dull and at the worst trivial. Our examination of the literature leads us to think that this is in marked contrast to the position in the U.S.A.... where there is a great deal of the work which is essential to sustained development."[25] Even the generally leading U.S.A. is considered backward in certain fields with respect to the Soviets. According to one observer, they are ahead in high energy and plasma physics and abreast of the West in space research, astronomy, and theoretical mathematics (Astrachan, 1973: 117).

By contrast the backwardness of the U.S.A. during its early years in basic research generally has been widely documented (Shyrock, 1962). The examples to make my point can be extended *ad libitum*.

What can be said concerning the thesis under investigation? I believe that I have demonstrated that one totalitarian system did not 'necessarily' restrict development in the natural sciences. For the reasons indicated at the outset, conclusions concerning relationships between totalitarianism and science in general are hazardous. The early demise of the Third Reich was untimely from the standpoint of a more conclusive investigation – it hardly need be said that it came none too soon. But we still have the opportunity to compare developments in the Soviet Union and China. It may well be the case that the concepts of 'totalitarian' and 'democratic' social structure are too gross to do justice to specific cases. As J. S. Salomon has pointed out,

> [In the case of the Soviet Union] if disciplines, theories, individuals and even teams may suffer – to the point of liquidation, as in the instance of Vavilov and his geneticist colleagues – from this interference with the affairs of science, other fields of research are not prevented from developing with as much success as under a liberal regime; the progress of physics in the USSR was not affected by the genetics crisis or even by the Stalinist purges of which some of the best physicists were victims. Conversely, the affairs of science are no less subject under liberal regimes than under totalitarian regimes to decisions imposed from without, and to interventions – more or less authoritarian, according to the nature of the regime, affecting the allocation of resources, the priorities assigned to certain sectors and therefore the general orientation of the research effort, but nevertheless unquestionable interventions, which cannot leave science indifferent.

He then goes on to say,

> Nowhere can political power dictate to science its precedures, the laws of its activity or its substance; power cannot determine the form or content of the scientific approach. The road to truth escapes political decision as much as truth itself; at a deeper level, truth – whether it be scientific or not, but especially if it is scientific – has its own authority which the authority of power cannot gainsay. Power may possibly hold it in check (by constraint or persuasion) by denying it access to public discussion, by screening part of its results or distorting its meaning, but no constraint or persuasion can change

what it has established, save the authority of the scientific process itself. (Salomon, 1973: 92).

He thus reiterates the positivist position taken by Lundberg and others. I question validity of this position, unless one takes an exceedingly long view of the course of history – a view which may be appropriate for philosophers, but not for those who are concerned with the state of science during the foreseeable future. In fact, one may put it to Salomon that he eliminates his own *raison d'être* as a 'statesman of science' by this assertion.

I am well aware of the deficiencies of this analysis. But "Any historian who returns from the past and claims to have discovered all of its secretes should be immediately suspected of not telling the truth, or all of it" (Peterson, 1969: 428).

I hope that others will fill the existing gaps. If history has any lessons to teach, present and future generations of scientists may learn something from the experience of their fellows – as scientists, citizens, and human beings.

Purdue University,
West Lafayette, Indiana

NOTES

* I thank the following for advice and information: E. Akeley, G. Benl, A. Beyerchen, E. Brueche, U. Clever, V. Foley, K. Freudenberg, W. Gerlach, J. Haberer, F. Hund, W. Koester, A. Lang, T. von Laue, K. Lemcke, G. Melchers, A. Peterson, H. Piepho, F. Saupe, B. Schroeder-Gudehus, E. A. Shils, E. Telschow, H. Weisert. The Purdue Research Foundation gave travel grants to defray part of the cost of this research.

[1] Merton (1957): Chs. 15 and 16; Barber (1952): Chapt. 3; Barnes and Doby (1970); Rothman (1972); Blissett (1972).

[2] For a general description and evaluation of the early stage of the Nazi takeover, see Haberer (1969): Chs. 7 and 8; Maier (1966).

[3] Hartshorne (1939); Ferber (1956). The losses, forced or voluntary, in different universities and disciplines, are described in greater detail in these sources.

[4] Freudenberg, K., personal communication.

[5] Gerlach, W. and Koester, W., personal communication.

[6] In addition to the regular funds Menzel had at his disposition as much as eight million RM in 1943 and apparently in subsequent years. This came from a fund for 'special military missions'. The actual use of this money could not be ascertained. Zierold (1968): 351–52.

[7] Library of Congress, Washington, D.C. microfilm PB 13347.

[8] Library of Congress, microfilm PB 18749.

[9] *Archives of the Max-Planck-Gesellschaft*, Munich, letter of 12 Sept. 1939.

[10] Telschow, E., personal communication.

[11] Kaiser-Wilhelm-Gesellschaft, Annual Report, 1941.
[12] Telschow, E., personal communication.
[13] Telschow, E., personal communication.
[14] Koester, W., personal communication.
[15] Melchers, G., personal communication.
[16] The political liberalism of physicists has been noted before. Cf. Meier (1951); Ladd and Lipset (1972).
[17] Laue, Letter to Planck, 22 December 1936, Deutsches Museum, Munich.
[18] 13 June 42, Laue correspondence in Deutsches Museum.
[19] Scherzer (1965); Brueche (1946). For Stark's apologia, in which he tries to show that he was trying to fight for freedom of research and comments by Laue, see *Physikalische Blaetter* (1945): 271–73. The conflict flared up once more in 1942 and was settled at a similar conference at Seefeld in the Tyrol, where another compromise was reached, apparently in order for the SS to save face. In any case there is no evidence that the ideology affected developments in theoretical physics. On the Seefeld conference see Goudsmith (1947): 153. The present paper does not deal with applied nuclear physics, since this area has been covered in great detail by others.
[20] Vol. 1, May–June 1944. Brueche, E., personal communication.
[21] August 1944, p. 118.
[22] Benl, G., personal communication.
[23] Seller, Cf., (1961). The author mentions a goodly number of anthropologists and others who were outspoken racists during the Nazi regime and remained in their chairs and received professional and civic honors after the fall of the Nazi's.
[24] The available comparative data for several countries are given in Tables II–V. They provide some indication of the impact of the war and general economic trends.
[25] *Report of the Working Group on Molecular Biology* 'Kendrew Report', Cmnd 3675, London, HMSO, 1968, p. 4.

BIBLIOGRAPHY

Archives of the Max-Planck-Gesellschaft: 1939, Munich, Letter of 12 Sept.

Astrachan, A.: 1973, Review of Loren R. Graham's *Science and Philosophy in the Soviet Union*, in *The New Yorker* (24 Sept.), 117.

Barber, B.: 1952, *Science and the Social Order*, The Free Press, Glencoe.

Barnes, S. B. and Doby, R. G. A.: 1970, 'The Scientific Ethos: A Deviant Viewpoint', *European Journal of Sociology*, **23** (March), 102–108.

Ben-David, J.: 1962, 'Scientific Productivity and Academic Organization in Nineteenth Century Medicine', in B. Barber and W. Hirsch (eds.), *The Sociology of Science*, The Free Press, Glencoe, pp. 305–28.

Ben-David, J.: 1968, *Fundamental Research and the Universities*, OECD, Paris.

Blissett, M.: 1972, *Politics in Science*, Little Brown, Boston.

Brueche, E.: 1946, 'Deutsche Physik und die deutschen Physiker', *Physikalische Blaetter* **9**, 232–36.

Ferber, C.: 1956, *Die Entwicklung des Lehrkoerpers der Deutschen Universitaeten und Hochschulen*, Vandenhoeck and Ruprecht, Goettingen.

Fest, J. C.: 1963, *Das Gesicht des Dritten Reichs*, Piper, Munich.

Glum, F.: 1964, *Zwischen Wissenschaft, Wirtschaft und Politik*, Bouvier, Bonn.

Goudsmith, S.: 1947, *Alsos*, H. Schuman, New York.

Haberer, J.: 1969, *Politics and the Community of Science*, Van Nostrand-Reinhold, New York.

Hartshorne, E. Y., Jr.: 1939, *The German Universities and National Socialism*, Harvard University Press, Cambridge, Mass.
Hirsch, W.: 1961, 'The Autonomy of Science in Totalitarian Societies', *Social Forces* **40** (October), 15–22.
Kaiser-Wilhelm-Gesellschaft: 1941, Annual Report.
Ladd, C., Jr., and Lipset, S. M.: 1972, 'Politics of Academic Natural Scientists and Engineers', *Science* **176** (9 June), 1091–1100.
Library of Congress, microfilm PB 13347, Washington, D.C.
Library of Congress, microfilm PB 18749, Washington, D.C.
Lundberg, G.: 1947a, *Can Science Save Us*?, Longmans, Green, New York.
Lundberg, G.: 1947b, 'Sociology vs. Dialectical Immaterialism', *American Journal of Sociology* **53** (Sept.), 85–105.
Lynd, R. S.: 1939, *Knowledge for What*?, Princeton University Press, Princeton, N. J.
Maier, H. (ed.): 1966, *Die Deutsche Universitaet im Dritten Reich*, Piper Munich.
Max-Planck-Gesellschaft: 1961, *50 Jahre Kaiser-Wilhelm-Gesellschaft und Max-Planck-Gesellschaft zur Foerderung der Wissenschaften, 1911–1961*, Max Planck-Gesellschaft, Goettingen.
Meier, R. L.: 1951, 'The Origin of the Scientific Species', *Bulletin of the Atomic Scientists* **8** (June), 169–173.
Merton, R. K.: 1957, *Social Theory and Social Structure*, The Free Press, Glencoe.
Niebuhr, R.: 1957, *New Leader* **40** (25 Nov.), 7–8.
Pascher, J.: 1966, 'An drei deutschen Universitaeten', in H. Maier (ed.), *Die Deutsche Universitaet im Dritten Reich*, Piper, Munich.
Peterson, E. N.: 1969, *The Limits of Hitler's Power*, Princeton University Press, Princeton.
Price, D.: 1963, *Little Science, Big Science*, Columbia University Press, New York.
Physikalische Blaetter: 1944, 1 (January-February).
Report of the Working Group on Molecular Biology: 1968, 'Kendrew Report', Cmnd 3675, HMSO, London.
Richter, M. N., Jr.: 1972, *Science as a Cultural Process*, Schenkman, Cambridge, Mass.
Ringer, K.: 1969, *The Decline of the German Mandarins*, Harvard University Press, Cambridge, Mass.
Roegele, O. B.: 1966, 'Student im Dritten Reich', in H. Maier (ed.), *Die Deutsche Universitaet im Dritten Reich*, Piper, Munich.
Rothman, R. A.: 1972, 'A Dissenting View on the Scientific Ethos', *British Journal of Sociology* **23** (March), 102–108.
Salomon, J.-J.: 1973, *Science and Politics*, MIT Press, Cambridge, Mass.
Scherzer, O.: 1965, 'Physik im totalitaeren Staat', in A. Flitner (ed.), *Deutsches Geistesleben und Nationalsozialismus*, Thuebingen, Wunderlich.
Schroeder-Gudehus, B.: 1972, 'The Argument for the Self-Government and Public Support of Science in Weimar Germany', *Minerva* **10** (October), 537–70.
Seller, K.: 1961, *Die Rassenlehre des Nationalsozialismus*, Progress Verlag, Darmstadt.
Shyrock, R. R.: 1962, 'American Indifference to Basic Science During the Nineteenth Century', in B. Barber and W. Hirsch (eds.), *The Sociology of Science*, The Free Press, Glencoe, pp. 98–110.
Speer, A.: 1970, *Inside the Third Reich*, Macmillan, New York.
Spiro, H. J.: 1968, 'Totalitarianism', *International Encyclopedia of the Social Sciences*, Macmillan, New York, Vol. 16, pp. 106–113.
Zierold, K.: 1968, *Forschungsfoerderung in drei Epochen*, Steiner, Wiesbaden.

DAVID E. MORRISON

PHILANTHROPIC FOUNDATIONS AND THE PRODUCTION OF KNOWLEDGE – A CASE STUDY

1. Introduction

Much of the work written on philanthropic foundations has been produced by persons who have at some time been connected with such organisations, and usually at a very senior level. Whilst furnishing valuable information with regard to the running of foundations, such writings have tended towards the anodyne, offering little critical perspective. Works by non-foundation men, on the other hand, have usually attempted an overview of foundations' activity, and have suffered from a surfeit of description at the expense of analysis. An outcome of such work has been the neglect of an important aspect of foundation activity, namely their role in the production of knowledge. Certainly, all works on foundations note at some point the support that these institutions have given to knowledge production and often they appraise trends in funding, but rarely have the mechanics of this relationship been analysed in any terms other than a crude cash nexus. A notable exception is the perspective developed by Coser (1970). Taking this as a starting point, the present essay sets out to explore further the role of the foundations as 'gatekeepers of contemporary intellectual life'.

This paper examines the manner in which changing social and political climates have influenced the operation of foundations, and through them, the production of knowledge. In attempting to offer an explanation of this process it is necessary to fully understand the position occupied by foundations within American Society, and particularly their relationship to the universities. When Lazarsfeld and Thielens (1958) surveyed social scientists for the possible effect of social pressure in a 'time of crisis', they concentrated on the actual point of production and knowledge transference. The present paper focuses on a prior point of the chain, and attempts to determine the consequences of social and political pressure at this point for intellectual activity. The position to be developed is that foundations are particularly responsive to social and political pressure

Knorr, Strasser, and Zilian (eds.), Determinants and Controls of Scientific Development, 367–402.

and that an important key to understanding why this should be so is provided by the concept of 'legitimation'. It is argued that the American philanthropic foundations have failed to legitimate themselves. Having 'explained' the reasons for this lack of legitimacy historically, and illustrated its consequences for the openness of foundations to social pressure, the paper focuses on a concrete case study of the way in which external pressure blocked a particular project which had important implications for a whole field of work.

2. The Lack of Legitimation

The vulnerability of philanthropic foundations to changing social and political climates is in part due to their failure to legitimate themselves. The concept of legitimation is used here in Lipset's (1960:46) sense of "the degree to which institutions are valued for themselves and are considered right and proper." It is argued that the foundations as institutions have failed to become 'accepted', in the sense of being taken as part of the natural ordering of things, having an unquestioned right to exist and operate. Further, such is the lack of acceptance that when controversies do break, in contrast to the universities which have legitimated themselves, the whole existence of the foundations can be called into question. This lack of basic acceptance is illustrated by the fact that the perceived indiscretions of one foundation have often been generalised to foundations *per se*. The fault is not seen as confined to the foundation responsible for the indiscretion but tends to be taken as symptomatic of foundations as a whole.

To understand this lack of legitimation two important factors must be taken into account. Firstly, the circumstances deriving from the historical and cultural situation surrounding the establishment of many of the large foundations, and secondly, the way in which their operations have developed over time. For, despite gestation difficulties, or labour pains, confronting the formation of many foundations, it is likely that they could, in time, have legitimated themselves had it not been for certain operating principles which led them into the controversial area of the social sciences. Once foundation funds began to flow into the social sciences it proved a quagmire from which there was little or no possibility of extraction into acceptance. The foundations were sucked into a vortex of controversies which they were ill-suited to withstand. Thus, one has two stages in the barrier to legitimation: the early suspicion surrounding the foundations' establishment which they gradually began to erode, followed by their entry into the social sciences when suspicion once more engulfed them.

3. THE BIRTH OF FOUNDATIONS – LABOUR PAINS AND LABOUR PROBLEMS

Despite antecedents in other periods and in other cultures, the modern philanthropic foundation is a peculiarly American, and essentially twentieth century, institution.[1] For as William Whyte Jr. (1955a: iii) notes:

> The concept of a private foundation for public progress is a very American development and a fairly recent one. Up until the late nineteenth century, businessmen used to give largely for the alleviation of misfortune, local usually, rather than prevention of it. But capitalism kept piling up more and more personal surpluses. Even with Newport chateaux, yachts and heirs, there was more left to give than they could give intelligently.

Even leaving aside the detailed psychological motivations behind this philanthropy, which are in themselves difficult to discern,[2] it is true as Lindeman (1936) notes, that:

> The distinction between ordering private charity and large scale philanthropy is the difference between a small and a large surplus. The former may remain on a personal level, but the latter involves organisation.

Philanthropy was put on the same entrepreneurial footing as productive enterprise. However, if the foundations provided their benefactors with an organisational solution to the problem of rationalising the distribution of large scale personal wealth, they were not always greeted with universal applause.[3] This lack of acclaim stemmed in part from recognition of the dubious nature of the source and methods which had facilitated such endowments. Fosdick (1952:19) in his history of the Rockefeller Foundation notes that such wealth was made "under conditions unique in the history of the country and not infrequently by methods which, if permissible at that time, no longer accord with social conscience or the requirements of the law."

The 'tainted money' spectre is a familiar theme in the distrust of foundations, early evidence of which is the establishment of Guys Hospital in London out of profits made by Sir Thomas Guy from the South Sea Bubble adventure.[4] A more recent example is the refusal of Mayor Hylam of New York to accept foundation assistance to build a hospital complaining that the money was tainted.[5] Another suspicion hanging over the early foundations was that their founders would use them to resist change and support their own economic views and practices.[6] There were therefore two strands to the early suspicion of foundations, one associated with the sources of endowment, and the other with the uses to which such monies might be put.

Although the cry of tainted money was a factor to be overcome if the foundations were to legitimate themselves, it has in recent years been raised less and less. Not so, however, the concern over the use to which foundation funds would be put. That fear has been persistent but, at the same time, it must be stressed that the focus of this suspicion has varied over time. For example, whilst the original criticism tended to cluster around fears that funds would be used to restrict social change, during the 1950's, there was a volte-face and criticism focused on fears that the foundations were encouraging radical social change. The suspicion that donors would direct funds to enhance their own industrial interests has, however, remained a constant source of criticism. Harrison and Andrews (1946:42), in discussing the concern generated by the establishment of foundations, note that:

> ... the foundations would use their establishments to resist social change and to support their own economic views and practices. Much of the criticism crystallised in the hearings before the United States Industrial Relations Commission [1915].

Turning to the Commission itself, one can clearly see concern over both the purpose to which such funds were being put, and the power of foundations to exert influence over recipients. Thus the Report (Manly Report of U.S. Senate Commission on Industrial Relations, 1916:83) reads:

> ... the so-called investigation of industrial relations has not, as it is claimed, either a scientific or a social basis, but originated to promote the industrial interests of Mr. Rockefeller.

In addition, the Report outlines as an example of foundation power:

> The abandonment by several colleges and universities of sectarian affiliations and charter clauses relating to religion in order to secure from the Carnegie Corporation pensions for the professors and for the advancement of teaching. It would seem conclu-

sive that if an institution will willingly abandon its religious affiliations through the influence of foundations, it will even more easily conform to their will in any other part of its organisation of teaching.

This investigation into foundation activity made headlines for weeks and generated great concern and apprehension as to what might follow from the foundations' power.[7]

The suspicions surrounding the foundations' activity and the purposes of their operations have not been easily shaken. Charles Dollard (Reece Report, 1955: 950), replying as Head of the Carnegie Corporation to such charges before the Reece Enquiry, well illustrates the nature of the criticisms. Part of his evidence reads as follows:

It is quite true that in recent years the development of foundations has been facilitated by tax provision; but it is a grave injustice to the American philanthropist to say that they are chiefly moved by considerations of tax avoidance. Both the Rockefeller and Carnegie Foundations were set up at a time when there was no federal income or estate tax.

Dollard is quite correct on this point. Except for a brief period during the Civil War, personal income tax was not levied in America until 1913, and then at the modest rate of 1% on income exceeding 20000 dollars, rising thereafter to a maximum of 6%. Furthermore, charitable contributions were not allowed a deduction until 1917, so the donors of older foundations, which includes many of the larger ones, had no possibility of personal tax savings.[8]

Although it is difficult to substantiate the charge that foundations were created primarily for tax gains, the accusations of ulterior motive remain with them. The Texas Democrat, Wright Patman, still launches attacks upon the foundations in this vein. He considers that foundations are often used firstly as tax dodges, and secondly to maintain control of industrial concerns by having stocks held by one of the foundations associated with the industrial enterprise. Although it is certainly true that in some cases foundations, such as the Howard Hughes Foundation, have at times not operated entirely within the spirit of the law,[9] it is a misreading of history to consider that foundations were created and continue to operate primarily as a tax benefit. At the same time, it is true that benefits can accrue to the donor's family in the course of operations, and that foundations have been loathe to sponsor work that may run counter to the donor's industrial interest,[10] but that would appear to be the limit to self-aggrandisement.

However, the suspicion surrounding the motivations behind donation has not entirely evaporated; remnants still exist today. For example, George Wallace, campaigning in the South, can still gain some political mileage from tapping hostilities towards the foundations but, in general, such antagonisms, at least on the surface level, have largely faded away. If, in a natural history of foundations' development, one could reasonably have expected an acquiescence towards the foundations to have emerged, to be later exchanged to acceptance through the emollient of time, this was not to be, for no sooner had the foundations begun to 'establish' themselves in American culture than a rupturing caused by the foundations themselves, occurred. That is, propelled by their own principle of 'improving Man's estate', they moved into the social field which, when allied to the professional ideology of 'risk capital philanthropy', created problems from which there was little respite.

4. Entry into the Social Sciences – the Paradox of Involvement

... whenever foundations get into trouble, as they do from time to time, it is practically always because of something they have done in the social field, or in the application of questioned social theories to business or industry (Hollis, 1938: 245).

As long as foundations operate within the area covered by the social sciences then the possibility of them extricating themselves from controversy is extremely limited. It is therefore valuable to plot briefly the manner in which foundations came to find themselves in this uncomfortable situation.

Whereas the very early American foundations were usually established through endowments for limited purposes, later foundations were established for general purposes.[11] The generalist nature of the modern foundation is reflected in the vagueness of their establishing charters. 'Education' is the term that occurs most frequently in the bequests and legal instruments which established foundations; in fact, nearly all the donors appear to have had education in mind when making their bequests.[12] This orientation is explicable firstly by reference to the symbolic role that education occupied as the "sovereign remedy for social ill" (Hollis, 1938: 116), and secondly to the unfavourable public estimates of the elder John D. Rockefeller and of Andrew Carnegie which militated against direct reformist grants in favour of general endowments.[13] This is not to say that many of these early general non-controversial grants did not have far-reaching reforms skilfully embedded in them – they did.[14] For example, the system of 'Carnegie Units' by which educational institutions could qualify for professorial pensions only after certain conditions had been met had far-reaching consequences within the American education system.[15] Such large-scale across-the-board grants eventually came to a halt however, partly through the sheer economics of the situation, and partly because of the increasing redundancy of such gifts, as government assumed responsibility for such areas. In 1921, the Carnegie Corporation gave 67% of the number of grants made, and 77% of the total sum – outside the appropriations to other Carnegie enterprises – in the nature of contributions to campaigns for general endowment or for the general support of organisations. Subsequently, the tendency of the Corporation has been more and more to support specific projects.[16] Hollis (1938: 42) underscores this point:

> By 1925 the cost of all education had mounted still higher and foundations, to maintain their effectiveness, gave up special endowments and concentrated on current support and research grants.

The economics of the situation are perhaps best summarised by the President of the Rockefeller Foundation. Reflecting on the situation in the 1920's, he noted that:

> ... there was a sense therefore in which 1928 [the amalgamation of many of the foundations in the Rockefeller group] marked the turning point in the thinking of the Trustees. This was in part due to the fact that they had come to the end of an era in philanthropy, an era that was reflected in many other foundations as well. Huge sums had been spent in the endowment of medicine, public health institutes and programs in higher education. Apart from the foundation's contribution, the General Education Board had given 50 million dollars on a matching basis to raise the endowment of colleges and universities, and over 90 million dollars for American medical schools. This type of giving could not continue without involving the rapid liquidation of the Rockefeller Boards (Fosdick, 1952: 159).

President Fosdick (1952) adds further comment on the overall economic situation stating that, after the Depression of 1929 and the decline in interest rates, it became increasingly difficult to induce other donors to match foundation funds, making appropriation for endowment an uncertain and dubious technique. He further notes that the interest rates of 4% and 5%, which in the 1920's could be relied upon, were cut and it became increasingly difficult to predict in terms of purchasing power; thus, "a new orientation of target, programme and technique became a vital necessity as the twenties drew to a close." (Fosdick, 1952: 159).

In addition to the basic economics of the situation there was the fact that in the early part of the century dramatic opportunities for giving on a massive scale had existed: there were colleges and libraries to be built, obvious social reforms to be aided and, since government contributions to such areas was small, foundation money represented a large proportion of the whole.[17] But, as government funding began to catch up, foundations were forced to search for pivotal points with which to maintain the effectiveness of philanthropy, and hence, they increasingly diverted funds to what Horowitz (1972: 245) calls "the growing edge of knowledge." This withdrawal from areas, when other agencies are prepared to take on the responsibility, is a common feature of foundations' activity; a good example is the Rockefeller Health Program.[18] The move away from general grant giving in the field of education and towards a concentration on the "growing edge of knowledge", that is, into innovation and growth within the knowledge institutions, ensured a greater cost effectiveness for the foundations.

Before the First World War, apart from such endowments as the Smithsonian and Carnegie Institutions, foundations in general had not been engaged very actively in supporting research in any of the sciences.[19] Even by 1920, only the physical sciences and natural sciences were receiving any appreciable foundation support.[20] The entry into support of research as such can be seen as a natural outgrowth of foundations' remedial and preventive health programmes. Like education, health was an obvious contender and target for grand demonstrations of human concern on the part of foundations. The support for the natural and physical sciences was to a large extent an outgrowth of these health programmes. Funds were given to the recognised medical sciences such as anatomy, physiology and pathology in the belief that they would provide the needed remedies but, when these were not forthcoming, the foundations spread their net of support to include the developing sciences of bacteriology, embryology and genetics, which in turn led back into fundamental study in the basic fields of biology, chemistry and physics.[21]

Support for the social and humanistic sciences was virtually nonexistent before the First World War and there was little enough shortly afterwards but Hollis (1938: 116) can write that since 1921 "the natural and physical sciences have continued to receive about the same proportion of the expanding philanthropic budget but the social sciences have received an ever-increasing proportion and are clearly the dominant foundation interest for the 1921–30 decade, as well as for the succeeding half decade." The First World War was undoubtedly largely responsible for stimulating foundation interest in support for the social sciences. The climate of opinion produced by the war began to run in favour of intervention in areas that had previously been closed.[22] Furthermore, the benefits attendant upon social science research were more clearly visible after the war for, as Lyons (1969: 31) notes:

> ... the wartime uses of intelligence testing set a precedent elsewhere in government and especially in private industry. Indeed, the research work that psychologists had performed for the military was called 'a war gift to industry'; private companies began to use psychologists, consulting firms were organised and new techniques invented.

It was within this improved atmosphere that Beardsley Ruml, as head to the Laura Spelman (Rockefeller) Memorial, spent fourteen million dollars of the foundation's money between 1922–29 "in order to put the social sciences on the academic map" (MacDonald, 1956: 24). However, if the aim was to put the social sciences "on the academic map" then the cartography was certainly that of empiricism. In fact, the earliest support for the social sciences in any significant amount was provided by the foundations,[23] the impact of which is witnessed by Donald Young's (1969: xv) statement that:

> In the years prior to World War Two it was practically impossible to write a text in any of the social sciences without relying on findings from research financed by the private foundations.

But, seen in a broader historical perspective, the importance of the foundations' contribution to the social sciences rests not so much in any individual piece or pieces of work that they supported but rather in the strain or style of academic endeavour that they encouraged and gave emphasis to. For, if America was not the home of quantification but the recipient of a European export, then it certainly adopted those techniques on a large scale.[24] A not insignificant part in aiding this adoption was the willingness of the foundations to provide the necessary funds for such expensive research. Not only were they prepared to support such quantitative work but, perhaps more importantly from the point of view of propagation, they supported and encouraged the requisite institutional forms within which such work could flourish and expand.[25] That is, research institutes, centres, bureaux and other similar offices. In fact the driving idea behind Ruml's giving of the Spelman Memorial money was "in order to develop major centres of research"[26] and as Ogg (1928: 156) notes:

> The single fact that, doubtless for good and sufficient reason, foundations and other patrons and benefactors will rarely place money directly at the disposal of an individual, but will turn it to the support of a council or bureau or other continuing, responsible, cooperative organisation, would alone account for the emergence of many of our present research groups or bodies.

The 'good and sufficient reason' mentioned by Ogg for the foundations' support of such institutions and the associated quantitative work rests on the imperatives of the foundations' own bureaucratic structuring (see p. 395) and the high value placed upon quantitative work – in particular "the fact finding non-controversial areas of the social and humanistic sciences"[27] which afforded protection against criticism. To understand the attraction of the 'scientific', associated with quantitative work, it must be remembered that the foundations' interest in serious research developed first in the natural and physical sciences and that their attitude towards the social sciences was to some extent coloured by that experience: it led easily to comparisons with 'the superior sciences'.[28] However, it is difficult to separate out such influences from the general developing pattern of American social science that fed into and gave support for such a comparison; nevertheless, the outcome tended to be an abstraction from the growing social sciences of that part which in *their view* most closely resembled the natural sciences and which at the same time gelled with their own organisational needs and problem solving approach to society. For example, Hollis (1938: 255) lodges the complaint against the foundations for having foisted natural science methods upon the social sciences and writes:

> The point to be established is that the social sciences are undergoing a profound and rapid development, and that more than any other supporting agency philanthropic foundations are pointing and guiding this phase of American higher education.

And further:

... They have I fear been the chief offenders in forcing the techniques of research which developed in the natural sciences onto the social sciences and humanities.

Yet ten years later, Donald Young (1948: 330), Director General of the Russell Sage Foundation comments:

The appreciable measure of success which has attended the increasing concentration on making social science more scientific gives promise that further effort in this direction will be rewarding.

Of course, it would be wrong, in the sense of offering an incomplete picture to the complex phenomena of the adoption of empirical methods and associated forms by American social science, to abstract the influence of the foundations from other influences that helped shape and structure its development. To do so would be to render violence to a complex phenomenon, but any such discussion of other influences must remain outside the possibilities of this paper. Although foundations were only one of the influences operating to encourage the empirical strain within American social science, it was undoubtedly a very important one. According to Berelson (1960: 8):

The foundations contributed to the predominant tone of the social sciences in these (early) decades, namely the empirical rather than say the reflective. This was a distinguishing feature of the 'new social sciences' and one hardly to be discarded or even modified much. But the foundations did go along with this trend and even encouraged it, with all that meant by way of natural observation, segmentalisation of problems, quantification and the rest.* Certainly the foundations did not try to halt or delay the movement towards empirical analysis of human problems ... quite the contrary.

* One might include research institutes among 'the rest'.

5. Operating in the Social Sciences – The Continuing Problems of Legitimation

Movements in society can force movements in the foundations, and can corner or hold them to their principles which, because of the foundations' weak position, paradoxically become inoperable. It is in this manner that the social sciences have provided foundations with their most controversial area of operation. As social problems have become 'visible', and as a body of knowledge has developed laying claim to the possibility of solution or amelioration, the foundations have been impelled by their principles of improving man's welfare[29] to engage the services of this knowledge to discharge their philanthropic duty.[30] This is the developing paradox for, in order to fulfil their philanthropic obligations, the larger foundations have been obliged to engage in activity that requires the help of social scientists. But to enter such a world is to enter the world of values and controversy at a level so direct that it has thrown into sharp relief the weakness of their position. This paradox of affirming their principles at the possible cost of denying their existence has within it a further paradox. Recognising the possible consequences of entering such controversial areas they have entered cautiously and at times defensively. Adopting this response of withdrawal in the face of criticism has meant that the necessary structural support within the foundation world itself has not been forthcoming. This failure to develop internal support systems in turn means that they constantly risk attacks which threaten their existence when controversial areas are entered. However, the increased support given to the Foundation Council[31] following the 1969 Tax Reform Act does show a certain collective recognition of this common situation. It is possible that from such a development a stronger position may emerge through which the paradox can be overcome.

Before leaving the enmeshment in controversy that entry into the social sciences entailed, it is worth noting that foundations have not been completely unprepared in the face of criticism. Quite apart from the fact that, as some commentators have noted, foundations tend to support safe and non-controversial projects, they have erected certain protective procedures. For example, nearly all foundations take the precaution of inserting statements of disengagement in published reports, either through an introductory note or else in a separate statement. But even here, the practical effectiveness is not all that might be hoped for: as Emerson Andrews (1956: 164) points out, the findings "will inevitably be attributed to the foundation itself." A much more effective defence, and one that has major importance for the type of work that foundations are willing to support, is that of 'objectivity'. As the foundations moved into the social sciences and followed the practice of 'target giving' their values were exposed in a much more observable manner than had ever been the case with general across-the-board giving. Clearly, one cannot raise the defence of objectivity for the area of research chosen for support, but the defence of objectivity in the practice of such research has been held to very closely indeed. This is one of the few defensive barriers that the foundations have managed to erect and shelter behind with some impunity. It is difficult, of course, to separate the foundation officer's insistence upon objective methodology as a product of his situation from insistence based upon intellectual training. Nevertheless, there is good reason to suppose that within foundations the emphasis upon the 'objective and scientific' is promoted far more than it is in the university by considerations of defence from criticism. It may be true, as Gouldner (1973: Ch. 3) argues, that insistence upon objectivity was one process by which sociology legitimated itself but within the foundations there is also a conscious awareness of its protective value which can be set aside from its perceived intellectual merits. This was revealed very clearly to the writer during conversations with several foundation officials, and is well expressed in the following conversational abstract:

> Where foundations have got into trouble is where they have done something political and not scientific. It is rare that a really good solid research project with good scientific methodology which is supposed to produce objective results is going to get criticised.[32]

Clearly, a defence resting upon adherence to 'scientific methodology' can only have persuasive power within a culture which regards this as the correct method, and what is considered correct procedure at one time can change and turn into its opposite at some other time. Such a situation occurred during the 1950's when the Reece Investigation stridently attacked the foundations for their scientific procedure.[33]

6. Changing times – changing confidence

> ... both in the past and in the present, fear of legislative retaliation has accelerated the tendency of some foundations to avoid controversial activities.

This statement by Richard Colvard (1964: 737), whilst being fundamentally correct, glosses over the interesting subtleties of foundation existence and operation. In the final analysis it is correct since it pinpoints the living reality against which foundations operate. Although foundations do exist by the acquiescence of the State, and very few could continue to operate should the tax concession be removed, a more productive insight into foundation granting policies is afforded by an understanding of the climate of opinion within which they must function. It is crucial to understand the atmosphere and social pressure surrounding the foundations since it is this which generates or fails to generate confidence, and thereby sets the parameters of operation. Whilst not denying that the possibility of "legislative retaliation" forms an ever-present backcloth, the reality of which has all too often been forced to the forefront, it would be wrong to see it as in any way capable of furnishing an understanding of overall operations. The intention now, therefore, is to examine the impact of the changing social climate on foundation confidence and to indicate the repercussions for funding policies. Following this general outline, I will focus in depth on one particular case of an unfavourable social climate undermining a foundation's confidence which then blocked a particular line of academic enquiry. The purpose of first giving an overview of fluctuations in foundation confidence is to provide a more total understanding of the structural situation of foundations and to situate the case study within a more general context.

Reference has already been made to the 1915 United States Commission on Industrial Relations which was occasioned by Rockefeller's entry into that area. What is now of interest for the understanding of foundation operations is not so much the public suspicion created by their entry into that field but rather their response to that suspicion and hostility.

Rockefeller Foundation was moving towards a more direct operation of its own programmes, especially in the fields of health and medicine, but when it attempted an application of the same principles to the volatile social world of industrial relations it was overwhelmed in controversy. The already delicate balance of the situation was tipped by the eruption of a strike in one of the Rockefeller-controlled companies – the Colorado Fuel and Iron Company – which resulted in the savage 'Ludlow Massacre' of strikers by the State Militia in 1914:

> The Rockefeller Trustees were given an unforgettable lesson about the hazards of becoming involved in social and economic issues, especially if the foundation itself undertook to carry out the work. As a matter of policy they decided thereafter to restrict the foundation's direct operation to scientific areas such as public health, medicine and agriculture. If the foundation supported work in controversial social fields, it would do so through grants to other independent institutions (Nielsen, 1972:53–54).

Not only did this incident result in concern among the Rockefeller Trustees but the shock-waves hit other foundations, demonstrating the vulnerability of their position.[34] The Rockefeller Foundation did not move into suppoıt of the social sciences in any definite manner again until the 1920's.[35]

Between the "unforgettable lesson" of the 'Ludlow Massacre' and

Rockefeller's second attempt at entry into the social arena the social climate had changed considerably. Not only had the social sciences become more acceptable through their war services but agencies such as the Department of Agriculture were engaged in more systematic research and in a more sustained utilisation of social science research.[36] This increased 'visibility' of the social sciences and, more particularly, their growing acceptability, facilitated by the 'war gift to industry' provided by psychological testing, allied with the growing public toleration of trusts, both commercial and philanthropic,[37] lessened the possibility of a repeat disaster. In the words of Fosdick (1952: 215), President of the Rockefeller Foundation, writing of the grand entry into the social sciences by the Laura Spelman Memorial and commenting on the favourable social atmosphere of the times, "it must be admitted that for this activity the environment of 1922–23 was propitious."

If the First World War eased the foundations' path of entry into the social sciences, a further impetus was provided by the employment of social scientists in the New Deal Administration[38] and even greater encouragement was given by the Second World War. Nevertheless, criticism is a constant reminder of the foundations' tenuous position and, even though the social sciences may have been receiving increased support from the foundations,[39] particular areas within these disciplines remained dangerous places to venture. The awareness of this situation is well illustrated by Carnegie's trembling involvement with the area of race relations. Despite the increased confidence built upon the increased acceptance of the social sciences, the foreword to Gunnar Myrdal's *An American Dilemma*, written in 1942, is pregnant with worry and the bated expectation of punitive criticism. The whole of the four page foreword is a justification for attempting such work, even to the extent of extolling the past works of the Corporation and urging the reader "that he make every effort to read these statements intellectually and not emotionally." Despite engaging in a search for a scholar from a country "with no background or tradition of imperialism," Keppel (1944: v), President of the Carnegie Corporation, still felt it necessary to exonerate the foundation:

> Provided the foundation limits itself to its proper function, namely to make the facts available and let them speak for themselves, and does not undertake to instruct the public as to what to do about them, studies of this kind provide a wholly proper, as experience has shown, sometimes a highly important use of their funds.

The atmosphere which allowed this tentative study into such a sensitive area as race began to ebb with the end of the Second World War and with the build-up of Cold War tensions, especially after the outbreak of the Korean War in 1950. The ensuing tensions resulted in suspicion of the foundations which was later translated into two Congressional Enquiries – the Cox Commission of 1952 and the Reece Commission of 1953. The cloud which formed over the foundations, blown by the winds of a more general suspicion, resulted in accusations of communist infiltration and misfeance in allocating grants to communists or to communist sympathisers. In a recent conversation with an official at the Carnegie Corporation who had worked through that period, the atmosphere inside the foundation was described as "oppressive"[40] and John Marshall, former Head of the Rockefeller Humanities Division – which actually engaged in much sociological work: it had sponsored Lazarsfeld's early Radio Research Projects – described the situation as follows:

> ... from that time on we were virtually required to consider the kind of suspicion that the committee had manifested. I say suspicion because that is all it was. I don't think we were ever afraid of criticism but one had to be aware that suspicion of that kind could be handicapping to us.[41]

In summary, John Marshall considered the atmosphere within the foundation to be more one of 'circumspection' than of fear. Be that as it may, the externals of the situation, so far as granting policies were concerned, were less circumspect and more direct. Hutchins (1956: 207), Head of the Fund for the Republic, writes:

> Congressman Reece was scoffed at. It was agreed that his investigation was a farce. I think he had good reason to be satisfied with himself. Without firing a single serious shot, without saying a single intelligent word, he accomplished his purpose, which was to harass the foundations and to subdue such stirrings of courage, or even of imagination, as could be found in them. As I have said, there were not many there when he came on the scene. Congressman Cox had been there before him. If there was a foundation that was willing to be controversial, that was willing to take risks, it learned its lesson by the time Cox and Reece got through. Who will venture now?

Reading the Reece Report affords a glimpse into the working of a true conspiratorial mind, ranging, as it does, from attacks upon the Rockefeller, Carnegie and Russell Sage Foundations for initially sponsoring the Encyclopaedia of the Social Sciences[42] to attacks upon the "dubious staff of the Ford Foundation" (Reece Report, 1954: 36) in the personage of Bernard Berelson. Nevertheless, despite the farfetchedness of some of the charges, Hutchins' comment "Who will venture now?" is apposite given the reaction of the foundation world.

Although foundations in general in the early- and mid-1950's were held under suspicion, the Ford Foundation was singled out for particular attention. According to MacDonald (1956: 25), the pillorying of the Ford Foundation got off the ground in 1951, when the Chicago Tribune ran a news story under the headline 'Leftist Slant Begins To Show In Ford'. This supposed left slant was an allusion to the participation in various Ford programmes of individuals such as Paul Hoffman who, as Head of the Marshall Plan, "had given away 10 billion dollars to foreign countries." This lead was taken up by the Hearst columnists such as Westbrook Pegler, George Sokolsky and Fulton Lewis Jr. Hoffman was described by Pegler as "a hoax without rival in the history of mankind" and the Marshall Plan was described as "the fabulous Roosevelt-Truman overseas squanderbund" (Reeves, 1969: 15).

But, if that was the periodic tone of the attacks, the impact was much more serious. The criticism of the foundation's activities, besides being extremely vocal, reached a wide and devoted audience. Time and again, Fulton Lewis Jr. had called upon his audience to exert pressure on behalf of some personality or cause, the result of which was that his broadcasts had been responsible for starting over a dozen Congressional Enquiries (MacDonald, 1956: 25). Lewis reached over 16 million people each weekday evening on radio, many watched him on one of 50 television stations, while millions read his syndicated columns in the Hearst press (Reeves, 1969: 124). Attacks were also delivered from groups such as the right-wing Constitutional Educational League which produced pamphlets against the Ford Motor Company linking it with communism, the reasoning behind such tirades being that, through ownership of a Ford motor car, one unwittingly gave sustenance to communism, since the Ford Motor Company's profits were spent by the leftist Ford Foundation (MacDonald, 1956: 27).

The singling out of the Ford Foundation for such attention was the result of two main factors. The first was the newly expanded programme of the foundation which included heavy support for the social sciences, both for its application to social problems and for its scientific development. In fact, Division Five of the Foundation, which was responsible for the establishment of the Center for Advanced Studies in the Behavioural Sciences at Palo Alto, received special adverse comment in the Reece Report (Reece Report, 1955: 36). The second reason, and perhaps the key to Ford's unpopularity, was the granting of 15 million dollars for the establishment of the Fund for the Republic. This radical foundation, whose purpose was the defence of civil liberties and which sponsored such works as Stouffer's communism studies (Stouffer, 1955) and Lazarsfeld's *Academic Mind* (Lazarsfeld and Thielens, 1958), time and again stung the foundation's critics to fury over its practices. Although the Fund was independent, the Ford Foundation was obliged to assume parental responsibility for its wayward child. Not only were the Fund's practices controversial but, in addition, its birth in 1953 came at a particularly defensive time for the foundations. This unfortunate timing – although the Fund had in fact been conceived some eighteen months before – resulted from the decision to postpone the founding until after the Presidential Election of November 1952.

The attacks upon the Ford Foundation left their marks, both at the personal level and at the level of granting policies. MacDonald (1956: 27) quotes Ford as saying:

> The dealers send us in letters from customers accusing the foundation of being communist and warning that they'll never buy another Ford. But I can't bother much with that sort of mail. Why should I?

The above statement by Ford is strictly for the 'official history' of the foundation, since the reality is somewhat different. He was worried, in fact he showed some of his mail to his speech writer and "expressed deep concern" (Reeves, 1969: 15). Neither does Hoffman's view of Ford lend support to the above nonchalant statement.

> I told him [Ford] that I wanted to experiment... to change things and that change always means trouble. But every time we got a dozen letters objecting to something we had done... a radio show of an overseas program or what-not... I'd have to spend hours reassuring the Board (Reeves, 1969: 15).

Henry Ford's behaviour under such pressure is also evidenced by his denunciation of the Fund for the Republic in 1955. This denunciation, circulated as it was in every major newspaper in the country, was an unexpected blow to the Fund (Reeves, 1969: 177). Why had Ford done it? It is not attributable to immediate economic considerations since, although boycotts of Ford products had been threatened, the company records reveal that sales and profits were breaking records (Reeves, 1969: 177). The reason would appear to be the amount and intensity of general criticism, plus criticism from dealers and customers. In an interview between Elmo Roper and Reeves, Reeves comments that:

> [Roper] knew that there had been heavy pressure from officials at the Ford Motor Company; Roper was given hints that his own business relations might be cut if he failed to resign from the Board, or to support the removal of the Fund's outspoken President. At one point, Ford himself had asked Roper to leave the Fund's Board (Reeves, 1969: 177).

If the pressure was too much for the Ford Foundation, then it was also too much for the 'stronger' Fund for the Republic which had a man such as Hutchins at the helm. Following a particularly blistering attack by Fulton Lewis Jr., after his discovery of the Fund's intention to study the American Legion, the Board Meeting of September 1955 was extremely worried (Reeves, 1969: 138). The outcome was the decision by the Fund to drop certain controversial programmes (Reeves, 1969: 138).

It must be remembered that the social class background of both foundation officers and trustees[43] had not prepared them for the unsophisticated and libellous nature of the attacks which were delivered. Nor are the Henry Fords of the world accustomed to give account of themselves before Congressional Enquiries, as indeed Ford had to do in the autumn of 1952. The effects of these onslaughts upon the funding policy were remarkable – although Lyons (1969) disagrees, especially in relation to the Congressional Enquiries. Lyons (1969: 278) considers that these "had no appreciable effect on foundation policies" and that whatever the effects were they were largely "attributable to the general anti-communist hysteria and anti-intellectualism of which they were a part." In support of this contention, he cites the fact that, in the early 1950's, the Ford Foundation increased support for the social sciences. Admittedly, the Ford Foundation did expand its programme in the social sciences in the early 1950's, but that begs the question, for this activity was in part responsible for the attacks. The interesting point is, what happened to granting policies once the attacks began in earnest?

At the Quarterly Meeting of March 1955, the trustees approved three big programmes on which they expected to spend 85 million dollars, all, or most of it, within the next three years. At that current rate of spending at that time those three programmes would together have accounted for nearly half of the regular annual budget through until 1957 (MacDonald, 1956: 166). In addition to the 20 million dollar National Merit Scholarship, there was a 50 million dollar programme to raise the salaries of college teachers, and a 15 million dollar fund for research into mental illness during the following five to ten years. They were good works according to MacDonald (1956: 167), but their attraction for the foundation was precisely that they "offered the maximum possibility for spending money and the minimum danger of getting into controversial territory." MacDonald (1956: 170) further adds the biting comment that "even this program was not large enough for the well-Gaitherised foundation and at the end of 1955 another was announced that was even safer and a great deal bigger". This refers to the grant of ½ billion dollars for privately supported institutions within America.

> The trouble probably was more lack of daring than of sophistication; those philanthropists at Ford are reasonable intelligent men, and it is unlikely that it did not occur to them that a more productive use could have been made of half a billion dollars than just giving it out pro rata to everybody. They were scared, or more accurately, Henry Ford, Donald Young and other trustees were scared, and their fear communicated itself, through channels to the philanthropists who ran the foundation for them. (MacDonald, 1956: 170).

The above outline provides some indications of the unfavourable climate within which foundations had to operate during the 1950's. Even though by the 1960's this climate of distrust had largely evaporated in the buoyancy generated by the 'New Frontier' in Washington and the emergence of many foundation officials in high places within the White House and State Departments,[44] the possibility that the foundations' own actions might stir up controversy remained as a reminder of their delicate position. The action of the Danforth Foundation in the sensitive area of race illustrates the price of 'risk capital philanthropy'. Following through its 1968 programme of aid to young persons from poor neighbourhoods, a young black militant, Percy Green, was awarded a scholarship in 1970. This resulted in pressure upon the foundation, even though the award had not been made by the foundation itself but by a committee of prominent private citizens. In response to this pressure, several members of the foundation's Board wanted to rescind the scholarship. In the event it was allowed to stand, but a price was exacted. In June 1970, in a "wide departure from its general pattern of grants", the foundation awarded a

15 million dollar support gift to Washington University, and another one of $1\frac{1}{2}$ million dollars to St. Louis University for similar purposes. According to the "privately expressed view of the trustees," this grant was a way of placating the "foundation's critics in the St. Louis area for having given educational assistance to a black radical" (Nielsen, 1972: 106).

Before turning to the case study, it is interesting to note that the recent 1969 Tax Reform Act, prompted to some extent by the support given by some foundations to the voter registration movement, has again undermined foundation confidence. Barbara Shenfield (1972: 11), in her summary of the debates of a recent Anglo-American conference on philanthropy, suggests that "the new legislation has encouraged a climate of suspicion towards foundations generally which in turn has caused many of them to be more cautious in their grant making than the restrictions in the legislation actually require."

7. Social pressure, television research and foundation collapse[45]

One of the problems for those working in the field of mass communication research who wish to confront their own history, is to explain the absence of any major sociological work on television in America during the 1950's. Although it is true that scholars such as the Maccobys and the Rileys were engaged in research in that area, it was not until 1961 with the publication of Schramm's (Schramm *et al.*, 1961) book *Television in the Lives of our Children* that the first major American work was produced.[46] The first major work on television originated not in America but England with Himmelweit's (Himmelweit *et al.*, 1958) study of *Television and the Child.* Commenting on this situation, Charles Wright (1956: 83), in his UNESCO report on American mass communications research for the decade 1945–55, wites:

> In reviewing this recent history, one is struck by the absence of material on what is perhaps the most salient development on the mass media in the United States during this period; that is, Television. There has been no *major* sociological study of the new medium to date.

Whilst not wishing to become enmeshed in the Berelson controversy over the 'withering away of mass communication research' (Berelson, 1959), the absence of both work on television and foundation support for this area of research is a problem requiring explanation, particularly as it runs counter to what may have been expected, given the foundations' past record in the area of mass communication research. For example, in the 1920's, when motion pictures were a cause of concern, the Payne Fund supported Charters' studies (1933); then, in the late 1930's, with the spread of radio, the Rockefeller Foundation provided support for a study of its impact upon American society.[47] Yet, the situation arose in the 1950's, when a new medium of enormous potential influence was developing, and no foundation came forward to offer support for an in depth examination of its impact.

The Rockefeller Foundation, following the familiar path of support and withdrawal, had already left the field of mass communication research and could therefore not really have been expected to re-engage itself in such work. Further, although the Rockefeller Foundation had given generous support to the social sciences, especially through the activities of the Laura Spelman Memorial, its influence was steadily being eclipsed by the entry of the Ford Foundation. The question then arises: why did the Ford Foundation never see fit to lend its support to television research? Of course, a possible explanation would be that such research never held any intellectual interest for the foundation but such an explanation would be mistaken since the foundation was very interested indeed. During the winter of 1951–52 the Ford Foundation began to discuss seriously the feasibility of studying the impact of television on American society. Several preliminary seminars and conferences were held to explore the possibilities of such work, and leading figures in the academic world and in the media were consulted. The outcome of this preliminary work was the establishment of an Advisory Committee on Television. The Committee met for the first time in New York on August 21 and 22, 1952, with Paul Lazarsfeld as its chairman. The Committee's brief was to develop proposals for something akin to a commission to study television and to make policy suggestions. The Report was finally submitted in the summer of 1953 with expectations that the proposals would be acted upon, as the Committee's members had been led to believe they would. In the event, however, nothing ever became of the Committee's recommendations.

A point of entry into the reasons surrounding the still-birth of the commission is provided by a reading of the Proceedings of the Kefauver Committee on Television and Juvenile Delinquency, Spring 1955. In discussing the lack of funds for a study of television, Paul Lazarsfeld (Kefauver, 1955: 54), who had been called as an expert witness, stated:

Unfortunately, the chances for such a turn of affairs [foundation support] are limited at this moment, because of the kind of criticism which has been levelled against foundations in recent years. A Congressional Committee [Reece] has criticised the foundation boards for certain action in other areas. The boards are frequently cautious in making funds available for new areas of study. When radio appeared on the scene, the Rockefeller Foundation was still quite willing to finance large studies on the effects the new medium might have on American life. Now that television is here, with presumably more intensive effect, no foundation has yet seen fit to sponsor the necessary research.

In the Report of the evidence given to the Committee, part of Lazarsfeld's original testimony was omitted. The missing part[48] reads:

Just as our Committee [refers to the Ford Advisory Committee on T.V.] submitted a detailed plan endorsed by industry as well as by critical reform groups, the attacks on foundations began and the sponsoring organisation [Ford] decided to drop the whole matter.

The reason for this omission is that a high-ranking official at the Ford Foundation wrote to Kefauver objecting to Lazarsfeld's interpretation of the reasons for 'dropping' the proposed project. Understandably, the suggestions put forward by Lazarsfeld did not make pleasant reading for the official and resulted in correspondence to 'clear the matter up'. Far from clearing the matter up, however, the correspondence provides further evidence of the uncertainty surrounding the whole situation and the necessity for discovering what really did happen to the proposed project. For example, correspondence between the foundation and Professor Lazarsfeld shows that the foundation considered it should have been obvious at the time why the project was never funded. If it was obvious to the foundation, then it was certainly not obvious to Lazarsfeld (1955a) for he writes to Vice-President McPeak:

I am, of course, very eager to avoid any misunderstanding as to what happened with the Citizens' Committee on T.V. Your letter came too late for changes to be made in the Congressional record. I am sure you realise where the difficulty lies. I have never had any communication from the foundation as to what disposition was made of our proposal. Whenever I was in New York some member of the Citizens' Group was likely to ask me about it. The most reasonable interpretation was that the matter had got forgotten in the turmoil created by the Reece Committee. This formulation saved me from embarrassment with my colleagues who had worked so hard on this assignment and I thought it also saved the foundation from the reproach of being discourteous to this distinguished group of men. Let me assure you that I will be equally cooperative in adjusting to your way of looking at the matter once you have explained it to me.

Considering the amount of work that Lazarsfeld had expended on behalf of the foundation, the above is a remarkably restrained and polite letter, but it is shortly followed by another and somewhat sterner letter (1955b) asking for clarification. He writes:

... at the time I rendered my report every member of the group got a thank-you letter and was told that in due time we would learn of the disposition the foundation made of our recommendations. Since then, neither I nor any other member of the group as far as I know has heard from you. We of course took it for granted that the foundation had decided to drop the matter but because we were left without any information most of us developed the theory that the Reece episode accounted for all of it... including the silence. Therefore, there are two things I will want to clear up with I think can be done in a few lines from you to me.

This request by Lazarsfeld for an explanation as to why the proposal was refused funds received a reply, but no explanation. It was not even a case of receiving 'no adequate explanation'; there was simply no explanation at all, except to inform Lazarsfeld of what he already knew – that the project had been refused funds. The whole matter was raised with Professor Lazarsfeld in some recent interviews that the writer had with him in New York.[49] The following extracts give some insight into Lazarsfeld's understandable confusion over the whole situation:

D. Morrison: In your evidence to the Kefauver Committee on T.V. and Juvenile Delinquency you mention that the foundations were reluctant to support T.V. research because they were feeling the pressure of the times.

P. Lazarsfeld: Oh yes, the foundations – there was a congressional investigation of all the foundations.

D. M.: Well, there was Cox and Reece.

P. L.: Yes, that's right, they were badly affected – not financially – there was nothing you could do – the atmosphere was very unpleasant. This Reece was really just like McCarthy.

D. M.: I wanted to ask you about some information I discovered on the T.V. Commission while I was going through the Ford Foundation files.

P. L.: Oh yes, that I wanted to make you aware of – that is something we never touched on before because it never had the slightest consequence.

D. M.: Yes, that's exactly what I wanted to ask you about. I've got my own interpretations.

P. L.: So have I. Well, I have a definite theory about it, but I have no evidence really. There was one man on my committee, or whatever you want to call it, who was very famous for many reasons but also completely erratic – that was Ruml. I think Ruml took a very definite dislike to me. I always had a theory that Ruml torpedoed the whole project and torpedoed it in the following way. There was a Vice-President who has since died – McPeak – who in my opinion was a very typical foundation bureaucrat, not in the slightest a John Marshall. And you know the reason. This committee just disappeared and my theory was always that Ruml, who had a great influence on McPeak – and McPeak had no reason to like me either [sic]. I had hardly any contact with him. But between these two men it was killed. It's a very funny episode. One day, Hutchins asks me to take it over. I think he had asked Stanton's advice, and he suggested me. You know there were eight or ten monographs written around it. Siepman and others and I worked. We worked endlessly and this was a very distinguished group. After all, these were top people on this board. I was very proud. One day, out of the blue, it was all ended.

From the position that Lazarsfeld occupied in the affair, the situation may well have appeared that way – and his supposition may well be correct with regard to the actual mechanics of the ending of the project. But the concern in the present context is more with the situation within which the mechanics operated – how it came to be that the project could be jettisoned.

Lazarsfeld and his colleagues had good reason to consider that the 'rug was being pulled out from underneath the project' since, in the trade paper *Variety* (May 9, 1952), the front page headline read 'Ford Foundation Maps 1000000 Dollars Survey to Cure the Woes of TV'. No authorisation had been given for such a release, in fact, so far as can be gathered, it remained a mystery just who had leaked the information. The release caused consternation among those who were committed to the project since, in the early days of negotiations, such publicity could only jeopardize the chances of the project. However, too much must not be read into this incident, except to note that it could have been responsible for the expressed feelings that the project was being sabotaged. There is no evidence that the project was actually 'torpedoed', and it is reasonable to suppose that, because of the prestige of the groups involved, plus the size of the enquiry, and the foundation's obvious commitment to it, any machinations powerful enough to sink the project would somewhere become visible. That is not to discount behind-the-scenes manoeuvering, and it may well be that, in the final demise of the project, personal antagonisms did come into play – they were not, however, responsible for it. There is other evidence available to account for its failure.

Hutchins was a moving force behind the whole project, but at the same time, Henry Ford opposed it on the grounds that it was too controversial. From the start, Henry Ford had insisted on the inclusion of the media industries in the project as a protection against criticism, despite early resolutions to exclude them on the grounds of 'vested interests'. It is true that in the early days of the Committee, certain sectors of the media industry were hostile, but they were placated by Lazarsfeld's diplomacy and became supporters. As Lazarsfeld (1953) informs the President of the foundation:

> I am very eager indeed to talk with you about the progress of the preparatory committee on the T.V. Commission. We have made good progess intellectually as well as in our efforts to secure the support of the industry.

Indeed, in personal conversation with the writer,[50] Lazarsfeld has underscored this point:

D. M.: I know that Ford insisted that the project should only go ahead if there were representation from the media industry involved, and, if my reading of the foundation's files is correct, the media people seemed quite responsive to the idea.

P. L.: Of course. Look, I had all sorts – I remember big station managers from the mid-West, and I was fairly skilled to reconcile them. The list of station managers is very impressive – the station representatives liked the idea. At least, they never created any trouble. It was one of the strangest episodes.

Unfortunately for the future of the project the support of the media personnel came too late. Ford's uneasiness probably stemmed from the fact that he had phoned Frank Stanton of CBS in the early days of setting up the committee to ask him to take part in it. Stanton agreed, and was then criticised by the industry. This got back to Ford, and made him unsure about the wisdom of such work, given the atmosphere of the times. Further evidence of the existence of 'nervousness' in relation to the project is the insistence that an outside group approach the foundation for support of the project when, in fact, the proposal had originated within the confines of the foundation itself. A certain amount of subterfuge followed to cover this deception.

The death knell of the proposed commission was sounded by two events – the first was the departure of Hutchins from the foundation to help establish the Fund for the Republic; the second was the establishment by Reece of his Congressional Enquiry into foundations. Both of these events occurred in 1953, the year that Lazarsfeld submitted the proposal for the Television Commission. Hence, the already tense atmosphere[51] within the foundation, further intensified by the proposed Reece Enquiry; in addition, with Hutchins' departure, the project's principal supporter was removed. Once this lockgate was opened, the waters of caution flooded in. However, the proposed Television Commission was submerged and not actually drowned, a fact that, to some extent, accounts for the absence of communication to Lazarsfeld regarding the foundation's disposition towards his submitted report. They could not decide what to do with it. It floated for some time in uncertainty, and even looked as if it might surface, but finally, excuses were given as to why it should not be supported. These excuses look very much like post hoc reasoning for dropping the proposal. They included the lack of financial support from the media industry and from citizens' groups, and the fact that the original release of money for Lazarsfeld's Advisory Committee had made no mention of guaranteeing support for any proposals made. Both of these can be discounted as serious reasons for ignoring the proposal since alternative financial support had never been part of the committee's brief in the first place and it is doubtful that the Ford Foundation actually needed the financial contributions that a citizens' group could make – unless, once more, it wished to spread responsibility for the commission. In any case, such reasons were never communicated to Lazarsfeld as excuses for the refusal to support the commission. With regard to the fact that there was never any promise to support suggestions that the committee produced, this is not true, in the sense that Lazarsfeld and his colleagues were in no doubt that their report was going to be accepted. Although it is true that the official release of the original money for the Advisory Committee contained no explicit promise of support for the proposals produced, there was nonetheless a definite tacit understanding. Lazarsfeld is adamant on this point[52] and other evidence leaves no room for doubt. Such an understanding was not a mis-reading of the situation by Lazarsfeld. Indeed, as Lazarsfeld repeatedly states, "it was a peculiar episode," reflective in many ways of the peculiarity of the times and the uncertainty existing within the foundations. But it would be a mistake to consider it as an isolated incident. Indeed, in order to understand the meaning of this particular episode, it is necessary to situate it within the kind of analysis of the foundations' social position and their development over time which is outlined in this paper.

8. Wider implications for the production of knowledge

In looking at the foundations' role in the production of knowledge, one can adopt varying points of entry, depending upon the level under consideration. For example, one can see them as 'brokers of ideas' (Coser, 1970: 337) or as 'standards of excellence' to which scholars can appeal for definitions of their work (Berelson, 1960). One can even widen the angle of vision to include foundations within the productive process itself, rather than as part of the apparatus of support. Indeed, it is difficult to draw the line between the 'creator of knowledge' and the 'support structure' necessary for this production, for both are part of the same process of knowledge production. It would be false to treat the academic world as the active partner and the foundations as somehow passive, merely giving over funds to 'worthwhile' studies, for both interrelate, shaping and informing the other. It is not sufficient to consider the foundation as 'out there', to be finally considered when all other aspects of the situation have been attended to and worked out. The foundations exist as part of the social world that the scholar must operate in and on. Their existence is an aspect of reality that must be confronted, an aspect that has its own qualities which need to be taken into account.

The purpose of this paper has been to indicate the way in which the foundations inform and shape the production of knowledge. Given the rapid expansion of empirical work in American social science, and its institutionalisation in research bureaux, research centres and institutes, the role of the foundations is of paramount importance. Many of these 'institutes', although linked to universities in a variety of ways, have little sound financial basis (cf. Lazarsfeld, 1962: 763) and have therefore been obliged to seek external sources of funds to support what is basically expensive work (cf. Zetterberg, 1956: 13). The relationships between such research institutes and foundations have wide implications for the production of knowledge. The relationship between the two is firmly cemented in the histories of both and, although one is not suggesting that foundations have been responsible for empiricism in any genitive sense, they have greatly facilitated its adoption and spread through their support of the institutional forms necessary for its advanced development. The relationship is a reciprocal one, however, for just as the organisation of empirical research required the financial support of the foundations so the foundations needed such organised research to help dispense their funds. The reason is that foundations, as bureaucratic organisations, have demands that do not necessarily spring from their stated purpose of existence but from the operational imperatives which have pushed them in the direction of dependence on other bureaucratic organisations. Relatively stated, the allocation of a small grant costs more than would the disbursement of a large grant (cf. Whyte, 1955b: 216). Furthermore, handing over large funds to research organisations relieves the foundation of an administrative headache. For instance, by allocating large funds to research institutes or departments attached to universities, they know that their money is 'safe' and consequently constant scrutiny for the possibility of misuse of funds is unnecessary. They can rest assured that, within reason, the money will be used for the purposes stated in the grant application. Its proper use is underwritten by the organisational structure within which the research is conducted. The money has been given from one bureaucracy to another with all the empathy of confidence that this implies. But, if the relationship is symbiotic, it may, as I have argued, have possible disadvantages for the production of knowledge. The interlocking nature of the relationship means that attacks upon the foundations leave the production of knowledge in a balance over which those producing the knowledge have little control. Whilst foundation funds are not the sole source, or even the largest source, of finance for research, other funds from other agencies such as the Federal Government, industry or commerce, are not always available for certain areas of work, and may have extra-academic conditions attached to their funds. The consequence is that foundation funds are often the most attractive to researchers.[53]

In the light of this, and by way of final comment, some possible solutions to the quandary may be suggested. Apart from abandoning a whole tradition of research or re-organising the institutes of higher learning to make the necessary finances available internally (thus lessening the necessity for operational dependence on foundation support), one is faced with trying to improve and strengthen the position of foundations to lessen their vulnerability to social pressure. It may well be that in asking for this one is asking the impossible. That is, one is here demanding the resolution of an inherent contradiction which, if successfully overcome, would necessarily entail the dissolution of the present foundation structure. One of the principal sources of the foundations' vulnerability to pressure lies in their lack of legitimation and, whereas one would normally expect an institution to legitimate itself by its actions or some other practices, such a path is denied the foundations by their continued operation within the controversial fields of social science. It must be remembered that foundations are not in any sense created by the State, or meaningfully ratified, but rather acquiesced to, and therefore cannot draw their legitimation from a higher authority.[54] Thus foundations are consequently thrown back in on themselves by appeals to their own actions for continued existence.

If fully-fledged legitimation is not possible so long as a foundation operates within the social sciences, then it may be possible to develop support structures around them. This could be achieved on one level by the foundations themselves giving greater support and energy to the development of the Foundation Council. Secondly, the universities themselves could offer more support when foundations find themselves under pressure. When called upon in the recent controversy surrounding the 1969 Tax Reform Act, not all universities lent their support. Of course, it may be that the academic world does not wish to see the continuation of such institutions but, if that is so, and this applies to other agencies as well, then alternative forms of research finance must be arranged. In the meantime, any weakening of the foundations' position should be viewed with concern by all those involved in the production of knowledge.

Centre for Mass Communication Research, University of Leicester

NOTES

[1] For a brief but comprehensive history of the evolution of philanthropic foundations see Hollis (1939).
[2] One of the factors producing suspicion against foundations is the difficulty in assessing the 'real' motives of the donors in setting up foundations. On the surface, such generosity simply did not fit with their other public behavior. For a pertinent description of the donors see Nielsen (1972: 10–11).
[3] For an elaboration of this argument and a description of the high-minded principles surrounding such large-scale giving see Harrison and Andrews (1946: 42).
[4] The South Sea Bubble adventure of 1720 was a commercial swindle and caused widespread alarm coming, as it did, among many other such swindles.
[5] For a discussion of this see MacDonald (1956: 23).
[6] For an argument against conspiratorial theories of foundation influence, in favour of a more subtle analysis of foundation power in protecting the value system in existence, see Lindeman (1936: 12). This is very similar to the position adopted by Whyte Jr. (1955b: 22) when he considers that foundations 'mirror the times'.
[7] It is interesting to note the emergence of the press as a factor in the generation of concern over foundation activities – see Nielsen (1972) for an example of this case.
[8] These figures support the assertion that tax advantages, in the main, were not an important consideration in the establishment of foundations – see Andrews (1956: 41) for evidence of the above and for examples of cases where such considerations did play a part.
[9] Cook (1963) provides a very good outline of Patman's grievances against foundations, singling out the Howard Hughes Foundation as an example of a 'tax dodge', the Lilly Endowment for supporting right wing political groups and the Rockefeller Foundation for commercial advantages.
[10] For example, the foundations connected with the chemical industry have been reluctant to take action on air and water pollution, and those connected with the automobile industry have been reluctant to finance research into car safety – Nielsen (1972: 319).
[11] For a discussion of the variety of foundations and the establishment of those for general purposes see Andrews (1956: 11–13).
[12] Lindeman (1936: 26) provides an interesting discussion of donors' interest in education, and their belief in it for making the democratic ideal workable.
[13] Hollis (1938: 128) comments on this, and notes that, except where there were scruples against accepting tainted money, there could be no objection to accepting grants for the raising of professorial salaries or for general building endowments.
[14] There is some confusion regarding the Carnegie Units, centred on the question of the deliberacy of the ensuing reforms and responsibility for them. President Pritchett (1935: 31) lays claim to the responsibility for implementing such reforms and absolves Andrew Carnegie of any responsibility – Hollis (1938: 128–129).
[15] Horowitz (1972) makes a scathing attack on foundations' role in shaping American higher education – he particularly criticises the Carnegie Units for institutionalising the PhD qualification. In addition, see Jencks and Riesman (1968: 240–242) for the importance attached to the PhD qualification, even within areas of teaching not considered to benefit directly from such a qualification.
[16] Hollis (1938: 43) quotes supportive evidence from the Carnegie Annual Reports for this point.

[17] Whyte Jr. (1955a: 112) suggests that, because foundation giving is not a large proportion of the whole, foundation influence has declined. However, Berelson (1960: 6) argues that the importance for influence is not the overall amount but the trend of support.

[18] "Now that public health in the United States has become largely a function for government... Rockefeller is getting out of it" – Whyte, Jr. (1955a: 258).

[19] A difficulty arises concerning what one considers 'support for the social sciences'. It is certainly true, as Oberschall (1972: 218) comments, that the flowering of the social survey movement in the 1910 decade owes much to the establishment of the Russell Sage Foundation in 1907. However, it is equally true that it was not until the late-1930's and World War Two that the survey technique was methodologically perfected and systematically applied in professionally directed scientific research projects.

[20] Hollis (1938: 240) discusses this trend over time.

[21] Such movements of foundation funds from area to area, as expectations are not met, is one of the more flexible features of foundation monies, and shows a mirroring of developments within the world of learning itself.

[22] Although the First World War, as far as the social sciences are concerned, is best considered as a dress rehearsal for the Second World War, it did, as times of crisis often do, force to the surface issues that more tranquil times tend to obscure.

[23] See Donald Young (1969: xiii) for this point and also for the importance of the Laura Spelman (Rockefeller) Memorial's contribution to the social sciences.

[24] Lazarsfeld (1961: 332) mentions that "Nothing is stranger than the idea often expressed by European colleagues that quantification is a United States export endangering their tradition. It is true that when this country [America] took over the European research technique it did so on a large scale."

[25] See Lazarsfeld (1962) for the importance of institutional organisation for the propagation of styles of work.

[26] See Fosdick (1952:217) also for a list of institutes that the Laura Spelman Memorial under Ruml helped support.

[27] Hollis (1938: 289) and also for public attitude towards science.

[28] See in particular Young (1948) for expression of this attitude.

[29] See Bremner (1960) for factual evidence of the principle of 'improving man's welfare', particularly with reference to the commitment of Rockefeller and Harkness to this principle. Bremner (1960: 193–194) cites, for example, the Rockefeller charter, which includes a commitment to "promote the well-being of mankind throughout the world".

[30] One of the problems with foundation activity is that, because the role is one of high affectivity, it is difficult for officials to gauge when their role has been discharged. One ironic index suggested to the writer during an interview with Rossant, Director of the Twentieth Century Fund, was that of 'criticism'. Rossant was possibly bolder than most foundation officials in suggesting that by using the principle of 'risk capital philanthropy', criticism was an indication of role fulfilment (Interview with Rossant in New York, 19 June 1973).

[31] The nature of the 'increased support' was expressed to the writer during an interview with a high-ranking official of the Carnegie Corporation in New York, 7 June 1973.

[32] Further supportive evidence was provided during an interview with Dr. Klein (Russell Sage Foundation) and four colleagues in New York, 24 May 1973.

[33] Norman Dodd, Research Director of the Reece Committe, in particular, strongly attacked the foundations in his report to the Committee.

[34] For example, although it was the Rockefeller Foundation's actions which precipitated the 1915 Industrial Relations Commission, the Carnegie Foundation was also brought to task.

[35] The continuation in a field of research depends greatly on the opportunities for expenditure that such fields offer (Fosdick, 1937). However, entry into a field greatly depends on the climate of acceptability surrounding that area. For instance, Dollard (1950: 18) considers that Kinsey's work demonstrated the freedom of academics to engage in research. Although this may be so, it does not demonstrate the freedom of foundations to support such work. For example, foundations were attacked during the Reece enquiry for supporting work such as Kinsey's – the Rockefeller Foundation had helped to fund Kinsey's studies.

[36] Some progress had already been made in this direction prior to the war – for example the Progressive Movement in the rural mid-West, which agricultural economists and rural sociologists were to continue during the New Deal (Lyons, 1969: 31).

[37] It is important to recognise the antagonism which was directed against philanthropic trusts as part of a more comprehensive antagonism directed against holders of great wealth.

[38] The interesting point is not so much the actual benefits that social scientists gave during the New Deal, but the recognition by the pragmatic administration of Roosevelt of the social scientist as adviser, a role traditionally reserved for businessmen and lawyers.

[39] This refers mainly to the large foundations.

[40] During this interview, mention was also made of the tremendous amount of work that both the Cox and Reece investigations involved. In addition, reference was also made to the necessity for considering the suspicions raised by these investigations resulting in the need for closer scrutiny of recipients' backgrounds. (Interview at Carnegie Corporation, New York, 7 June 1973.)

[41] John Marshall expressed sentiments similar to those of the Carnegie official in an interview with the writer in New York, 6 July 1973.

[42] Reece (1955: 31) objected to the fact that the articles dealing with subjects of a left wing nature were given to 'leftist' contributors and articles of a right wing nature were given to similar contributors.

[43] Big foundations are a microcosm of what has been called 'the establishment', 'the power elite' or 'the American ruling class' (Domhoff, 1967). Both Lindeman (1936) and Andrews (1956) show the trustees to be men of considerable affluence, having attended the 'best' universities and having been members of the 'best' clubs. Foundation officers closely resemble these trustees to a large extent, although not ranking so highly on affluence or age. The data provided by Lindeman and Andrews on their social background shows remarkable consistency over time, and Nielsen (1972) gives little evidence to suggest any major alterations in this picture.

[44] This is a point made by Nielsen (1972) but it should be mentioned that the relationship between high ranking foundation officials and government has not been altogether absent in the past – note the cases of John Foster Dulles and Dean Rusk of the Rockefeller Foundation.

[45] This section relies heavily on information gained whilst examining Ford Foundation files (PA 53-16) and although the Ford Foundation was very cooperative in allowing access to such material, they have not given permission for direct reference to this material. An exception is the use of letters written by Professor Lazarsfeld, since he gave the writer personal permission to quote.

[46] It is worth noting that Schramm in the preface to his book gently complains about the lack of money for his research and contrasts his situation with that of Himmelweit when he notes that "who knew in advance the amount of their support, and could therefore plan the entire study before they gathered any data."
[47] This refers to the Office of Radio Research established at Princeton in 1937 which was moved to Columbia University in 1940, eventually becoming the Bureau of Applied Social Research. Early notable examples of work include those of Lazarsfeld (1940), Lazarsfeld and Stanton (1941, 1944) with perhaps the most famous being that of Lazarsfeld *et al.* (1944).
[48] The missing part of Lazarsfeld's testimony was discovered in the Ford Foundation files (PA 53-16).
[49] This refers to an interview with Lazarsfeld in New York on 15 June 1973.
[50] This refers to an interview with Lazarsfeld in New York on 15 July 1973.
[51] This point was brought out in an interview with a high ranking official of the Ford Foundation who was active in the foundation during that period (Interview in New York on 12 July, 1973).
[52] This point was rechecked in particular during a recent conversation with Lazarsfeld on 28 March, 1974 in Cambridge, England.
[53] Certainly, in discussion with foundation officials, academics were perceived to prefer foundation funds to funds from other sources. Whyte Jr. (1955a: 141) also comments on the lack of 'strings' attached to foundation grants.
[54] Whitaker (1974: 235–242) in his very recent book on foundations questions the future of foundations after the inroads made by the 1969 Tax Reform Act.

BIBLIOGRAPHY

Andrews, F. E.: 1956, *Philantropic Foundations*, Russell Sage Foundation, New York.

Berelson, B.: 1959, 'The State of Communication Research', *Public Opinion Quarterly* **23** (1), 1–6.

Berelson, B.: 1960, 'The Place of Foundations', Paper delivered at American Sociological Association Meeting.

Bremner, R. H.: 1960, *American Philanthropy*, University of Chicago Press, Chicago.

Charters, W. W.: 1933, *Motion Pictures and Youth*, Macmillan, New York.

Colvard, R.: 1964, 'Risk Capital Philanthropy: The Ideological Defense of Innovation', pp. 728–749 in G. K. Zollschan and W. Hirsch (eds.), *Explorations in Social Change*, Routledge and Kegan Paul, London.

Cook, F. J.: 1963, 'Foundations as a Tax Dodge', *The Nation* **196** (No. 16, April 20); 321–325.

Coser, Lewis A.: 1970, *Men of Ideas: A Sociologist's View*, The Free Press, New York.

Dollard, Charles: 1950, 'A Middleman Looks at Social Science', *American Sociological Review* **15** (February), 16–20.

Domhoff, G. W.: 1967, *Who Rules America?*, Prentice-Hall, Englewood Cliffs, N.J.

Fosdick, R. B.: 1937, 'Trusting the Future', in Rockefeller Foundation, *Annual Report* pp. 5–61

Fosdick, R. B.: 1952, *The Story of the Rockefeller Foundation*, Odhams Press, London.

Gouldner, Alvin W.: 1973, *For Sociology: Renewal and Critique in Sociology Today*, Allen Lane, London.

Harrison, S. M. and Andrews, F. E.: 1946, *American Foundations for Social Welfare*, Russell Sage Foundation, New York.
Himmelweit, H. T. *et al*: 1958, *Television and the Child*, Oxford University Press, London.
Hollis, E. V.: 1938, *Philanthropic Foundations and Higher Education*, Columbia University Press, New York.
Hollis, E. V.: 1939, 'Evolution of the Philanthropic Foundations', *Educational Record* (October), 575–588.
Horowitz, D.: 1972, 'Billion Dollar Brains: How Wealth Puts Knowledge in Its Pockets', pp. 298–314 in Milton M. Mankoff, *The Poverty of Progress: The Political Economy of American Social Problems*, Holt, Rinehart and Winston, New York.
Hutchins, R. M.: 1956, *Freedom, Education and the Fund*, Meridian Books, New York.
Jencks, C. and Riesman, D.: 1968, *The Academic Revolution*, Doubleday and Co., New York.
Hearing Before the sub-committee to U.S. Senate Judiciary Committee: (Kilgore) The sub-committee to Investigate Juvenile Delinquency; (Kefauver) 83rd Congress 2nd Session Senate Resolution 89, 1954, Catalogue No. 3151-3154, 1955.
Keppel, F. P.: 1944, 'Foreword' in G. Myrdal (ed.), *An American Dilemma*: *The Negro Problem and Modern Democracy*, Harper and Bros, New York.
Lazarsfeld, P. F.: 1940, *Radio and the Printed Page*, Duell, Sloan and Pearce, New York.
Lazarsfeld, P. F.: 1953, Letter to President of Ford Foundation, Ford Foundation files, PA 53-16. (June 15)
Lazarsfeld, P. F.: 1955a, Letter to Vice-President of Ford Foundation, 16 August, Ford Foundation files PA 53-16.
Lazarsfeld, P. F.: 1955b, Letter to Ford Foundation, 27 September, Ford Foundation files, PA 53-16.
Lazarsfeld, P. F.: 1962, 'Sociology of Empirical Social Research', *American Sociological Review* **27** (No. 6); 767–797.
Lazarsfeld, P. F. and Stanton, F.: 1941, *Radio Research 1941*, Duell, Sloan and Pearce, New York.
Lazarsfeld, P. F. and Stanton, F.: 1944, *Radio Research 1942–3*, Duell, Solan and Pearce, New York.
Lazarsfeld, P. F. *et al.*: 1944, *The People's Choice: How the Voter Makes up His Mind in a Presidential Campaign*, Duell, Sloan and Pearce, New York.
Lazarsfeld, P. F. and Thielens, W.: 1958, *The Academic Mind: Social Scientists in a Time of Crisis*, The Free Press, Glencoe, Illinois.
Lazarsfeld, P. F.: 'Notes on the History of Quantification in Sociology: Trends, Sources and Problems', *Isis* **52** (June 1961), 277–333.
Lindeman, E. C.: 1936, *Wealth and Culture*, Harcourt, Brace and Co., New York.
Lipset, S. M.: 1960, *Political Man*, Doubleday, New York.
Lyons, G. M.: 1969, *The Uneasy Partnership: Social Science and the Federal Government in the Twentieth Century*, Russell Sage Foundation, New York.
MacDonald, D.: 1956, *The Ford Foundation: The Men and the Millions*, Reynal, New York.
Nielsen, W. A.: 1972, *The Big Foundations*, Columbia University Press, New York.
Oberschall, A. (ed.): 1972, *The Establishment of Empirical Sociology*, Harper and Row, New York, pp. 187–251.
Ogg, F.: 1928, *Research in the Humanistic and Social Sciences*, The Century Co., New York.

Pritchett, H.: 1935, *Thirtieth Annual Report*, Carnegie Foundation, quoted in E. V. Hollis (ed.), *Philanthropic Foundations and Higher Education*, Columbia University Press, New York.

Report of Special Committee to Congress on Tax-exempt Foundations and Comparable Organisations. House Report 2681 83rd Congress. House Resolution 217 1954, Catalogue No. 396 1955. [Reece Report]

Reeves, T. C.: 1969, *Freedom and the Foundation: The Fund for the Republic in the Era of McCarthyism*, A. Knopf, New York.

Schramm, W. *et al.*: 1961, *Television in the Lives of Our Children*, Oxford University Press, London.

Shenfield, Barbara: 1972, *Philanthropy in the 70's: An Anglo-American Discussion*, The Council on Foundations, New York.

Stouffer, S. A.: 1955, *Communism, Conformity and Civil Liberties: A Cross-Section of the Nation Speaks Its Mind*, Doubleday, New York.

Manly Report (Commission on Industrial Relations), Senate Document 415, 64th Congress 1st Session Congressional Document, serial no. 6929, 1916.

Whitaker, B.: 1974, *The Foundations: An Anatomy of Philanthropy and Society*, Eyre Methuen, London.

Whyte, W. H., Jr.: 1955a, 'What Are the Foundations up to'? *Fortune*, October.

Whyte, W. H., Jr.: 1955b, 'Where the Foundations Fall Down', *Fortune*, November.

Wright, C. R.: 1956, 'Sociology of Mass Communications, 1945–1955', in H. L. Zetterberg (ed.) *Sociology in the U.S.A.*, UNESCO, Paris, pp. 78–83.

Young, D.: 1948, 'Limiting Factors in the Development of the Social Sciences', *Proceedings of the American Philosophical Society* **92**, No. 5, 325–336.

Young, D.: 1969, 'Foreword' to G. M. Lyons, Paris (ed.), *The Uneasy Partnership*, Russell Sage Foundation, New York.

Zetterberg, H. L.: 1956, *Sociology in the U.S.A.*, UNESCO, Paris.

LEWIS A. COSER

COMMENT ON DAVID E. MORRISON'S PAPER ON PHILANTHROPIC FOUNDATIONS AND THE PRODUCTION OF KNOWLEDGE – A CASE STUDY

Agreeing, as I do, with the major thrust of this paper, I shall limit myself to a few remarks that are meant to supplement rather than to controvert David E. Morrison's analysis.

The author rightly stresses the insecurity on the part of Foundation executives who have no sense of the fundamental legitimacy of the enterprise they are engaged in. This seems to me an essential point worth emphasizing further.

As I have said earlier, "Democratic political thought generally agrees that where there is authority there ought to be accountability to those over whom authority is exercised. But the foundation is not held accountable to its grantees or to the public at large. Its grantees are simply a group of claimants, and foundation officials have no obligation to account to them for their decisions. Similarly, the public at large exercises no control over the foundations' decisions – beyond some very general legal regulations connected mainly with their tax exemptions. Foundation officials... make their decisions 'in a responsible manner' but they are in fact responsible to no one outside the foundations."[1] This very general sense of being responsible to no one, makes for a pervasive insecurity and hence, as David E. Morrison has shown, a proclivity to eschew 'risky' ventures which might offend powerful decision makers whether in the public or in the private sphere.

Lack of legitimacy and insecurity also explain the tendency of foundations to make grandiose public claims while in fact engaging in utterly safe and humdrum activities. Let me cite an example bordering on caricature. The 1969 report of the Irvine Foundation – located in Southern California – states: "[Foundations] are able to question the status quo, encourage experimentation, and provide 'seed money' for new institutions and new ideas. Foundations do the essential job which the government by nature cannot do. The Foundation is more than a mechanical alternative for government action.... The Foundation is an instrument for our citizens to transfer profit from the commercial sector and put it

Knorr, Strasser, and Zilian (eds.), Determinants and Controls of Scientific Development, 403–406.

directly to work as risk capital for the general betterment of society." It sounds impressive – until one compares this grandiose vision with the institutional reality. Here is a representative sample of the grants that the Irvine Foundation actually gave that year: $25000 to the Orange County Children's Hospital to furnish the third floor; $10000 to the Orange County Society for Crippled Children and Adults; $5000 to the Orange County Symphony Association; $5000 to the Boys Club of South San Francisco to help build a gym....[2]

The general sense of lack of legitimacy and insecurity of foundation officials is compounded by their institutional isolations and insulation and the fact that they usually lack evaluative information as to the effects of their own decisions. In what follows I shall rely heavily on a paper by the sociologist Orville G. Brim, Jr., a long-time foundation executive with a rather unusual capacity for self-conscious insight into his own role.[3] " Foundation executives " he writes, "operate with few, if any, reality checks. They are cut off from the natural flow of evaluative information that other institutions receive in American life. They do not know whether they are doing what they think they are doing – or whether what they are doing makes any difference to anyone or not. Institutional isolation breeds narcissism and illusory feelings of power, and separates administrators from the frontier of thought."

Brim shows that this isolation and insulation operates in five different ways:

(1) Foundations' officials are highly autonomous because, as I have said already, they are not really accountable to anyone.

(2) There is no readily available way to evaluate their performance. There is no way to use monetary standards or profit-and-loss statements as a measure of performance.

(3) Foundation executives are socially encapsulated, living among friends and colleagues of similar Eastern seaboard Waspish background, and have few challenging contacts with people of different backgrounds and stations in life.

(4) Their exchanges with grant applicants are usually of a sort that precludes critical appraisals from such applicants. Few people will bite the hand that might at one time or another feed them.

(5) "Foundations lack natural enemies in our society. This has two consequences. While there are some occasional critics, some muckraking,

some intelligent challenges, there is a virtual absence of a body of criticism compared with what is received by other American institutions, such as universities, hospitals, government, and corporations. There is no systematic continuous monitoring by journalists; no features, no foundation beat, no section on philanthropy, no continuing coverage. Nor is there any deadly competition from other institutions in American life."

Lacking institutionalized criticism as well as fully legitimized authority, foundation officials tend to be timid and insecure in their decision-making while being given to grandiose statements in their public pronouncements. Without securely institutionalized exchanges with stable role partners, be it in the world of the intellect or that of power, they but too often succumb to the lure of conventional wisdom. David E. Morrison has shown how they often set their sails according to the prevailing conservative winds in Washington or in the country at large. But at times, though this is much rarer an occurrence, they set aside secure intellectual standards in order to follow the latest fashion, to be *with it*, in order not to miss the lastest turn and twist of the *Zeitgeist*.

It is not easy, as Morrison says, to think of remedies for such ills. But, a need for the reduction of structural insulation and isolation, and, above all, for greater accountability, is plainly apparent. It could be met, at least in part, through the creation of supervisory boards on which, in addition to the present trustees, the 'consumers', i.e. present and future grantees as well as universities and research centers, would be represented. If foundations were more directly tied to 'consumers', they would be forced to develop more elaborate methods of evaluation in order to counter possible criticism. They would be forced to abandon the atmosphere of encapsulation and secrecy in which they now tend to operate.

What the foundation needs, or so it would seem to me, are new role partners with a measure of real power so that the dialogue with them would be more than the polite prattle that now characterizes its exchanges with the public at large. No doubt, this would from time to time create controversy and conflict where a kind of genteel harmony and consensus now operates. But that would be all to the good since, as John Dewey remarked long ago, conflict is the gadfly of thought.

State University of New York
at Stony Brook

NOTES

[1] Coser, L. A.: 1965, *Men of Ideas*, The Free Press, New York, p. 338.
[2] Quoted in Victor S. Navasky's review of Waldemar Nielsen, 'The Big Foundations', in *Partisan Review* **61**, 1(1974), 143–144.
[3] Brim, O. G., Jr.: 1973, 'Do We Know What We Are Doing', in F. Heiman (ed.), *The Future of Foundations*, Prentice-Hall, Englewood Cliffs, N.J., pp. 216–258.

HILARY ROSE

THE SOCIAL DETERMINANTS OF REPRODUCTION SCIENCE AND TECHNOLOGY*

In the sociology of science there is frequently – perhaps more than in any other sociological speciality – a sharing of concern, and sometimes of discourse, between the state, certain social movements and the speciality. I am not sure why this should be so but suspect that the relatively uninstitutionalised nature of the speciality means that ideas are considered from a much wider range of origins than is the case in a well established area. Thus, that we are meeting in Graz to discuss 'The Determinants and Controls of Scientific Development', reflects this convergence, echoing in its theme both the starting point of the theoretical debate between 'internalists' and 'externalists' in the 1930's and also the contemporary renewed interest in this question.

1. The development of 'pragmatic and externalism'

While the debate between 'internalism' and 'externalism' was conducted primarily between historians of science, a variant of internalism has also held sway in sociology. This I would characterise as the sociological preoccupation with science as a more or less autonomous social subsystem (Merton, 1957; Storer, 1960; Zuckerman, 1967; Hagstrom, 1965; Crane, 1972; Gaston, 1970; Cole and Cole, 1973).

At the same time, separate from sociology, although drawing on some of its concepts, developed a substantial if pragmatic literature of science and government, which implicitly and often explicitly took the view that it was possible through the funding of science to influence the direction (and speed) of scientific advance. By pragmatic I mean that the major preoccupation of this literature was with the administration of what were seen as immediate or short run problems. While rarely was this pragmatic externalism of the policy makers so crude as to suggest that the frontiers of knowledge could be extended by the mechanical application of resources to a particular problem, they were none the less seeking in practice to promote particular developments and lines of work. Perhaps the most

Knorr, Strasser, and Zilian (eds.), Determinants and Controls of Scientific Development, 407–430.

successful of the crude application stance was Kennedy's response to Sputnik – 'a man on the moon in ten years!' – while Nixon's crash programme on Cancer has been manifestly less successful. Similarly Greenberg (1969) documents some of the more notorious scandals in pure science, such as Project Mohole and the Linear Accelerator, while as a technological development we have the spectacle of the British and French Governments agonizing over the viability of Concorde, to say nothing of the political implications for the survival and technological level of the European aerospace industry vis-à-vis the Superpowers, and, not least the socio-economic implication for some 50000 skilled workers chiefly in Bristol and Toulouse.

These examples reflect a commonly made distinction between scientific and technological enterprises, but while to some extent this remains possible, within one important sector of science I would argue that this is an increasingly unsustainable distinction. The sector is Big Science, where the change is not simply a question of scale but of a qualitative change in the mode of production of scientific knowledge. This change can be seen most sharply in the context of high energy physics, where the convention remains that only 'scientists' sign the paper and therefore produce and own the knowledge, none the less increasingly there may be twenty signatures or even the whole laboratory (of scientists) signs the paper. Typically a paper produced by such collective work is accompanied by an acknowledgement to the technical staff – of the bubble chamber or whatever, saying that without their help the paper could not have been produced. This polite genuflection to reality attempts at the same time to mask reality, for it is clear that the producers of the new knowledge are a measurably larger number of persons than those recognised by the convention of signatures (Levy LeBlond, 1974).

In the case of Little Science, which is what most scientists are engaged in, the situation is rather different, and much of the traditional distinction remains. None the less, because Big Science is the most technologically advanced it is also where the interpenetration of science and politics is most complete. Like Salomon (1974) and Blume (1974) I would regard Robert Oppenheimer as the symbol of this interpenetation, not as a tragic deviant from the myth of pure science – and presumably pure scientists.

In this same vein all those many books entitled *Science and Society* (alas including one by authors who perhaps should have known better) are, as Levy LeBlond and Jaubert (1973) observe, sociologically incorrect, in that they juxtapose, as autonomous, scientific and social systems. It would be more accurate – if ponderous – to entitle such studies as 'Science in Society and Society in Science'. Indeed two recent books by Blume (1974) and Salomon (1974) do attempt to do precisely this. Salomon's book is divided into two main divisions: science in politics and politics in science. Blume's work attempts to integrate the sociology of science together with the science policy literature in order to develop a political sociology of science. As such, both authors reflect part of a general move among social scientists to locate science within its social context, and away from any of the various forms of internalism. This move away from sociological internalism is important on two scores. First, an adequate sociology of science cannot exclude technology because it is precisely the technological capacity of modern science which differentiates it from pre-modern science. Embedded within modern science is the intention of achieving the control and mastery over nature, and it is this achievement which sets modern science apart from that of Babylon, Greece or India. Secondly, the massive growth of science since the mid-20th century and the intrusion of scientific rationality into all forms of life itself posed important sociological questions which any variant of internalism excludes.

But I would like to return a little further to considering the activities of those elite scientists who are engaged in science policy making. It is worth emphasizing what a very small number of scientists the elite contains, for example it has been variously estimated that between 200–300 out of a possible two million qualified scientists and technologist play a key part in science policy making in the U.S.A. The language of this elite, *par excellence* the 'pragmatic externalists', reflects a theory of happy elision between the interest of science and those of the nation-state. At the height of a boom economy and boom spending in science this elision was seen to be particularly felicitous as "what's good for high energy physics, radio astronomy or molecular biology is good for the nation." Later when government spending slackened, the scientists still attempted to convince their governments of the automatic correlates between basic research spending, the maintainance of civilisation and economic growth.

Indeed when the British government in the late 60's refused the multi-million dollar extension to CERN, the response of the physicists was not only a vigorous protest that the development of fundamental knowledge about the physical world was to be crippled but that the root of modern civilisation itself was threatened. The correspondence in *Nature* and the *Times* suggested that while sociologists regard the relationship between cognitive and social structure as problematic, economists similarly view the relationship between science spending and economic growth, the most eminent physicists appeared to have no doubts (Rose and Rose, 1969). However, within the elite, the limitations to scientific growth had been recognised, the brashest both scientifically and financially of the research proposals had begun to be weeded out, and for our purposes the most interesting activity of the pragmatic externalists began, namely, how to select, direct and control scientific development.

One of the most interesting attempts to articulate the criteria by which specific research proposals were to be assessed was the work of Weinberg (1963). His 'Criteria for Scientific Choice', received with great interest by the scientific and science policy making community, attempted to manage science, embracing and balancing both 'internalist' and 'externalist' arguments. For example he proposed that major investments in a given field were correct if those developments would nourish other adjacent fields (incidentally an argument, which was to be used by the British against the CERN proposal, as the work was sufficiently arcane to offer little or nothing to adjacent fields).

Theoretically unassisted by the methodological stance of internalism of both sociology and history, the policy makers simultaneously proceeded with pragmatic externalism and at the same time set out to promote research which would hopefully provide them with a theoretical framework. Both at a ministerial and interstate level (such as UNESCO and OECD) research units were established, together with a proliferation to research units labelled, 'science studies', 'science of science', 'science policy', 'science and government' etc., according to the ideological perspectives within which the question were to be posed. This pattern held for both advanced capitalist and socialist countries. In both the character of the research produced was initially predominantly empirical and, because it was separate from the thrust of the necessity felt by the science policy makers, detached and abstract from the political situation, yielded for the most part neither theoretical explanation nor even facts very helpful to policy making. In America during the last few years there is evidence of withdrawal of funding from these research groups, which suggests that as far as their sponsors were concerned they failed to deliver the goods.

With the deepening economic crisis and the sharpening recognition in an increasingly corporate capitalism of the need to manage science, the lack of theoretical explanation becomes even more embarrassing. The desparate nadir of science policy studies was surely reflected in the meeting jointly sponsored by DGRST and OECD, where immense trouble had been taken to invite the radical critics of science in Europe, and to provide them with their normal work-shop style environment, with the presumable intention of receiving a theoretical blood transfusion into the sickly body of pragmatic science policy.[1]

Whether working in the tradition of Scheler (1926), Mannheim (1936), and Merton (1945) or in that of Hessen (1931) and Bernal (1939), the theoretical prospects are, while perhaps more encouraging than they were in the sixties, still sketchy. In order to approach the question of the interpenetration of science and politics we have to set aside or transcend the extremely well articulated, and empirically fruitful Mertonian paradigm. Equally in the effort to avoid the extreme (and often extremely dull) empiricism of the science policy studies there is presently a tremendous thrust towards theory. It seems to me that the present danger for the sociology of science is the enticement of metatheory, where concepts cease to be analytical tools becoming instead things in themselves. The difficulty experienced for example in the Whitley (1974) edited symposium of linking empirical work with theoretical work may well be indicative of this danger. In this situation I want to argue that we retain some capacity to work from the concrete, so that the work of theory building does not develop into some grand reification.

It is in this spirit that I want to present some material I have been working on with a colleague in the sociology of reproduction science and technology. Within the context of the long standing ideological debate on regulating reproduction, I shall first discuss some of the new technologies; then consider how the theoretical analysis of the position of women and the role of science under capitalism may be used to cast some light on the interpenetration of science and the position of women, and the patriarchal and class bound society into the science and technology of reproduction. Thus the rest of my paper divides into two parts which I will broadly characterise as (1) Labelling the problem and (2) Explaining the problem.

2. Labelling the Problem

2.1. *The Ideologies of Regulating Reproduction*

Regulating reproduction through late marriage, infanticide, abortion, some forms of circumcision, rudimentary chemical and physical intervention (vinegar on a sponge in the uterus, linen condoms) are, as Himes' (1936) classical text points out, not new in medical history. Historically the Church both opposed the idea of artificially regulating reproduction and also persecuted as witches the practitioners of abortion and birth control advice (Ehrenreich *et al.*, undated). However, over the last one hundred years science and technology have dramatically increased the efficacy of the means of regulating reproduction. Beginning with the advance in rubber technology in the 1870's, which produced more effective condoms and eventually the cap, research moved from physical methods of regulation to chemical methods. The triumph of the pill in the late 1950's marked a dramatic step forward and has been particularly adopted in the industrialised societies. A return to physical means was evinced by the introduction of the coil, a technique which has been widely employed in developing societies.

Both these existing and widely used science based technologies together with the more futuristic technologies discussed later limit if they do not mechanically determine the social relationships of men to women.[2] If we leave to one side the question of whole categories of people denied access to birth control technologies, for economic, political or religious reasons the particular technology itself is expressive of particular social relationships. The condom, for example, is a male controlled technology and thus affirms male domination in a way that the cap and the pill do not. (Although access to the technology is still primarily male controlled in that the majority of those with the power to prescribe are male). The accounts in *Coal is Our Life* (Dennis *et al.*, 1957) of the condoms being thrown on the fire because they diminish male satisfaction shows that where the technology is male controlled, when there are conflicting interests – satisfaction versus pregnancy, it is the woman's interests which can be sacrificed. Thus the cap and the pill represent for women a gain in securing control over their bodies. The social implications of the pill have been particularly dramatic in this respect, as the woman unable to shelter behind the acceptable fear of an unwanted pregnancy, has to accept or reject sexual invitations on other grounds. Ironically because women are socially inferior to men, the very freedom that the technology offers, may turn to a male advantage as unless the woman is sufficiently conscious of herself as a free agent her practice of subordination may yet again leave her as a sexual object (although unimpregnated) rather than (becoming) an agent. None the less the technology itself, and it has been received by many women in this way, offers a degree of personal autonomy and a greater equality in the sexual relationship.

To return to my brief history: once the efficacy of scientific birth control was established, the ideological debate over its employment was also enjoined. The eugenicist movement, partly as an ideological expression of imperialism abroad and ruling class interests at home, had long argued that population control of inferior people as defined by class and race was a necessary and desirable goal (Pickens, 1968). Naturally it saw in scientific birth control a more acceptable and efficacious way of securing its ends. The most brutal forms of eugenic control saw their apotheosis in Nazi Germany, beginning with sterilization and marriage laws, but leading finally to the gas chamber. There were ripples in Britain both in the proposals canvassed in the 1930's, including those of Lord Beveridge, the architect of the Welfare State, that the consistently poor and unemployed should be sterilised, and also in the practice of shutting away in subnormal institutions young girls who had evinced their feeblemindedness by having a baby without securing a husband first. Some such women were released only in 1973, having been shut away for 30 years for their 'crime'. In the United States where the eugenicist movement was a great deal more powerful, compulsory sterilisation was, and to a much lesser extent still is, practiced, particularly against poor blacks. By 1935 no less than 20000 sterilisations had been performed by law, nearly half of these in California, the heartland of the eugenicist movement (Ludmerer, 1972). In 1931, 30 states had sterilisation laws for the 'feebleminded, criminals and paupers'. Even in 1964 two young black sisters aged 12 and 14 had been sterilised involuntarily under the auspices of the American Poverty Program (NWRO, 1974). It is also the situation that in some states free Medicare for women giving birth to a baby is dependent upon her agreeing to subsequent sterilisation. A similar situation seems to be developing as hospital policy for abortions carried out by the British National Health Service.

Not surprisingly the eugenicist nature, both of internal programmes of birth control and also of those in developing countries (particularly when sponsored by imperialist countries), has been widely recognised. Birth control is seen by third world liberation movements as genocidic. However, some of the debates such as those conducted in *Science for the People* last year, which have taken place concerning population control programmes show signs of acknowledging both the eugenicist element and also the liberatory element for women themselves. Increasingly it seems that it is women who are recognising the positive aspect of birth control technology even while sharing with men an agreement as to its societal function.

To this traditional eugenicist argument has come a new ally, the ecologist movement preaching zero-population growth. Starting from an unimpeachable position concerning the finite nature of the earth and its resources, the new eco-eugenicists deftly move into an argument for retaining the developing world as a natural wilderness, and restraining economic growth within the industrialised world. As little redistribution is proposed, zero population and the provision of wilderness seem to be little more than the claims of the haves to preserve their advantages. Indeed the ecologist's interest in population control has mobilized a good deal of enthusiastic verbal support from the middle classes but ruling class allies are drawn from both old landed interests and new business syndicates holding land for sport. In so far that women are made to feel guilty for polluting the earth with their children, the ecology movement merely adds sexism to the traditional racism and class dominance to the eugenicist movement.

2.2. *Birth Control for Women's Freedom*

The alternative view of regulating reproduction as essential to female liberation was strongly held by women doctors such as Marie Stopes and by the feminist movement in the 19th and early 20th centuries. For that matter many men also recognised in birth control the potential for freeing women from an eternal round of pregnancy, childbearing and rearing. Whether and when to have a child become, *at the technological level*, a question of choice, while social class continued to determine the effective access of women to birth control technology. The conflict between the eugenicist strand and the liberatory strand has thus had a constant history.

2.3. *Social Engineering and Population Control*

While at present no society has the knowledge of the social and psychological factors and the level of technological competence needed to control population growth in any precise way, either to increase, stabilise or decrease it, there is a long history of state intervention. In feudal Japan, for example, the *samurai* in certain rural areas where infanticide had been carried out to such a level that the economic activity was threatened, provided family allowances to encourage the peasants to preserve their young. However while the case for intervention was for feudal Japan a matter of survival, intervention and estimates of a desired population level are rarely so straightforwardly rational. Exhortation, financial inducements, enlarging or restricting birth control facilities have all been used. France, for example, has had for many years a generous family allowance programme and limited birth control facilities but was still unable to increase her population to the levels deemed attractive by politicians.

Increasing the population through financial incentive is less efficient as a means of intervention than the technique of first providing an extensive abortion service and then withdrawing it. The Soviet Union in the 1930's, Rumania in 1966 and Bulgaria, Czechoslavakia and Hungary in 1973 have all used this with some success. A decision which previously belonged to the woman is withdrawn – via the medical profession – to become the decision of the State.

Various social techniques are being used to reduce population. Negative family allowance, for example, where the mother receives family allowances only if she produces the right number of children, were discussed for Mauritius by Abel-Smith and Titmuss, and have recently been adopted by Hong Kong. This, as studies of family budgeting under duress show, affects the fathers least because as the bread winners they have to have food, the children most sharply because of a young child's response to poor nutrition together with the mothers because they typically deny themselves food in order to feed first the husband then their children. Few administrative proposals for regulating population through economic incentives and disincentives have been so draconian, it is not without significance that such a proposal was made and carried out in the context of a colony. Other techniques dependent on the provision of services to limit population are a mix of social and technological engineering – epitomized perhaps by the highly publicized, though not very successful, transistor/male sterilization programme in India.

As Hawthorne (1970) points out, fertility depends on seemingly small things such as the age people decide to get married. In a country where individualism is highly valued the choices of 'if' and 'when' to marry can only be influenced indirectly. China untroubled by such individualism has developed a clear political line both on the need to marry and when. Women are expected to marry at about 23 and men at 25; this both limits population and extends the maximally socially productive years. After two children, spaced with the aid of pill or coil, women are encouraged to have themselves sterilised – a very final step for a woman who may be only 25 or 27. At present the brightness of the future they see for their two children compared with the almost certain distress and even death of all their children in old China is perhaps sufficient compensation. However, as China's social and economic wealth grows it is to be hoped that greater flexibility on questions of sexuality and reproduction will develop, as the present policy might in the future – if not now – harm the woman's position.

None the less, within the context of a capitalist society which places a premium on individualism, the intractability of human beings and therefore the immense complexity of the social engineering strategy (albeit with biological elements) leads under conditions of deepening economic crisis to one which is primarily biological engineering. It is, therefore, to some of the new technologies at least partially under development to which we now turn our attention.

2.4. *Biological Engineering and the Control of Reproduction*

Many of the proposals of the biological engineers stem from a body of scientific work in genetics and developmental biology which has been taking shape over the last forty years. Throughout this time various speculations have been made about its social implications from the vision and nightmare of Huxley's (1947) *Brave New World* to those of the geneticists Müller (1936) and Haldane (1925). Presently we approach a situation when fantasy becomes practical possibility.

It is now possible to fertilise the human egg outside the body and maintain it during the first, admittedly brief period of embryological development. This work of Edwards *et al.* (1971) is the first step to the extrauterine production of babies. At present they fertilize the egg in the test tube and then transfer the fertilized egg into the womb of a married woman otherwise unable to have babies. This work is conducted within a particular set of beliefs about women and their biologically determined social role. Edwards and his associates recognise the crucial significance of biological reproduction as determining the structure of the family, and their work is explicitly designed to enable women (otherwise biologically childless) to fulfill their biological destiny. They set to one side the possibility of adoption. Thus their work using the language of helping women is in fact deeply conservative in terms of preserving the woman's role. Their speculation that some women, who are better breeders, could carry the babies of other mothers, suggests a form of biological emancipation of a dominant class of women achieved only by the biological exploitation of another subordinate class. Updated and appropriately modernised by the scientific and technological revolution, it is at root uncommonly similar to wet nursing, whereby poor women nursed rich women's babies at the expense of their own children and bodies. Rent-a-tit gives way to rent-a-belly, and it is implied that this is progress.

One of the further possibilities which is well advanced technically concerns the choice of the sex of the unborn child. Despite Ogburn's (1964) cultural lag hypothesis of the time gap between technological change and societal adaptation, only Etzioni (1971) seems as a sociologist to have given any thought as to the kind of society this intervention in reproduction will produce. Sharing the assumption that Western civilisation proceeds by adjusting to new technologies, Etzioni considers the implication of choosing boy babies.

As Etzioni points out there will be no difficulty in finding a business investor to promote a product that can guarantee male children. People have measureable preferences about the sex of their children. More families stop having children after a boy is born than a girl, and men, in particular, state a preference for boys. Etzioni predicts that if it were actually possible to choose, many more than those who stated a preference when there was no means of definitely acquiring it would choose boys, "even if it were taboo or unpopular at first... could become quite widely practised once it became fashionable"; thus defining the role for advertising companies. In a capitalist society nothing is more sacrosanct than the "individual choice" of the consumer, thus making any move to not allow such techniques to be commercially developed much more difficult. While he acknowledges that a certain sexual imbalance would occur, that both homosexuality and prostitution would increase, Etzioni basically thinks that such a shift in the sex ratio would do little to transform society. Questions concerning the effects on the position of women, the family and the development of future technologies are firmly swept under the carpet.

The biologist Postgate (1973) is a bolder sociological prophet. He offers a male utopian conception of how sex selection would change male-female relations and deal with the population question. Postgate links the three eugenicist concerns – class, race and sex. He argues that overpopulation is the most important problem facing humanity today, and that population, starvation, and social instability are dependent upon it. Birth control is rejected as he claims it works best in countries that least need it, i.e. 'wealthy educated countries' but not in 'underdeveloped unenlightened ones'. Alternative forms of population control, such as war, disease, legalised infanticide, euthanasia, are rejected as they are not selective, acceptable, quickly effective, or permanent enough.

Breeding male, however, meets these criteria as "countless millions of people would leap at the opportunity to breed male" (particularly in the third world), and "no compulsion or even propaganda would be needed to encourage its use, only evidence of success by example." As fewer females than males would be born, the subsequent generation would produce fewer children, and a general population decline would set in. According to Postgate when the world population stabilised at a much lower number of people, the real benefits of developing changes in industrial production could be reaped.

Some people, of course, might object to a 'male child pill' because of the social consequences of the 'transitional phase', which he describes as a "matter of taste, rather than serious concern." During this period women would be kept in purdah, no longer able to work or travel freely, given as rewards for the most outstanding males, polyandry could be introduced, and women come to be treated as queen ants. While no woman could begin to equal a queen ant in production, the term indicates an intensified technological process where women bear children throughout their fertile years; the analogy is factory farming, though with battery, rather than free range hens. Thus the production of children would mirror existing industrial techniques; a realistic form once the absolute numbers of women are reduced. The logic of capitalism is to rationalise production, and this is one way it could be accomplished. Rowbotham's (1973) suggestion that the logical thrust of 'pure capitalism' is towards 'baby farms and state controlled breeding' is another.

While Postgate's utopia will come as he says – and I would agree with him – to resemble "a giant boys public school or huge male prison," the model the physicist Shockley offers us is one of friendly facism (Postgate's view can perhaps be compared with National Socialism before the Night of the Long Knives, while Shockley's is after).

Shockley, Nobel Laureate for his work on the transistor, and notorious for his views of the heritability of IQ genes is also concerned with the population question. Although his plan embraces both social and biological engineering, it turns on a new technology and is therefore included here (Shockley, 1971, 1972). He offers a population control plan that reinforces the monogamous unit, and strengthens the patriarchal system. There are five steps beginning by convincing people that population limitation is desirable and necessary for survival. The Census Bureau then calculates the number of children each woman may have (2.2 if $\frac{1}{3}$ of a percent increase is permitted each year). The Public Health Department then sterilises every girl as she enters puberty by a subcutaneous injection of a contraceptive time capsule which provides a slow seepage of contraceptive hormones until it is removed. When the girl marries she is issued 22 deci-child certificates, and her doctor will remove the contraceptive capsule on payment of 10 certificates which is replaced when the baby is born. After two babies, the couple may either sell their remaining two certificates (through the Stock Exchange), or try to buy eight more on the open market and have a third child. Those who do not have children have 22 to sell.

This technology which in principle is not impossible, transfers control not merely from the woman to the man – as in the case of the condom –

but directly to the state. Nor is it chance that Shockley's baby credit system would disfavour the low income groups, which in the United States would felicitously coincide with the blacks, a group he regards as eugenically inferior. Low income people are more likely to sell their credits and in this friendly model, the goal of limiting inferior children is achieved by the workings of the market. In addition power is not distributed evenly. Within the family unit, whether rich or poor the situation of all women, but particularly those with low incomes, who want children when their husbands do not, would be markedly worsened.

2.5. *Cloning*

Cloning is asexual reproduction of identical individuals. One technique in animals is to transplant the nucleus of any body cell into the de-nucleated infertile female egg cell. This has formed a central element in the proof that each body cell contains all the genetic information originally thought only to be found in the sex cells. The work originally carried out on frogs is thus central to the theories of development biology. What we have recently seen is the transfer of this work to human beings ostensibly because knowledge of foetal growth would aid the clinical treatment of infertility. However, to a public educated by Huxley's *Brave New World*, the social implications of cloning in the hands of a less than benign state represents a considerable threat.

What is interesting in this situation is that where in the 30's the Marxist geneticists, such as J. B. S. Haldane or H. Müller, considered human cloning to be an interesting and eventually practical possibility, today the orthodox left geneticists such as J. Lederburg, M. Pollock, W. Hayes or M. H. Wilkins seem to suggest that cloning is a science fiction, which, because it is unthinkable will therefore not happen. It is left to the liberals, and even more the right, to argue that cloning is practicable. J. D. Watson (like Wilkins – one of the D.N.A. model builders) argued recently in *Atlantic Monthly* (1971) that "we must therefore assume that techniques for the *in vitro* manipulation of human eggs are likely to be in general practice, capable of routine performance in many major countries, within some ten to twenty years."

The possibilities of cloning for genius nourish the strong eugenicist current which has emerged over the last five years or so within the scientific community. Mainly focussed around the IQ question, the scientific legitimation for contemporary eugenicism is provided by the work of W. Shockley, A. R. Jensen, H. J. Eysenck and R. Hernnstein. For the most part the response to this latest wave of eugenicism has raised questions concerning class and race: to this we would add the third dimension – sex.

If the discussion of cloning is carried out at a biological level, then because the woman has both the egg cells – to be denucleated – and the body cells from which the nucleus can be transferred to the egg cells, in principle a kind of virgin reproduction could take place. In this sense a world without men is biologically possible, and further, so long as female egg cells are necessary, a world without women is not. (Although as each woman has several hundred egg cells it would be possible to make do with very few women.)

Discussions of cloning tend however to be carried out in a social vacuum (a form of internalism) and thus mystify the character of science and technology under capitalism. This character is demonstrated when we inspect the proposals of the would be biological engineers for those who should contribute the critical body nuclei. Nor despite the fact that these lists have been drawn up from time to time over a period of forty years have they altered in one respect. Writing in 1936 the Marxist geneticist Müller suggested: Lenin, Leonardo, Pasteur, Beethoven, Omar Khayyam, Pushkin, Sun Yat Sen and Marx. He added "I purposely mention *men* [my emphasis] of different fields and races."[3] Indeed, present proposals initiated by contemporary and distinctly more conservative biologists are not dissimilar in their equation of cognitive ability and sex. The only two women I have ever been able to find listed are Elizabeth Taylor and Brigitte Bardot, which would suggest that rather different criteria have been employed.

3. EXPLAINING THE PROBLEM

Now while I have had a lot of fun *labelling* this particular area within science as sexist, this is not the same as offering an explanation of how this happens. There are clearly two analytically distinct but interrelated theoretical tasks: the first is the analysis of the position of women in capitalist society, the second is the analysis of science and technology development.

The richest theoretical writings on the position of women in recent years have not come from academic sociology but from the women's movement. There are two main theoretical strands within the movement, the radical feminists and the Marxist feminists. Despite the temptation I shall limit myself and discuss the ideas only of two of the most distinguished theorists of these groupings, namely Firestone (1971) and Mitchell (1966). Firestone is particularly relevant for our purpose in that she is concerned with both the position of women and science. Mitchell, though less immediately concerned with science, provides us with a method of analysis, which is potentially fruitful for linking women with science.

Firestone's basic premise is that it is women's biology which is the fundamental cause of their social inferiority. Like the early American sociologist Lester Ward (1914), she argues that it is because women bear children that it has been possible for men to gain ascendency over them, for the subjugation of women is rooted in the division of labour, which begins with the differing biological roles males and females have in the reproduction of the species. This division of labour is institutionalized in the family. Therefore, to free women it is necessary to eradicate the family through at first developing alternative life styles and social institutions and eventually reproducing people artificially. The solution is thus to eliminate the female reproductive function.

She argues that certain current social, economic and technological trends support the changes she is advocating. The family is under a death sentence from two sources. First, there is no longer a need for universal reproduction, and secondly, the coming cybernetic mode of production will make the family as a unit of reproduction and production obsolete. Cybernetics alters human beings' relations to work and even the need to work and this change in production will eventually "strip the division of labour at the root of the family of any remaining practical value." Thus, the women will no longer be needed to service the paid male worker (and his children) as in present day assembly line technology. Simultaneously with these changes in the social and economic organization of society which will modify the dominance of males, Firestone ultimately looks to biologists to generate the technological fix of artificial reproduction which will achieve equality for women.

Firestone having analysed the core of woman's oppression as residing in her biological destiny invokes the aid of science and technology to liberate her. For the social problems which confront women (albeit with a biological base) a technological fix is prescribed. Now, while Firestone's conception of science and technology is more sophisticated than those within the women's movement who, denied all access to science and technology, identify it with mending cars and understanding television sets, her conception is almost 19th century with its belief in the progressive nature of science and technology. Whatever doubts there were before, since Nagasaki and Hiroshima science has unequivocally lost this innocence. As Oppenheimer said "the physicists have known sin" (Haberer, 1969). Whereas in the writing of left and right scientists up to the late 30's, science was portrayed either as neutral or as progressive, contributing to human liberation, following 1945 the previous naive juxtaposition of science and progress settled down to a more cautious view of science as a two-edged sword, either cutting for or against humanity (Rose, 1971). Eventually a third attitude arose highly critical of science and of the inhumanity of its methodology; this view, whether promulgated by the social theorists of the neo-marxist Frankfurt school (Marcuse, 1964; Habermas, 1971; Schmidt, 1971), the conservative Ellul (1965) or the anarchist writings of Roszak (1970), contributed towards a view of science as repressive.

For that matter if we examine the social structure of science it is difficult to understand why science and technology should appear to be the allies of women. They are historically a stronghold of men, and, since the end of the first world war, increasingly locked into the state and industry. Whereas the pursuit of scientific knowledge in the past was the activity of relatively disinterested amateurs, scientists are now increasingly controlled by state and industry. The major items of the scientific research and development budget continue to be directed towards the needs of military, space, atomic, economic goals with 'welfare' occupying a residual role. The contrast in the organization of British Science, for example, between the Haldane Report of 1918 on the Machinery of Government, which defended the principle of autonomy for science, and the recent Rothschild Report (1972) which seeks to strengthen the customer/contractor relationship in scientific research, epistomises this trend. In a situation where not only the allocation of research funds between research areas substantially determines the structure of knowledge, but when within this overall funding pattern Rothschild seeks even greater control, where can Firestone's expectations of science seriously be placed?

My view is not only that such hopes are unrealistic but that also by their naivete about science may make it more difficult for women to see the dangers flowing from some new technologies of reproduction.

Firestone, as one of the most influential of the radical feminists, has, in her misplaced advocacy of asexual modes of reproduction, ignored the actual character of science and technology under contemporary capitalism. Essentially her argument is that of the *deus ex machina*, whereby the god in a white laboratory coat resolves the biologically founded contradictions of the women's situation. The male biological engineers are thus to create the feminist utopia.

Despite the feminist passion of her writing and the wide influence of these ideas upon women within the movement, they clearly fail in both major theoretical tasks. The view that the position of women is fundamentally determined on biological grounds displays the same reductionist perspective shared by the most male chauvinist of the biologists. Sociological analysis cannot ignore the biological parameters; if it does so it tends to exhaust itself in a semantic pursuit of the phenomena, but to *reduce* sociological problems to a biological explanation rapidly generates its own theoretical – and rapidly ideological – difficulties.

The alternative theoretical accounting of the position of women is that being developed by feminist Marxists. I would suggest that the quite unconscious sexism of, for example, Müller (1936) as a Marxist biologist was not some fearful aberration on the part of a natural scientist (anyway, there were exceptions like Haldane (1925) and at a theoretical level Bernal (1939)), but was part of a general failure within Marxist theory to articulate the relationship between the liberation of social classes and the liberation of women.

Orthodox Marxism sees the class struggle as primary, a concern for the position of women as secondary: an issue which will be resolved – and depending on the perspective possible only thought about – after the revolution. This is no recent development, it arose prior to the degeneration in Marxist thinking associated with Stalin (although with his medals for fecund Stakhanovite mothers and the restoration of the family Stalin played his part). It is, for example, expressed quite unambiguously in

Lenin's exchanges with Zetkin and Kollontai during the early years of the revolution. Whether it was the practical issue of famine and civil war or a more profoundly embedded view of Lenin's is not material to the argument. Lenin clearly put to one side the ideas of free love and the withering away of the family as expressed in the *Communist Manifesto* as a bourgeois preoccupation. Instead, like Engels, he saw the solution in terms of women entering the productive process as workers; liberation would, in some unspecified way, flow from this. The limitation of this 'solution' is illustrated by the present situation whereby women in the Soviet Union, despite some necessary assistence with child care in nurseries and creches, merely carry out double work – in the workplace *and* in the home.

The theory and experience of the Chinese revolution has been different. At a theoretical level the question of sexual (and generational) repression within the family and for the need for the struggle to be antipatriarchal has been discussed since a very early period within Mao's writings. In addition within the history of the revolution itself, the part played by the profoundly oppressed peasant women was critical to the success of the struggle for liberation (Belden, 1949).

For Marxism developed in the context of industrialised societies with a substantial urban proletariat, something of an invisible hand of socialism was, and is, thought to operate on the question of women. Apart from the *Holy Family*, there is a long silence, the question itself deemed of little strategic significance for the revolution and therefore even the discussion is relegated to the nursery.

It is therefore the feminist Marxists sustained by the experiential analysis of the movement to whom we should look for theoretical advances. Gardiner's (1974) work on the political economy of women, Dalla Costa and James' (1972) or Lonzi's (1970) attack on Hegelianism – to suggest but a few – are clearly part of this advance.

Mitchell's (1966) work that draws extensively on Althusser (1969), pioneers such an analysis, which represents both an advance on the orthodox view and seems potentially fruitful for examining the relationship between the position of women and the science and technology of reproduction. Instead of discussing women's condition in the conventional way as a monolithic entity, Mitchell seeks to specify the separate structures, which together form the complex whole. Each separate sector is to some considerable extent autonomous, and each has its own momentum, therefore the complexity created by the synthesis of all the structures is itself continuously shifting. Mitchell then distinguishes the four basic structures as: Production, Reproduction, Sexuality and Socialisation. Each is autonomous and therefore requires discrete analysis yet at the same time all four are linked in the complex continuously changing totality, which is the women's situation. Sometimes the movement in the separate structures cancel each other out, but at the moment when the separate structures reinforce one another and intensify the contradictions the conditions for revolutionary change exist.

In this distinction between the four sectors, biology is neither ignored nor made a fetish of, and is therefore not permitted to dominate all sectors. The science and technology of reproduction control therefore relates most immediately to the sectors of sexuality and reproduction and only indirectly with production and socialisation. The gain over the shortcomings of Firestone's monolithic perspective is marked.

Mitchell has little to say about science and technology as such, although together with most other feminists writing in a Marxist tradition, she sees science as far from neutral being mainly concerned with profit and social control.[4]

By and large sociology finds it difficult enough to examine *existing* science and technology and its interpenetration with society and has tended to ignore the discussion of future technologies. I think this is a pity as in some of these areas of current rapid development, such as in military technology, neurobiology or reproduction control the relationship between science and social control is very graphically visible. Least, for example, Rowbotham's (1973) observation be dismissed as purely speculative, and the examples I have presented here as unthinkable and therefore impossible, I think we have to recognise that between the old technologies of regulating reproduction and the new is a major shift from an individually controlled technology (whether male or female) to a potentially state controlled technology. Between using a condom or a pill and having a contraceptive agent in the drinking water is not merely a comparison of greater or lesser efficiency in birth control, but a whole series of complex issues concerning the changing nature of the state, the social relationships between men and women, and the potentiality for women achieving more or less control over, and in harmony with, their personal biology.

Of the new technologies of regulating reproduction discussed here, not only is there a qualitative shift between the individually based and the mass administered, but there is also a graduation between these technologies which are presently beginning to be practiced (test tube babies) which are immediately practicable under liberal capitalism (choosing the sex) and those which require a much more corporate form of capitalism to be practicable (cloning and mass application of test-tube baby rearing). Whereas in the part the Church as the major control agency insisted that women reproduce, now Science seems to be moving in the opposite direction, denying the woman natural reproduction, if not inviting her to commit voluntary 'femicide'.

Not only are these technologies linked but as part of a transitional phase towards their adoption, existing natural methods of reproduction are being increasingly rationalised to fit into the needs of a corporately planned medical care system. Induction initially used for reasons of the individual health of the mother or her as yet unborn child, has been increasingly used as a routine medical practice to ensure that babies arrive at administratively convenient time, now has been proposed as standard policy for one large region within the British Health Service. The logic of the rationalisation and mechanisation of reproduction is this not merely a possible future but an existing present.[5]

Indeed if women's liberation is to extend into these critical sectors of sex and reproduction, Dalla Costa and James (1972) are correct in arguing that women must enter and work to control and direct science and technology. At present doing science and technology is seen as something of a cop-out for both the women's, and also the black liberation movement, but as the understanding both theoretically and experientially of science in capitalism grows, the call such as Dalla Costa and James' to the women, and Anderson's (1974) to the blacks, reflects a determination to enter another arena of cultural struggle. Here I would remind those to whom action is seen as inimicable to theory, that the theory of social class came after the practice of class struggle, and that the moments where science and society touch each other most abrasively are politically and theoretically most revealing.

Lastly while science and technology continue to offer humanity the *possibility* of living harmoniously with nature whether at the level of sexuality and reproduction or the entire bio-system, and thus aid the long struggle from nature to a truly human culture, the alternative is grim. In this scenario science and technology dominate nature in the interest of the dominant class, race, and sex, turning for example reproduction itself into a mechanical process as part of the logic of an increasingly corporate society, grimly reflecting the interpenetration of science and society.

London School of Economics and Political Science

NOTES

* This paper draws extensively on some work carried out with a colleague, Jalna Hanmer. I am also grateful to Steven Rose for discussions and Hermann Strasser for editorial comment.

[1] Needless to say whether this was either the manifest or latent function of the meeting, it was farcical except that the British establisment captured one radical, however as he was shortly to declare himself a believer in divine light the loss or gain was equivocal.

[2] "Technology reveals the active relation of man to nature, the immediate process of production of his life, and thereby also his social life relationships and the cultural representations that arise out of them" (Marx, 1938: 367).

[3] But Müller's comment on artificial insemination elsewhere demonstrates that curious fallacy often evinced in human genetics that cognitive gifts descend solely (and satisfactorily) through the male genes. "How many women," Müller asks "in an enlightened community devoid of superstitious taboos and sex slavery, would be eager and proud to bear a child of Lenin or Darwin" (Müller, 1936: 153). The geneticist misses the elementary possibility seized on by the nongeneticist Shaw in response to the proposal of a very beautiful actress that they jointly bred a child: "Supposing madam the child had my looks and your brains."

[4] "This invention (the self-acting mule) confirms that the great doctrine already propounded, that when capital enlists science into her service, the refractory hand of labour will always be taught docility" (Marx, 1938: 477).

[5] I am indebted to Margaret Stacey for this observation.

BIBLIOGRAPHY

Althusser, L.: 1969, *For Marx*, Allen Lane, London.

Anderson, S.: 1974, 'Science Technology and Black Liberation', in *The Black Scholar*, 5(6), 2–8.

Belden, J.: 1949, *China Shakes the World*, reprinted (1973) Pelican, Harmondsworth.

Bernal, J. D.: 1929, *The World, The Flesh and The Devil*, Kegan Paul and Trench; reprinted (1971), Cape, London.

Bernal, J. D.: 1939, *The Social Functions of Science*, Routledge and Kegan Paul, London.

Blume, S.: 1974, *Towards a Political Sociology of Science*, Collier Macmillan, New York and London.

Cole, J. and Cole, S.: 1973, *Social Stratification in Science*, University of Chicago Press, Chicago.
Crane, D.: 1972, *Invisible Colleges, Diffusion of Knowledge in Scientific Communities*, University of Chicago Press, Chicago.
Dalla Costa, M. and James, S.: 1972, *The Power of Women and the Subversion of the Community*, Falling Wall Press, Bristol.
Dennis, N., Henriques, B., and Slaughter, C.: 1957, *Coal Is Our Life*, Eyre and Spottiswoode, London.
Edwards, R. G., Bannister, B. D., and Steptoe, P. C.: 1971, 'Early Stages of Fertilization in Vitro of Human Occytes Matured in Vitro', *Nature* **221**, 632–635.
Ehrenreich, B. and English, D.: undated, *Witches, Midwives and Nurses*, Glass Mountain Pamphlet.
Ellul, J.: 1965, *The Technological Society*, Cape, London.
Etzioni, A.: 1971, 'Sex Control, Science and Society', in A. Skolnick and J. Skolnick, (eds.), *Family in Transition*, Little Brown, Boston.
Firestone, S.: 1971, *The Dialectics of Sex*, Cape, London.
Gardiner, J.: 1974, 'The Political Economy of Women', mimeo.
Gaston, J.: 1970, 'The Reward System in British Science', *American Sociological Review* **35** (August), 718–732.
Greenberg, D.: 1969, *The Politics of Pure Science*, Penguin, Harmondsworth.
Haberer, J.: 1969, *Politics and the Community of Science*, Van Nostrand-Reinhold, New York.
Habermas, J.: 1971, *Towards a Rational Society*, Heinemann, London.
Hagstrom, W.: 1965, *The Scientific Community*, Basic Books, New York.
Haldane, J. B. S.: 1925, *Daedalus: Or Science and the Future*, Kegan Paul and Trench, London.
Haldane Report: 1918, *Machinery of Government Report*, H.M.S.O. Cmd, London.
Hawthorne, G.: 1973, *The Sociology of Fertility*, Collier, Macmillan, London.
Hessen, B.: 1931, *Science at the Cross Roads*, Kniga, London; reprinted (1973), Cass, London.
Himes, N. E.: 1936, *A Medical History of Contraception*, Williams and Wilkins, Baltimore.
Huxley, A.: 1947, *Brave New World*, Chatto and Windus, London.
Levy LeBlond, J. M.: 1974, 'L'Idéologie dans la Physique', in H. Rose and S. Rose (eds.), *Ideology in the Natural Sciences*, Penguin, Harmondsworth, in press.
Levy LeBlond, J. M. and Jaubert, A.: 1973, *Critique et Autocritique de la Science*, de Seuil, Paris.
Lonzi, C.: 1970, *Sputiamo su Hegel*, Rivolte Feminile, Rome; available in English *We spit on Hegel* (1973), Women's Workshop, mimeo, London.
Ludmerer, K.: 1972, *Genetics and American Society*, Johns Hopkins University Press, Baltimore and London.
Mannheim, K.: 1936, *Ideology and Utopia*, Routledge and Kegan Paul, London.
Marcuse, H.: 1964, *One Dimensional Man*, Routledge and Kegan Paul, London.
Marx, K.: 1938, *Capital*, Vol. 1, Allen and Unwin, London.
Merton, R. K.: 1957, 'Priorities in Scientific Discovery: A Chapter in the Sociology of Science', *American Sociological Review* **22** (December), 635–659.
Merton, R. K.: 1945, 'Sociology of Knowledge', in G. Gurvitch, and W. E. Moore (eds.), *Twentieth Century Sociology*, Philosophical Library, New York.
Mitchell, J.: 1966, 'Women: The Largest Revolution', *New Left Review* **40**, 11–37.

Mitchell, J.: 1970, *Women's Estate*, Penguin, Harmondsworth.
Müller, H.: 1936, *Out of the Night*, Gollancz, London.
National Welfare Rights Organisation (editorial): 1974, 'Forced Sterilization: Threat to Poor', *The Welfare Fighter* **4** (1).
Ogburn, W. F.: 1964, *On Culture and Social Change*, University of Chicago Press, Chicago.
Pickens, D. K.: 1968, *Eugenics and the Progressives*, Vanderbilt University Press, Nashville, Tennessee.
Postgate, F.: 1973, 'Bats Chance in Hell', *New Scientist* **58**, 12–16.
Rose, H.: 1971, 'Pangloss and Jeremiah', *Nature* **229**, 459–462.
Rose, H. and Rose, S.: 1969, *Science and Society*, Allen Lane, London.
Rothschild Report: 1972, *A Framework in Government Research and Development*, Cmnd 1272, HMSO, London.
Rowbotham, S.: 1973, *Woman's Consciousness, Man's World*, Penguin, Harmondsworth.
Roszak, T.: 1970, *The Making of a Counter Culture*, Faber, London.
Salomon, J. J.: 1974, *Science and Politics*, Macmillan, London.
Scheler, M.: 1926, *Die Wissensformen und die Gesellschaft*, Der Neue-Geist Verlag, Leipzig.
Schmidt, A.: 1971, *The Concept of Nature in Marx*, New Left Books, London.
Shockley, W.: 1971, 'Dysgenics. A Social Problem. Reality Eroded by Illusion of Infinite Plasticity', *Review Educational Research* **41**, 227.
Shockley, W.: 1972, 'Dysgenics – A Social Problem, Reality Evaded by Illusion of Infinite Plasticity of Human Intelligence', *Phi Delta Kappan* **1** (March), 291–295.
Storer, N. W.: 1966, *The Social System of Science*, Holt, Rinehart and Winston, New York.
Ward, L. F.: 1914, *Pure Sociology: A Treatise on the Origins and Spontaneous Development of Society*, Macmillan, New York.
Watson, J. D.: 1971, 'Moving toward the Clonal Man', *The Atlantic Monthly* **228**.
Weinberg, A.: 1963, 'Criteria for Scientific Choice', *Minerva* **1** (2), 159–171.
Whitley, R. (ed.): 1974, *Social Processes of Scientific Development*, Routledge and Kegan Paul, London.
Zuckerman, H.: 1967, 'Nobel Laureates in Science', *American Sociological Review* **32** (June), 391–403. Also *Scientific Elites, Noble Laureates in the United States*, Chicago University Press (in press).

HERMANN STRASSER

SOCIAL TECHNOLOGISTS AND SOCIAL EMANCIPISTS: FACTORS IN THE DEVELOPMENT OF SOCIOLOGY

1. Introduction

Eminent sociologists such as Small (1924), Parsons (1968, Vol. 1: Preface), Merton (1968: Ch. 1), and Gouldner (1965; 1970: Part 1) have followed Comte's (1863: 22) dictum that "a new science must be pursued historically." Nevertheless, students of social theories have been hesitant in applying the methods of sociological analysis to the development of their own discipline.[1] To know its past and present state "is the only firm basis for evaluating whether we have 'progressed', and, if so, how much and in what ways" (Gouldner, 1962: 8).

In order to find out how far sociology has travelled and in what direction, we shall apply the techniques of the sociology of knowledge and the sociology of science to the more recent history of social thought. Social thought is here defined as a body of ideas referring to the social situation as an integral part of the socio-cultural milieu, in which these ideas are produced. In this paper an attempt is made to separate the factual, hypothetical, and normative elements involved in the development of social thought, which should hopefully lead to an appraisal of the degree of its social-existential determination as distinguished from its more cognitive-internal elements.

We shall begin with presenting three working hypotheses, which, it is hoped, will subsequently be verified.[2] First, it is hypothesized that the sociological discipline originated from the dilemma of political involvement and opposition on the one hand, and of reflecting scientifically on the problem of how best to organize society on the other. There is no doubt that "the theory of social evolution and the evolution of social theories [are] related by more than a mere punning verbal tie" as they are both embedded in a common matrix of social change (House, 1929: 545). As Goldenweiser (1924; as quoted in Jensen, 1957: 53f.) has aptly shown, though, the continuity of social thought only approximates the continuity of history. The unity that characterizes the processes of social thought and of

Knorr, Strasser, and Zilian (eds.), Determinants and Controls of Scientific Development, 431–447.

history is in each case one of a different order of relations. While the unity of history refers to a relation of existential dependence of *events*, the unity of social thought appears as a relation of logical dependence between *ideas*.

The second hypothesis that is advanced concerns the early foundation of an organismic analogy and social system perspective that has had a decisive influence on the development of the sociological enterprise in the past. In their attempt to determine the meaning and the limits of social phenomena as well as sociological thinking, sociologists have continually participated to varying degrees in the cognitive orientations pervading a number of otherwise distinct scientific fields (cf. Martins, 1972: 21). Sociologists draw on these broad conceptual orientations such as analogies or ideals of natural order "because they refer to facts which are considered similar to those studied within an area of discourse" (Isajiw, 1968: 115). In other words, we do not assume, as Kuhn (1970) does, that scientific fields practically develop their paradigms or research programs by endogenous change only. Pioneering developments in prestigious scientific fields such as astronomy (17th century), physics (18th century), biology (19th century), and the new physics (20th century) had significant effects on the change potential of other areas of scientific inquiry including the scientific study of society.

And thirdly, we believe that there is evidence for an overall trend in social theory, which is guided by a conservative, i.e., social-technological research interest, and dominated by the structural-functional approach. Moreover, it is held that these conservative images of society tend to be continually challenged by progressive, radical perspectives. Consequently, we assume not only interaction of contemporaneous paradigms in *different* scientific fields but also mutual influence of the various components of existing schemes of explanation *within* the area of sociological inquiry.

2. Origins of Sociology: Highlights of Humanism and Progress

By linking the original intention of the scientific study of society to its relation with the established social order, we hope to be able to derive propositions regarding the nature of sociological theory, its past development and its current state.

Although there are various conceptions of the origins of sociology,[3] we have reason to believe that sociology evolved at a time when rationalism was conjoined with humanistic values in the newly forming nations. The rise of modern science was closely connected with the destruction of the metaphysical view of life as much as the fall of the social order of the Middle Ages had to do with the rise of capitalism, of the new nations, of the inventions and discoveries around the world. The emerging autonomy of the human intellect found in modern philosophy a medium through which it expressed itself. These humanistic perspectives on social and intellectual affairs arose in educated and politically powerful circles of the medieval and Renaissance cities. Humanism represented a *normative* orientation that was intended to justify and implement the new kinds of individuality and community (Martindale, 1964: 460).

As soon as a new ideal of knowledge based on rational proof was linked to a method of establishing 'truth' in the world of physical things, namely, empirical experimentalism, modern science was born. A functional conception of knowledge, for which knowledge represents an instrument of prediction on the basis of sense observation, began to replace a transcendental conception of knowledge whose criterion of truth depends on other sources than sense perception (Reichenbach, 1953: 252; Whitehead, 1933: 3). As a non-normative method, science arose in circles of artists and craftsmen who aimed at the acquisition of an exact knowledge of nature with the ultimate goal in mind to maximally improve man's ability to control the material world (Martindale, 1964; Bernal, 1970, Vol. 4: 970). It seems that the fundamental antagonism that characterizes presentday social science came then into existence: While humanism provided emotional energy, science was meant to establish facts; while humanism was dialectical, science claimed to be cumulative (cf. Horowitz, 1968b: 201).

The sociological perspective came into focus as the individual began to acquire intrinsic value vis-à-vis the collectivity. Social analysis developing out of a system of laws of human nature replaced the metaphysical construction of reality. The religious, national, and class struggles of the late sixteenth and the seventeenth century and subsequently the rising middle classes served the function of a socio-political midwife of the new science of man-in-society (Bernal, 1970, Vol. 4: 970ff.). In the century prior to the French Revolution, the Scottish moral philosophers were among the first to view man as a psycho-social entity and to place him in the center of social phenomena. Their sociology represented a natural history which aimed at providing an orientation for social and political action, that is, for a rationalization of civil society. Their sociology was conservative within the limits of this natural development; it was critical insofar as it dealt with the utility and the malfunctioning of existing institutions. Already in the seventeenth century Lilburne and Winstanley had presented a critical assessment of contemporary conceptions concerning the meaning and the purpose of society and, as a consequence, suggested practical reform measures. However, Hobbes' analysis of the coercive nature of society, which led him to postulate the establishment of a powerful government, the Leviathan, was more acceptable to his contemporaries (cf. Bernal, 1970, Vol. 4: 970). Giovanni Battista Vico followed at the beginning of the eighteenth century with the first explicit outline of a science of society (*Scienzia nuova*), in which he professed that human society had been created by man and could therefore be understood by man (Scheler, 1926: 239).

As the case of the Scottish Enlightenment demonstrates, Adam Smith, Adam Ferguson, John Millar, David Hume and others not only attempted to explain the emancipation of civil society, but also conceived of themselves as part of the very process of liberation. Their research interests were guided by the Whiggest principle to pursue enlightenment in the sense of naturally progressing civil society according to the standards of social uility (Forbes, 1954: 643ff.; Habermas, 1963: 219). Sociology thus paraded as a *science of progress* as long as the practical and theoretical connection between authority and utility in the development towards civil liberty was maintained. The French Revolution and its aftermath destroyed this context of authority and utility. Moreover, it brought about a change of the discipline's awareness of its scope and objectives. Sociology was claimed by both adversaries and advocates of the fall of the *ancien régime* as contributing to the solution of a crisis that had become permanent. The new design of the sociological discipline was that of a *science of crisis*. Saint-Simon wanted it to be an opposition discipline, while de Bonald attributed to it the function of stabilization. The spirit of revolution and of restoration seem to provide both material and perspective to early sociology: the revolutionary spirit aimed at (critically) dissolving traditional authority, while the restorative forces championed its conservation at any cost. Finally, Comte integrated the 'industrial' and 'traditional-constitutive' notions of society into a structural-functional outline of societal organization. In his attempt to harmonize order and progress under the rule of order, he eventually stripped the idea of progress of its emancipatory power and sociology of its critical mission. Sociology was on the road to become a science of justification (of the status quo).

3. Sociology established: competing interpretations of social existence

Thus far we have seen that a strong will to progress led to social and intellectual currents, the structuring elements of which consisted of a maturing nation-state, rising middle classes, and an unfolding science of society. These social and intellectual developments profoundly reflect the great metamorphosis of the rising middle classes, i.e., the bourgeoisie.

As indicated above, the advent of middle classes in the countries of Western Europe and in the United States drastically changed the value system by which individuals and social roles were now to be evaluated. All social strata came to be measured in terms of the middle-class standard of utility. The middle classes were useful because of the services they performed, and because what they produced, they held, was what others wanted (Gouldner, 1970: 62–64). Without overstating the case one can say that Western sociology may be understood as a response to the conditions of the utilitarian culture in which it evolved. By the time of the last decade of the eighteenth century the power of the bourgeoisie, in England at least, was securely anchored. In the course of the Industrial Revolution, the bourgeoisie gained an enormous amount of wealth. There were hardly any reasons to extend the existing liberties and privileges, though many for limiting them (Bernal 1970: 979).

The French Revolution did succeed in liberating man from his political confines. However, it stopped short in breaking his economic chains. Socialism and the laboring classes were next to claim to be the historical forces of the revolutionary, humanistic, ferment, which had brought the middle classes to power. However, it was not until the publication of the *Communist Manifesto* (1848) by Marx and Engels, the *Capital* (1867) by

Marx, and the *Anti-Dühring* (1878) by Engels that the basic outline of scientific socialism came into existence. The postrevolutionary societies of Western Europe witnessed the evolution of a different political program which also generated a distinctive mode of thinking.

Similar to the progressive-humanistic way of thinking, discussed earlier, its conservative counterpart also envisioned a total structure of the world, including a science of man-in-society. Mannheim (1964b: 423) attributes the rise of modern conservatism in many countries to the fact "that the modern world has become dynamic": a dynamic caused by an accelerating differentiation in social structure and function. In these societies particular events and phenomena were increasingly related to the problem of growth and development of the total societal complex with the result that homogeneously reacting social strata originated. Correspondingly, the intellectual world and its supporting sentiments split along the lines of the socially divided world. The basic intentions of the major social strata became the center that generated opposing world views and patterns of social thought. These bifurcations into progressive and conservative elements began to cluster most particularly in the political and economic spheres.

Besides the general factor of an accelerating social dynamic, one specific development in early industrial society must be mentioned in accounting for the formation of distinctive intellectual, and for that matter, sociological currents: the rise of ideological politics. As long as politics were not the vehicle for the realization of the right social order but rather the mere maintenance of order, there was no room for ideological politics (Shils, 1972: 49). Then, politics only concerned the rulers and those who aspired (or better: were destined) to become rulers. Since the sixteenth century, however, a body of intellectuals emerged, whose imagination carried them beyond the requirements of everyday life, and who were no longer forced to depend exclusively on ecclesiastical, governmental, aristocratic, or mercantile patronage for their existence (Shils, 1972; Mannheim, 1960: Ch. 1). The turning point has been located at the time of the French Revolution, that is:

> with the transformation of human society from the relatively simple conditions of traditional agrarian communities to the vast complexity of the conditions of industrial and urban organization, a new body of knowledge has been, and is, *necessary* – if men are going to exercise any effective control whatsoever over the social forces which these new conditions have unleased (Fletcher, 1971: 156).[4]

What we are saying is that the new science of society was felt to be important as a means of control and was to take on a subject matter dominated by the bourgeois value of utility, a utilitarian morality, and a particular way of thinking. It is not held that social science merely reflects class interests. What we do maintain is that the development of the sociological discipline shows a decisive influence of antagonistic social groupings, each of which emits a more or less distinctive *Weltanschauung* and clings to a particular pattern of social thought.

The works by Auguste Comte, Frédéric Le Play, and Lorenz von Stein exemplify that sociology was molded by a conservative philosophy, which meant to come to grips with the given reality in its relativity. Conservative thought is concrete; it rests on the factual substratum when it deals with the normative realm. Progressive thinking, by comparison, tends to see empirical reality through the lenses of the possible, the normative, some utopia. In progressive thought, aspects of social reality take on meaning through their relation to a utopia or normative code which contrasts with the factual substratum. Conservatism, on the other hand, derives the meaning of empirical items from the positive acceptance of the beginning, i.e., from the past. For the conservative, the past partakes of the present. To him, the present represents the last stage of the past, while the progressive experiences the present time as the beginning of the future in that he seeks to re-enact the enunciation of Reason in things (cf. Mannheim, 1960: 211f.; 1964b: 435–439; Scheler, 1926: 204ff.; Stark, 1958: 77f.; Blum, 1974). In other words, the conservative thinker views things, people, and institutions as 'organically grown', as having necessarily come into existence, while the progressive deals with them under the aspect of some kind of 'ought-to-be'. While the conservative orientation likely accepts the given circumstances, also enforced by the 'realistic' assumption that social experiments tend to fail, the progressive orientation tends to neglect the immediate reality of the current world (e.g., human selfishness rooted in organic and cultural properties and its manifestation in society) and to surpass it by negation.

Conservative philosophers and sociologists such as G. W. F. Hegel, Edmund Burke, and Auguste Comte pointed to the indispensable value of the sacred, nonrational, and, in part, nonutilitarian elements of man's existence in society. Society was seen as an organic whole that was greater than the sum total of its members. Idealistic philosophies, including the assumption of a pre-established state of order or the insistence on the primacy of social controls, eventually gave rise to predominantly organismic analogies and finally to a structural-functionalist outline of social explanation. To be sure, conservatism provided both idealistic and positivistic approaches to social theory with a stable ground of social anchorage (Martindale, 1960: 76).

In conclusion, we may state that doing sociology, that is, investigating and explaining social phenomena, is grounded in normative objectives. The history of sociology is characterized by a dialectical development, above all with respect to the public interpretation of human existence in society. We have found that four structural conditions of social life produce this dialectical element in social analyses: First of all, our point of departure is the insight, most cogently formulated by Durkheim (1965: 21, 485–488) and Mannheim (1960: Chs. 1 and 2), that as individuals act in differently organized groups with and against each other they think in accordance with, or in contradiction to, others. Considering the structure and position of the respective group, their membership will either try to change or to maintain nature and society that environ them. We can thus assume that the 'essential' ideas dominating our intellectual life represent the work of the *group*. Secondly, as a consequence we have seen that social groupings with rather *distinctive world views* persist. Thirdly, following the second assumption and as we shall further elaborate, the phenomenon of *competition* pervades intellectual life. Fourthly, also to be dealt with later in the paper, the succession of *generations* significantly influences the development of the sociological discipline. We shall subsequently try to demonstrate further the validity of these assumptions.

4. Guiding research interests and vocabularies of social explanation

Before we can discuss the role competition and generations play in shaping the development of sociology, we should briefly outline its normative and conceptual structure,[5] within which the competitive and generational factors show their effects.

So far we have looked at key issues accounting for the origin of sociology and found that it was guided by certain interests of cognition.

Specifically, we conceive of scientific activity as involving prescriptive attitudes concerning the expectations held by members of a disciplinary reference group that consists of the group(s) one feels attached to, the wider public, and one's fellow (social) scientists. Underlying such a conception of social science is a specific relationship between knowledge and interest:

> We know from our everyday experience that often ideas serve well enough to substitute for the true motives of our actions. What on this level we call rationalisation, on the level of collective actions we call ideology. In both cases the manifest content of statements is falsified through the unreflective attachment of an only seemingly autonomous consciousness to its interests. It is therefore right that the discipline of trained thought should aim at precluding all such interests....
>
> Because science has to achieve objectivity for its statements notwithstanding the pressure and seduction of particular interests, it mistakenly disregards fundamental interests to which it owes not only its impulse, but also the very condition that makes objectivity possible.
>
> The attitudes towards technical control, towards an understanding of life's activities and towards emancipation from the quasi-natural forces of history and society determine the specific points of view from which we are first able to conceive reality as such (Habermas, 1970: 47f.).

We now formulate the proposition that a socio-logical position is decisively guided either by a social-technological interest (as exemplified by the works of Auguste Comte, Herbert Spencer, Ludwig Gumplowicz, Talcott Parsons, Lewis A. Coser, and Ralf Dahrendorf) or by a social-emancipatory interest (e.g., Adam Ferguson, Henri de Saint-Simon, Karl Marx, C. Wright Mills, and Alvin W. Gouldner). The normative perspective adopted by a sociologist is determined by his relation to the *scope* of social reality, particularly in view of the rules according to which he applies methods and theories to reality.

Sociologists with a social-technological interest assume, often implicitly, that the structure of the subject matter under consideration exists independently of those in search for scientific knowledge. They purport to produce nomological knowledge. In contrast, social emancipists are not content with that. They assume that the process of social research cannot be separated from the creation of society and hence does not serve only as a means for reproducing social structures, but also to establish the very definitions of social life. Thus, social scientific knowledge potentially transcends mere self-preservation. Therefore, sociologists interested in social emancipation also try to uncover theoretical statements that express unalterable laws of social action, which are factually and principally subject to change (Habermas, 1970: 46, 49). In contrast to social technologists, their aim is not saving theory for the benefit of reality, but rather to conceptualize unbearable social circumstances for the sake of changing them. Their prime concern is the materialization of theory and not its confirmation. Obviously, "The relation of theory and experience changes with that of theory to its subject matter" (Habermas, 1969: 159).

Although the sociologists' guiding research interests, as expressed in value-sentences, are relatively vague and abstract ('freedom', 'autonomy', 'order', 'social stability', 'equality', 'justice according one's achievements/needs', 'happiness', etc.), the discipline was bound to develop into conservative and radical schools in correspondence with either a social-technological or a social-emancipatory intellectual setting. It is important to realize that neither the normative universe nor the explanatory universe represents a statement about reality but rather is created by man. The universe in which man lives is factually created by him and not discovered. In science we discover, and insofar we also re-create, the factual order; we set it up conceptually. In sociological terms, this is to say that the theory of society becomes a function of *the problem* of social order (cf. Meadows, 1967: 77).[6] However, to actually formulate a social theory we need to connect the normative perspective, expressed in value-sentences, with empirical observations, expressed in terms of data-sentences. The vocabulary of social explanation or model provides this mediating function.

A model consists of a set of concepts and their composites, i.e., propositions, which are logically connected with each other. In that it focuses attention on relevant problems, it suggests *important* data that should be methodologically produced. Generally speaking, a vocabulary of social explanation must deal with three important aspects of social life, each of which may render the explanation of social order problematic: (1) the aspect of societal development (i.e., the genesis of new social structures); (2) the aspect of societal organization in the sense of regulation and self-regulation (i.e., the equilibrium of social forces); and (3) the aspect of societal exchange between the social system and its environment and between subsystems (i.e., the autonomy dimension) (Piaget, 1970: 471f.). A model usually imposes a judgment of what is important on the respective subject matter. It thus provides a selective focus and structures the multiplicity of social facts (Emmet, 1945: 196f.).

It is assumed that sociologists have adopted two models more regularly than any other modeling device when looking for a clue to understand the social world, namely, the order and conflict models of society. These models are considered to be adequate if they are capable of generating credible interpretations of given facts in terms of their respective categories. Such a vocabulary of social explanation must therefore be capable of reasonably interpreting the errors of the other system of explanation. If for no other reason than mutual comparison and correction of interpretative bias we need alternative models, granted that they are adequate.[7] A scheme of major types of sociological theory constituted by

TABLE I

Scheme of major types of sociological theory

		Guiding Research Interest	
		Social-emancipatory (progressive)	Social-technological (conservative)
Vocabulary of social explanation (Model)	Order	(1) Transitional system theory: – Past-oriented (de Bonald) – Future-oriented (Smith)	(2) Structural-functional theory: – Positivistic organicism (Comte) – Western functionalism (Parsons) – Eastern functionalism
Vocabulary of social explanation (Model)	Conflict	(4) Radical conflict theory: – Humanistic rationalism (Ferguson) – Scientific socialism (Marx) – Critical social theory (Gouldner)	(3) Conservative conflict theory: – Early industrial conflict theory (Stein) – Darwinistic conflict theory (Gumplowicz) – Formalistic conflict theory (Simmel) – Industrial conflict theory (Coser)

those elements on the basis of which sociological theories develop their informational power, can thus be developed (see Table I).

Since we have dealt with the elements of these major types of sociological theory elsewhere at some length (Strasser, 1975), we suggest to move on to the discussion of the impact which the factors of competition and the persistence of generations have had on the development of sociological theory.

5. Remarks on competition and generations in sociology

There is no doubt that competition and the succession of generations among scientists did always have an important influence on the development of any scientific discipline. Ever since intellectuals had freed themselves from the rigid organization of the church and had destroyed its social monopoly to produce interpretations of reality, the competition between different modes of mental production began to predominate the intellectual sphere. It should be pointed out that, unlike the clergy, the intellectuals had to compete for the approval of the wider public. Social scientists, in particular, began early to devote themselves to rather different modes of thinking and of cultural as well as material experience then prevalent in society (cf. Mannheim, 1960: Ch. 1).

Earlier in this paper, we showed how sociological knowledge has become embedded in, and has been borne by, the drive for power and privilege of basic social groupings, which, consequently, competed for the 'right' public interpretation of social existence. We have come to realize that the public interpretation of the world, and of the social world in particular, is never given but always struggled for. Insofar we are in agreement with Kuhn's (1970) central thesis that the life of the scientific community shows considerable affinity to the life cycle of the political community.

The close relationship between sociology and the interests of major social groupings, as Bernal (1970, Vol. 4: 945) observes, is mostly responsible for the fact that it (as well as other social sciences) did not take on a relatively independent posture as the 'exact' sciences did. While natural sciences are concerned with the *forces* of production of a society as their subject matter, social sciences investigate the *relations* of production and the ide(ologic)al superstructure, which also means to legitimize, to maintain, and to scrutinize these relations of production.

There is some evidence that the polarization of types of social theory, as effected by the selective process among views and needs of struggling groups, generated to some extent theoretical syntheses. The simple law of efficiency contest, it is argued (Mannheim, 1964a: 604), may bring about the adoption of opposite theory elements. For example, the great concern of sociologists such as Emile Durkheim, Vilfredo Pareto, Max Weber, Karl Mannheim, and Talcott Parsons with values, socialism, stratification, socialization, and economics was to a considerable extent due to the ever-present challenge of Marxian sociology. More obvious is the change in Parsons' work over the last forty years or so. These changes are clearly related to the shifting problems of legitimization and control, with which American capitalism had to deal, say, in the 1930's as compared with the 1950's. Although his conceptual scheme changed to some extent, his disciplinary reference group did not change.

These comments should not obscure the fact that there is not only a competition between conservative and radical approaches on normative grounds, but also between Structural-Functionalism (and its precursor) and the various types of Conservative Conflict Theory. Inspite of the predominance of either conservative types of sociological theory in Western sociology, Radical Conflict Theories did gain momentum in recent years. They continually represent viable theoretical alternatives. Only recently the radical caucus in the U.S. and West Germany succeeded in penetrating, to some extent, the ranks of 'establishment sociology', even if it were only for the fact that certain topics were discussed in official meetings or that the chances of radical sociologists to be employed by universities and other institutions were improved. This kind of competition often takes the form of a (part-)paradigm syndrome in that some sociologists are occupied with a theoretical rhetoric directed at weakening the explanatory power of rival theories. More generally, that so many discussions among fellow sociologists turn into status management games seems to be part of the working of the social system of a scientific discipline: "The need to persuade others and to answer objections from others in the discipline imperceptibly transforms the *ex*ponent of a theory into a *pro*ponent of it" (Weigert, 1970: 113).

Closely connected with competition is the factor concerning the succession of generations of sociologists. According to Gouldner (1970:

7), in the 1960's "a younger generation has emerged with a sharply different structure of sentiments, with collective feelings that are not resonated by the very different kinds of sentiments that have been historically deposited in older theories... ." The idea is that the career and (often only sensed) validity of a theory depends, to a considerable degree, on whether or not those who offer and those who listen to the theory share in the personal experiences and the sentiments to which these experiences give rise. In other words, parts of the new generation of sociologists were increasingly characterized by a lack of congruence between new sentiments and old theories. In an earlier study of American sociologists, Gouldner and Sprehe (1965) found that, at the time of the survey (1964), younger sociologists were more likely to express unfavorable views of Functionalism than did their older colleagues.

The preceeding discussion suggests that the competition between scientists, on the one hand, and between theoretical approaches, on the other, is mediated by the activities of the respective groups of scientists, which imply differential development and acceptance of (theoretical) innovations by cohorts of different ages. It is not sufficient to consider competition under the aspect of particular constructions of social reality as they are related to social commitments and identities. We must also take into account that the degree of institutionalization of the discipline, of specialities, and research areas tends to generate different types of competition, which, in turn, have supposedly different consequences for the further development of the discipline as a whole (cf. Whitley, 1974).[8]

In conclusion, we may state that the persistence of theoretical alternatives in sociology forces the dominant type of sociological theory to adjust to these challenges. We have presented some evidence that the sociological discipline is poly- or part-paradigmatic, whose (part-) paradigms lack individual self-sufficiency and are, to some extent, complementary. Kuhn's emphasis on the autonomy of paradigm succession does not apply to sociology. A total change of the dominant paradigm would not only mean that it alters its vocabulary of explanation but also its normative elements. It is hard to believe that all important cognitive norms of a discipline or speciality are paradigm-bound. Rather, they appear to be common to a multiplicity of contemporaneous and successive (part-)paradigms: The vocabulary of social explanation, which was prevalent in sociology during the past 150 years, has often been modified (organic analogy; evolution; structure-function; interaction; system; exchange), while the normative component, stemming from the scientist's guiding research interest, remained essentially the same, that is, kept its conservative bias. To analyze those variables, most particularly the groups competing for public interpretation of social existence, that contributed to the emergence of normative perspectives and their subsequent adoption by social scientists was the main purpose of this paper.

Institute for Advanced Studies, Vienna

NOTES

[1] Several reasons have been suggested to explain the general neglect of early social thought in showing "the dynamic interrelationships of personality and culture and the emergence of social thought as the joint product of its author and its age" (Jensen, 1957: 37).

[2] The present paper is based on some of the ideas I have expressed in Strasser (1975).

[3] Gitler's (1941: 3f.) classification of various conceptions of the origins of the sociological discipline into five groups may very well be cited to represent current views on the subject (except, perhaps, that today a few more names could be added to the fifth category): "... (1) the theory that sociology dates back to ancient times and to the Greeks in particular (this view is upheld by Spann, Bogardus, Lichtenberger, Menzel, Ellwood, and Hankins); (2) the conception of the origin of sociology in the seventeenth and eighteenth centuries (this is supported by such men as Sombart, Brinkmann, and Tönnies); (3) the view that it begins in the romantic period of about 1800 (Small, von Below, and Baxa believed in this conception); (4) the idea that the nineteenth century is the period which gave rise to sociology (this theory has such adherents as Barth, Oppenheimer, M. Weber, Gothein, Freyer, Squillace, Stein, and Mohl); (5) and the theory which places the origins of sociology in the twentieth century (Wiese strongly supports this theory)." The present author considers himself to adhere to the second group (cf. Strasser, 1975: Chs. 1–3).

[4] The view that ideological politics originated in the French Revolution is also supported by Aron (1957: 42), Shils (1972), and Brogan (1947). However, Horowitz (1968a: 22) takes the view that "The solution of a primary level of economic needs, the transparent realization of economic freedom through the creation of a stratum of society absolved from pure scarcity, commences the production of psychic types that play a critical role in the formation of both science and ideology. Aristotle may have been correct in noting that in leisure begins the production of ideologies geared to justify the bifurcation of society into leisure groups and laboring groups."

[5] For a presentation in greater detail the interested reader is referred to Strasser (1975).

[6] The term 'problem' refers here primarily to the object of problematization rather than to the discrepancy between a vision of what society (or an aspect of it) should be like and the reality of social institutions under consideration. It seems to cover the whole range of possible definitions between Hobbes' 'problem of order' (Parsons, 1968, Vol. 1:89ff.) and C. W. Mills' 'public issues' (Mills, 1959:8ff.).

[7] Following our conception of sociological theories as focusing on some social phenomenon with social order implications, we presuppose that the order-conflict alternative will also deal more efficiently with the problem of displacement of theoretical scope (cf. Wagner, 1964). It is thus hypothesized that structural theories as well as interactional theories are characterized by steady processes of fading in and out of micro and macro explanations respectively (Strasser, 1975: Ch. 1).

[8] This point needs more systematic study. The sociological discipline has hardly ever been studied under this aspect. At any rate, an extension of the concept of competition, which may then include the idea of the development of generations, is suggested. For an attempt in this direction, see N. C. Mullins' contribution to this volume and Mullins (1973).

BIBLIOGRAPHY

Aron, R.: 1957, *The Opium of the Intellectuals*, Doubleday, New York.

Bernal, J. D.: 1970, *Wissenschaft*, Vol. 4: 'Die Gesellschaftswissenschaften', Rowohlt, Reinbek.

Blum, A. F.: 1974, 'Positive Thinking', *Theory and Society* **1**, 245–269.

Brogan, D. W.: 1947, 'Was the French Revolution a Mistake?', *The Cambridge Journal* **1** (October), 43–55.

Durkheim, E.: 1965, *The Elementary Forms of Religious Life*, Free Press, New York.

Fletcher, R.: 1971, *The Making of Sociology: A Study of Sociological Theory.* Vol. 1: 'Beginnings and Foundations', Michael Joseph, London.

Forbes, D.: 1954, ' 'Scientific Whiggism': Adam Smith and John Millar', *The Cambridge Journal* **7** (Augusts), 643–670.

Gitler, J. B.: 1941, *Social Thought Among the Early Greeks*, University of Georgia Press, Athens, Georgia.

Goldenweiser, A. E.: 1924, 'Psychology and Culture', *PASS* **19**, 15–23.

Gouldner, A. W.: 1962, 'Introduction', to E. Durkheim, *Socialism*, Collier Books, New York.

Gouldner, A. W.: 1965, *Enter Plato: Classical Greece and the Origins of Social Theory*, Basic Books, New York.

Gouldner, A. W.: 1970, *The Coming Crisis of Western Sociology*, Basic Books, New York.

Gouldner, A. W. and Sprehe, T.: 1965, 'The Study of Man, 4: Sociologists Look at Themselves', *Trans-Action* **2** (May and June), 42–44.

Habermas, J.: 1969, 'Analytische Wissenschaftstheorie und Dialektik', in Theodor W. Adorno *et al.*, *Der Positivismusstreit in der deutschen Soziologie*, Luchterhand, Neuwied, pp. 155–191.

Habermas, J.: 1970, 'Knowledge and Interest', in D. Emmet and A. MacIntyre (eds.), *Sociological Theory and Philosophical Analysis*, Macmillan, New York, pp. 36–54.

Horowitz, I. L.: 1968a, 'Scientific Criticism and the Sociology of Knowledge', in I. L. Horowitz (ed.), *Professing Sociology*, Aldine, Chicago, Ill., pp. 17–29.

Horowitz, I. L.: 1968b, 'Mainliners and Marginals: The Human Shape of Sociological Theory', in I. L. Horowitz (ed.), *Professing Sociology*, Aldine, Chicago, Ill., pp. 195–220.

House, F. N.: 1929, *The Range of Social Theory*, Henry Holt and Company, New York.

Isajiw, W. W.: 1968, *Causation and Functionalism in Sociology*, Schocken Books, New York.

Jensen, H. E.: 1957, 'Developments in Analysis of Social Thought', in H. Becker and A. Boskoff (eds.), *Modern Sociological Theory*, Holt, Rinehart and Winston, New York, pp. 35–59.
Kuhn, T. S.: 1970, *The Structure of Scientific Revolutions*, 2nd ed., University of Chicago Press, Chicago, Ill.
Mannheim, K.: 1960, *Ideology and Utopia*, Routledge & Kegan Paul, London.
Mannheim, K.: 1964a, 'Die Bedeutung der Konkurrenz im Gebiete des Geistigen', in K. H. Wolf (ed.), *Wissenssoziologie: Auswahl aus dem Werk*, Luchterhand, Neuwied, pp. 566–613.
Mannheim, K.: 1964b, 'Das konservative Denken', in K. Mannheim, *Wissenssoziologie: Auswahl aus dem Werk* (ed. by K. H. Wolff), Luchterhand, Neuwied, pp. 408–508.
Martindale, D.: 1960, *The Nature and Types of Sociological Theory*, Houghton Mifflin, Boston, Mass.
Martindale, D.: 1964, 'The Roles of Humanism and Scientism in the Evolution of Sociology', in G. K. Zollschan and W. Hirsch (eds.), *Explorations in Social Change*, Houghton Mifflin, Boston, Mass., pp. 452–490.
Martins, H.: 1972, 'The Kuhnian 'Revolution' and Its Implications for Sociology', in A. H. Hanson, T. Nossiter, and S. Rokkan (eds.), *Imagination and Precision in Political Analysis*, Faber & Faber, London, pp. 13–58.
Meadows, P.: 1967, 'The Metaphors of Order: Toward a Taxonomy of Organization Theory', in L. Gross (ed.), *Sociological Theory: Inquiries and Paradigms*, Harper & Row, New York, pp. 77–103.
Merton, R. K.: 1968, *Social Theory and Social Structure*, Enlarged Edition, Free Press, New York.
Mills, C. W.: 1959, *The Sociological Imagination*, Oxford University Press, New York.
Mullins, N. C.: 1973, *Theories and Theory Groups in Contemporary American Sociology*, Harper & Row, New York.
Parsons, T.: 1968, *The Structure of Social Action*, 2 vols., Free Press, New York.
Reichenbach, H.: 1953, *The Rise of Scientific Philosophy*, University of California Press, Berkeley, Cal.
Scheler, M.: 1926, *Die Wissensformen und die Gesellschaft*, Der Neue-Geist-Verlag, Leipzig.
Shils, E. A.: 1972, 'Ideology and Civility', in E. A. Shils (eds.), *The Intellectuals and the Powers and Other Essays*, University of Chicago Press, Chicago, Ill., pp. 42–70.
Small, A. W.: 1924, *Origins of Sociology*, University of Chicago Press, Chicago, Ill.
Stark, W.: 1958, *The Sociology of Knowledge*, Free Press, Glencoe, Ill.
Strasser, H.: 1975, *The Normative Structure of Sociology: Conservative and Emancipatory Themes in Social Thought*, Routledge & Kegan Paul, London.
Wagner, H. R.: 1964, 'Displacement of Scope: A Problem of the Relationship between Small-Scale and Large-Scale Sociological Theories', *American Journal of Sociology* **69** (May), 571–584.
Weigert, A. J.: 1970, 'The Immoral Thetoric of Scientific Sociology', *The American Sociologist* **5** (May), 111–119.
Whitehead, A. N.: 1933, *Science and the Modern World*, Cambridge University Press, Cambridge.
Whitley, R.: 1974, 'Types of Competition, Autonomy and Modes of Development in Scientific Specialties'. Unpublished Paper, Manchester.

INDEX OF NAMES

INDEX OF SUBJECTS

THEORY AND DECISION LIBRARY

An International Series in the Philosophy and Methodology of the Social and Behavioral Sciences

1. GÜNTER MENGES (ed.), *Information, Inference, and Decision.* 1974, viii+195 pp.
2. ANATOL RAPOPORT (ed.), *Game Theory as a Theory of Conflict Resolution.* 1974, v+283 pp.
3. MARIO BUNGE (ed.), *The Methodological Unity of Science.* 1973. viii+264 pp.
4. COLIN CHERRY (ed.), *Pragmatic Aspects of Human Communication.* 1974, ix+178 pp.
5. FRIEDRICH RAPP (ed.), *Contributions to a Philosophy of Technology. Studies in the Structure of Thinking in the Technological Sciences.* 1974, xv+228 pp.
6. WERNER LEINFELLNER and ECKEHART KÖHLER (eds.), *Developments in the Methodology of Social Science.* 1974, x+430 pp.
7. JACOB MARSCHAK, *Economic Information, Decision, and Prediction. Selected Essays.* 1974, three volumes, xviii+389 pp.; xii+362 pp.; x+399 pp.
8. CARL-AXEL S. STAËL VON HOLSTEIN (ed.), *The Concept of Probability in Psychological Experiments.* 1974, xi+153 pp.
9. H. J. SKALA, *Non-Archimedean Utility Theory.* 1975, xii+138 pp.